AF615164

Relativity and gravitation

Relativity and gravitation

Edited by
CHARLES G. KUPER
and
ASHER PERES
Technion – Israel Institute of Technology
Haifa, Israel

GORDON AND BREACH SCIENCE PUBLISHERS
New York London Paris

Gordon and Breach, Science Publishers, Inc.
150 Fifth Avenue
New York, N.Y. 10011

Editorial office for the United Kingdom

Gordon and Breach, Science Publishers Ltd.
12 Bloomsbury Way
London W.C. 1

Editorial office for France

Gordon & Breach
7–9 rue Emile Dubois
Paris 14e

Library of Congress catalog card number 72–118689. ISBN 0 677 14300 1.

To **Nathan Rosen**
teacher and friend

Foreword

This volume is based on the proceedings of an International Seminar on Relativity and Gravitation, which was held in July 1969, at the Einstein Institute of Physics, Technion City, Israel, on the occasion of the 60th birthday of Professor Nathan Rosen. This seminar was held under the auspices of the International Committee on General Relativity and Gravitation. A number of articles from distinguished relativists, who were unable to participate personally in the symposium, are also included.

The editors wish to express their gratitude to the Technion–Israel Institute of Technology, and to the Israel Academy of Sciences and Humanities, for their generous financial support which contributed so much to the success of the symposium.

C.G.K.
A.P.

Contents

PAPER 1

A class of exact solutions for the motion of a particle in a monopole-prolate quadrupole field*

ANGELO ARMENTI, JR. and PETER HAVAS
Temple University, Philadelphia, Pa. U.S.A.

ABSTRACT

We consider the motion of a particle in the combined gravitational field of a monopole and a prolate quadrupole, both in Newtonian mechanics and in general relativity (using the metric found by Erez and Rosen). In Newtonian mechanics there exists a class of exact solutions, corresponding to circular motion with constant angular velocity in planes parallel to the plane of symmetry of the quadrupole; similar solutions exist in general relativity if the quadrupole moment is not too small. The motion is stable in a wide region, up to a maximum value of the radius, reached at the plane of symmetry; this value coincides exactly with the minimum value of the radius for stable circular orbits within the plane of symmetry for all Newtonian solutions, as well as for the general relativistic solutions. Possible generalizations of these results are indicated.

1 INTRODUCTION

The complete solution for the motion of a particle in a spherically symmetric gravitational field (the "Kepler problem") has been known in classical mechanics since Newton, and was found in general relativity within a few years of the creation of the theory[1]. On the other hand, exact solutions for the classical motion in a combined monopole-quadrupole field were found

* Research supported by the *National Science Foundation*.

only recently, and only for motion in the plane of symmetry of the field[2]; no exact solutions at all are to be found in the literature for the general relativistic case. Indeed, the problem of the exact form of the field itself, trivial in Newtonian mechanics, poses serious difficulties in general relativity. Solutions of Einstein's field equations which can be interpreted as corresponding to multipoles of various orders were first given by Erez and Rosen[3] on the basis of Weyl's work on axially symmetric fields[4]; other solutions were given later by Zipoy[5], which can be shown to be equivalent to Erez and Rosen's, however.

The usual classical treatment of the motion outside the plane of symmetry is based on considering the effect of the quadrupole as a small perturbation of the Kepler problem[6]. All Keplerian orbits lie in planes passing through the center of symmetry, and the effect of the addition of a small quadrupole field to that of the monopole is mainly a slow rotation of these planes around the axis of symmetry of the quadrupole. The calculations are quite involved, and not suitable for studying the effect of a large quadrupole moment.

It seems to have been completely overlooked that in the case of a prolate quadrupole there exists a class of very simple exact solutions for all values of the quadrupole moment, corresponding to circular motion with constant angular velocity in planes parallel to the plane of symmetry. These solutions can be obtained by elementary methods.

A similar class of solutions exists for motion in the general relativistic field of a monopole-quadrupole as given by Erez and Rosen if the quadrupole moment is not too small; very small moments, as well as very small radii, are excluded because they would require an orbital velocity exceeding the local velocity of light. Both the Newtonian and the general relativistic solutions are stable in a wide region.

2 NEWTONIAN SOLUTIONS

The Newtonian gravitational potential of a monopole and quadrupole in spherical coordinates is[6]

$$V = -\frac{GM}{r} + \frac{GQ}{2r^3}(1 - 3\cos^2\theta), \tag{1}$$

where G is the constant of gravitation, and M and Q are the mass and quadrupole moment of the source. The r and θ components of the force per unit mass on a particle are

$$f_r = -\frac{\partial V}{\partial r} = -\frac{GM}{r^2} + \frac{3GQ}{2r^4}(1 - 3\cos^2\theta), \tag{2}$$

$$f_\theta = -\frac{1}{r}\frac{\partial V}{\partial \theta} = -\frac{3GQ}{r^4}\sin\theta\cos\theta\,. \tag{3}$$

We now investigate the possibility of steady motion of the particle in a circle, parallel to the plane of symmetry, with constant angular velocity ω around the z-axis. Here and in the following we will identify equations which hold only for this special motion by a subscript o on the number of the equation.

For such a motion the components of the acceleration are

$$a_r = -\omega^2 r\sin^2\theta, \tag{4$_o$}$$

$$a_\theta = -\omega^2 r\sin\theta\cos\theta, \tag{5$_o$}$$

and thus such a motion is possible provided

$$\omega^2 = \frac{G}{r^3\sin^2\theta}\left[M - \frac{3Q}{2r^2}(1 - 3\cos^2\theta)\right], \tag{6$_o$}$$

$$\omega^2 = \frac{3GQ}{r^5}\,. \tag{7$_o$}$$

The last equation shows that Q must be positive, which for an extended body corresponds to a prolate mass distribution.

Eqs. (6_o) and (7_o) imply

$$Mr^2 = 3Q\left(\tfrac{5}{2}\sin^2\theta - 1\right). \tag{8$_o$}$$

For M to be positive, we must therefore have

$$\sin^2\theta > \tfrac{2}{5}. \tag{9$_o$}$$

For a given M and Q, Eq. (8_o) determines a real r for any θ satisfying (9_o). The motion takes place in a plane at a distance $d = r\cos\theta$ from the plane of symmetry, which from (8_o) equals

$$d = \pm r\left[\frac{3}{5} - \frac{2Mr^2}{15Q}\right]^{1/2}. \tag{10$_o$}$$

The maximum value of this is reached at $r = (3/2)\,(Q/M)^{1/2}$, and equals

$$d_{\max} = \pm \frac{3}{2} \left[\frac{3Q}{10M} \right]^{1/2} . \tag{11$_o$}$$

We now investigate the stability of the circular motion following the standard method, which is based on the Lagrangian formalism[7]. We form the Routhian

$$R = L - \frac{\partial L}{\partial \dot{\varphi}} \dot{\varphi} = \frac{1}{2} \dot{r}^2 + \frac{1}{2} r^2 \dot{\theta}^2 - V - \frac{C^2}{2r^2 \sin^2 \theta}, \tag{12}$$

where dots denote differentiation with respect to t, V is given by (1), and

$$C = \omega r^2 \sin^2 \theta \tag{13}$$

is the angular momentum per unit mass. From (7_o) we have

$$C^2 = \frac{3GQ \sin^4 \theta}{r}. \tag{14$_o$}$$

We now put

$$r = r_0 + \varepsilon, \quad \theta = \theta_0 + \eta, \tag{15}$$

where r_o and θ_o are the values of r and θ for the steady motion, and neglect terms of higher power than the second in ε and η. The motion is stable provided that, keeping C constant, the modified potential

$$U = V + \frac{1}{2} \frac{C^2}{r^2 \sin^2 \theta} \tag{16}$$

is a positive quadratic form in ε and η. A simple calculation shows that this is the case provided

$$\sin^2 \theta > \tfrac{8}{15}. \tag{17$_o$}$$

Then it follows from Eq. (8_o) that there is no stable motion for $M = 0$, and that stable orbits satisfy

$$\frac{Q}{M} < r_0^2 \leqslant \frac{9Q}{2M}, \tag{18$_o$}$$

where r_o takes its maximum value in the plane of symmetry. However, within this plane r_o is not restricted by (18_o), since Eq. (7_o) and thus (8_o) and (14_o) do not hold there. Furthermore, since Eq. (7_o) does not hold, Q is no longer required to be positive; however, we only give the results for positive Q, since we are only interested in comparing them with those valid outside the plane of symmetry. Then, from Eq. (6_o), motion in a circle is possible for $\theta = \pi/2$ provided $r_0^2 > (3/2)\, Q/M$; an investigation of the condition for stability by the same method as above shows that this motion is stable for any $r_0^2 > (9/2)\, Q/M$.

3 RELATIVISTIC SOLUTIONS

We now consider the general relativistic metric corresponding to a monopole-quadrupole following from the solutions given by Erez and Rosen[3]. In Schwarzschild coordinates $x^0 = t$, $x^1 = r$, $x^2 = \theta$, $x^3 = \varphi$ (with units such that $c = 1$), it equals

$$\mathrm{d}s^2 = \mathrm{e}^{2\psi}\, \mathrm{d}t^2 - \mathrm{e}^{2\gamma - 2\psi} \left\{ \left(1 + \frac{m^2 \sin^2 \theta}{r^2 - 2mr} \right) \mathrm{d}r^2 + (r^2 - 2mr + m^2 \sin^2 \theta)\, \mathrm{d}\theta^2 \right\}$$
$$- \mathrm{e}^{-2\psi} (r^2 - 2mr) \sin^2 \theta\, \mathrm{d}\varphi^2 , \tag{19}$$

where

$$\psi \equiv \frac{1}{2} \left\{ \left[1 + \frac{1}{4} q \left(\frac{3r^2}{m^2} - \frac{6r}{m} + 2 \right) (3 \cos^2 \theta - 1) \right] \ln \left(1 - \frac{2m}{r} \right) \right.$$
$$\left. + \frac{3}{2} q \left(\frac{r}{m} - 1 \right) (3 \cos^2 \theta - 1) \right\}, \tag{20}$$

and[8]

$$\gamma \equiv \frac{9}{64} q^2 \left\{ \left(\frac{9r^4}{m^4} - \frac{36r^3}{m^3} + \frac{44r^2}{m^2} - \frac{16r}{m} \right) \ln^2 \left(1 - \frac{2m}{r} \right) \right.$$
$$\left. + \left(\frac{36r^3}{m^3} - \frac{108r^2}{m^2} + \frac{80r}{m} - 8 \right) \ln \left(1 - \frac{2m}{r} \right) + \frac{36r^2}{m^2} - \frac{72r}{m} + 20 \right\} \times$$
$$\times \cos^4 \theta + \left\{ \frac{9}{32} q^2 \left(- \frac{5r^4}{m^4} + \frac{20r^3}{m^3} - \frac{24r^2}{m^2} + \frac{8r}{m} \right) \ln^2 \left(1 - \frac{2m}{r} \right) \right.$$
$$+ \left[\frac{3}{2} q + \frac{9}{32} q^2 \left(- \frac{20r^2}{m^2} + \frac{40r}{m} - \frac{8}{3} \right) \right] \left(\frac{r}{m} - 1 \right) \ln \left(1 - \frac{2m}{r} \right)$$
$$\left. + 3q + \frac{9}{32} q^2 \left(- \frac{20r^2}{m^2} + \frac{40r}{m} - \frac{28}{3} \right) \right\} \cos^2 \theta - \frac{1}{2} (1 + 2q + q^2) \times$$
$$\times \ln \left[1 + \frac{m^2 \sin^2 \theta}{r^2 - 2mr} \right] + \frac{9}{64} q^2 \frac{r^4}{m^4} \left(1 - \frac{2m}{r} \right)^2 \ln^2 \left(1 - \frac{2m}{r} \right)$$

$$+ \left[\frac{1}{16} q^2 \left(\frac{9r^2}{m^2} - \frac{18r}{m} - 6\right) - \frac{3}{2} q\right]\left(\frac{r}{m} - 1\right) \ln\left(1 - \frac{2m}{r}\right)$$

$$+ \frac{9}{16} q^2 \left(\frac{r^2}{m^2} - \frac{2r}{m} - \frac{1}{3}\right) - 3q. \tag{21}$$

The Erez–Rosen mass and quadrupole parameters m and q are related to the corresponding Newtonian parameters M and Q by

$$m = GM, \qquad q = \frac{15GQ}{2m^3}. \tag{22}$$

The geodesic equations describing the motion of a test particle[9]

$$\frac{d}{d\tau}\left[g_{\mu\nu} \frac{dx^\nu}{d\tau}\right] = \frac{1}{2} \frac{\partial g_{\varrho\sigma}}{\partial x^\mu} \frac{dx^\varrho}{d\tau} \frac{dx^\sigma}{d\tau} \tag{23}$$

for the metric (19) are

$$\frac{d}{d\tau} [e^{2\psi} \dot{t}] = 0, \tag{24}$$

$$\frac{d}{d\tau}\left[e^{2\gamma - 2\psi}\left(1 + \frac{m^2 \sin^2\theta}{r^2 - 2mr}\right)\dot{r}\right]$$

$$= -\frac{1}{2}\left[\frac{\partial g_{00}}{\partial r} \dot{t}^2 + \frac{\partial g_{11}}{\partial r} \dot{r}^2 + \frac{\partial g_{22}}{\partial r} \dot{\theta}^2 + \frac{\partial g_{33}}{\partial r} \dot{\varphi}^2\right], \tag{25}$$

$$\frac{d}{d\tau} [e^{2\gamma - 2\psi} (r^2 - 2mr + m^2 \sin^2\theta)\, \dot{\theta}]$$

$$= -\frac{1}{2}\left[\frac{\partial g_{00}}{\partial \theta} \dot{t}^2 + \frac{\partial g_{11}}{\partial \theta} \dot{r}^2 + \frac{\partial g_{22}}{\partial \theta} \dot{\theta}^2 + \frac{\partial g_{33}}{\partial \theta} \dot{\varphi}^2\right], \tag{26}$$

$$\frac{d}{d\tau} [e^{-2\psi} (r^2 - 2mr) \sin^2\theta\, \dot{\varphi}] = 0, \tag{27}$$

where dots now denote differentiation with respect to the proper time τ, with $d\tau = (g_{\mu\nu}\, dx^\mu\, dx^\nu)^{1/2}$. Eqs. (24) and (27) have the first integrals

$$E = e^{2\psi} \dot{t}, \quad C = e^{-2\psi} (r^2 - 2mr) \sin^2\theta\, \dot{\varphi}, \tag{28}$$

where the constants E and C are relativistic analogues of the Newtonian energy and angular momentum per unit mass.

We now investigate the possibility of steady motion of the type considered in Section 2, restricting ourselves to the region $r > 2m$. Clearly

$\dot{r} = \dot{\theta} = 0$ will be first integrals of Eqs. (25) and (26) provided

$$\frac{\partial g_{33}}{\partial r} \dot{\varphi}^2 = - \frac{\partial g_{00}}{\partial r} \dot{t}^2 \tag{29$_o$}$$

and

$$\frac{\partial g_{33}}{\partial \theta} \dot{\varphi}^2 = - \frac{\partial g_{00}}{\partial \theta} \dot{t}^2, \tag{30$_o$}$$

where $\dot{\varphi}$ and $\dot{t}$ are now constant by Eq. (28). Eliminating $\dot{\varphi}$ and $\dot{t}$ from (30$_o$) by means of Eq. (28) we easily find

$$q = \frac{C^2 (r - m) e^{4\psi}}{[E^2 (r^2 - 2mr) \sin^2 \theta + C^2 e^{4\psi}] g \sin^2 \theta}, \tag{31$_o$}$$

where

$$g \equiv - \frac{3}{2} (r - m) \left\{ \frac{1}{2} \left(\frac{3r^2}{m^2} - \frac{6r}{m} + 2 \right) \ln \left(1 - \frac{2m}{r} \right) + \frac{3}{m} (r - m) \right\}. \tag{32}$$

It is easy to verify that $g > 0$ for all $r > 2m$ and thus q must be positive as in Newtonian mechanics.

From Eqs. (29$_o$) and (30$_o$) we get, using (19),

$$m = q [(g + 3h) \sin^2 \theta - 2h], \tag{33$_o$}$$

where

$$h \equiv \frac{1}{2} (r^2 - 2mr) \times$$

$$\times \left\{ \frac{3}{2m^2} (r - m) \ln \left(1 - \frac{2m}{r} \right) + \frac{m}{2 (r^2 - 2mr)} \left(\frac{3r^2}{m^2} - \frac{6r}{m} + 2 \right) + \frac{3}{2m} \right\}. \tag{34}$$

Thus we must have

$$\sin^2 \theta > \frac{2h}{g + 3h} \tag{35$_o$}$$

for m to be positive. From (32) and (34), this expression approaches the Newtonian value 2/5 as $r \to \infty$ and vanishes as $r \to 2m$. In addition, it may be shown[10] from Eqs. (32)–(34) that, for given m and q, $\sin^2 \theta$ approaches a minimum value zero with infinite slope as $r \to 2m$. For the Newtonian case, on the other hand, it follows from (8$_o$) that $\sin^2 \theta$ approaches its minimum value 2/5 with vanishing slope as $r \to 0$.

The functions g and h are everywhere positive, go to zero for $r \to \infty$, and approach their single maximum value for $r \to 2m$. This maximum equals $(1/2)\, m$ for h; for g it diverges as $-(3/2)\, m \ln (1 - 2m/r)$. It follows from Eq. ($33_o$) that for $q \ll 1$ we can have a real θ only for $r/m - 2 \ll 1$.

From (32) and (34) it follows that

$$g = -(r - m) \frac{\mathrm{d}h}{\mathrm{d}r}, \tag{36}$$

while the Taylor expansions for g and h (valid for $2m/r < 1$) show that for sufficiently large r

$$g = 2h. \tag{37}$$

Eqs. (36) and (37), together with the Taylor expansion for h, imply

$$h = \frac{m^3}{5\,(r - m)^2}. \tag{38}$$

The accuracy of the approximate expressions (37) and (38) improves rapidly with increasing r; at $r = 4m$, they are good to about five percent, while at $r = 10m$ they are already accurate to one-half of one percent.

For a given m and q, Eq. (33_o) determines a real r for any θ which satisfies (35_o). The motion takes place in a (coordinate) plane, a "distance" $d = r \cos \theta$ from the plane of symmetry, and with constant angular velocity $\omega \equiv \mathrm{d}\varphi/\mathrm{d}t$, where from ($33_o$), (28), and ($30_o$), and subject to ($35_o$),

$$d = \pm r \left[\frac{g + h - m/q}{g + 3h}\right]^{1/2}, \tag{39_o}$$

and

$$\omega = \left[\frac{g\,(g + 3h)\, \mathrm{e}^{4\psi}}{q\,(r^2 - 2mr)\,[(r - m)\,(g + 3h) - g\,(m + 2qh)]}\right]^{1/2}. \tag{40_o}$$

The maximum value of d equals

$$\mathrm{d}_{\max} = \pm \frac{2\,(g + 3h)^{1/2}\,(g + h - m/q)^{3/2}}{h'\,(2g - 3m/q) - g'\,(2h + m/q)}, \tag{41_o}$$

where primes denote differentiation with respect to r.

From (19), (28), (30_o), and (33_o) it follows that for circular motion in a plane parallel to (but not in) the plane of symmetry

$$E^2 = \frac{[(r - m)\,(g + 3h) - g\,(m + 2qh)]\, \mathrm{e}^{2\psi}}{(r - m)\,(g + 3h) - 2g\,(m + 2qh)} \tag{42_o}$$

and

$$C^2 = \frac{g\,(r^2 - 2mr)\,(m + 2qh)^2\,e^{-2\psi}}{q\,(g + 3h)\,[(r - m)\,(g + 3h) - 2g\,(m + 2qh)]}. \tag{43$_o$}$$

Thus, in order for both E and C to be real and finite, we must have

$$r - m > 2g\,(m + 2qh)\,(g + 3h)^{-1}. \tag{44$_o$}$$

It may be shown that, for a given m and q, this equation determines a unique minimum value of r_o within which circular motion is not possible, and whose value increases with q. It follows easily from Eqs. (44$_o$) and (38) that for small q this value must be close to $3m$. For $q \ll 1$, on the other hand, Eq. (33$_o$) demands that r_o be close to $2m$, as discussed before. Therefore, no circular motions parallel to the plane of symmetry are possible for very small q. Numerical evaluation of the equations shows that the limiting values of q and of the corresponding r are 2.2544 and 2.4481 m, respectively. It also follows from the limitation imposed on r that the value of $\sin^2\theta$ given by (33$_o$) cannot drop below its Newtonian minimum value 2/5.

For $q \gg 1$ we get from Eqs. (44$_o$) and (33$_o$), using (38),

$$r_{\min} = \tfrac{2}{5}\,(5q)^{1/3}\,m \tag{45$_o$}$$

and

$$\sin^2\theta_{\min} = \tfrac{2}{5}\,[1 + 2\,(5q)^{-1/3}]. \tag{46$_o$}$$

From Eq. (33$_o$), the radius of the orbit corresponding to $\theta = \pi/2$ is determined from

$$m = q\,(g + h). \tag{47$_o$}$$

Substituting Eqs. (32) and (34), we obtain

$$\frac{1}{3q} = \sum_{n=1}^{\infty} \frac{[(r/m) - 1]^{-2n}}{3 + 2n}, \tag{48$_o$}$$

from which we get

$$(r - m)^2 = \frac{3}{5}\,qm^2\left(1 + \frac{5^2}{3\times 7q} + \frac{5^3\times 4}{3^4\times 7^2 q^2} + \cdots\right), \tag{49$_o$}$$

where from (22) we have $(3/5)\,qm^2 = (9/2)\,Q/M$, which is the Newtonian value (18$_o$) for r^2. All radii corresponding to $\theta < \pi/2$ are smaller than the value obtained from (49$_o$). However, just as in the Newtonian case, within the plane of symmetry the radius is not restricted by this condition.

For comparison we now investigate the possibility of circular motion within the plane of symmetry $\theta = \pi/2$. We shall identify all equations for this motion by a subscript s on the number of the equation. From (19), (25), and (28) we find now

$$E^2 = \frac{(r - 2m + qh)\,e^{2\psi}}{r - 3m + 2qh} \tag{50$_s$}$$

and

$$C^2 = \frac{(m - qh)\,(r^2 - 2mr)\,e^{-2\psi}}{r - 3m + 2qh}. \tag{51$_s$}$$

Just as in the Newtonian case, solutions for circular motion within the plane of symmetry exist for both signs of q, but we shall discuss only those of positive q. Then since $r - 2m + qh > 0$ for all $r > 2m$ and $q > 0$, E and C can be real and finite only if

$$r - 3m + 2qh > 0 \tag{52$_s$}$$

and

$$m - qh > 0. \tag{53$_s$}$$

We note that for $q = 0$ we get $r > 3m$, which is precisely the condition obtained for the existence of circular motion in the Schwarzschild field[11]. For $q \ll 1$, Eqs. (52_s) and (38) give

$$r > 3m\left(1 - \frac{q}{30}\right), \tag{54$_s$}$$

and Eq. (53_s) is always satisfied. For $q \gg 1$, on the other hand, Eqs. (53_s) and (38) require

$$r > m\left[1 + \left(\frac{q}{5}\right)^{1/2}\right], \tag{55$_s$}$$

and (52_s) is always satisfied. Furthermore, for any value of $r \geqslant 2m$ one can find a range of values of q such that Eqs. (52_s) and (53_s) are satisfied; e.g. for $r = 2m$ we must have $1 < q < 2$.

The stability of the circular motions may be investigated exactly as in the Newtonian case, since the geodesic equations (23) follow from a variational principle with the Lagrangian[9]

$$L = -\frac{1}{2}\, g_{\mu\nu}\,\frac{dx^\mu}{d\tau}\,\frac{dx^\nu}{d\tau}. \tag{56}$$

We remove the cyclic coordinates φ and t by forming the Routhian

$$R = L - \frac{\partial L}{\partial \dot{t}}\,\dot{t} - \frac{\partial L}{\partial \dot{\varphi}}\,\dot{\varphi} = -\frac{1}{2}\left[g_{11}\dot{r}^2 + g_{22}\dot{\theta}^2 - \frac{E^2}{g_{00}} - \frac{C^2}{g_{33}}\right], \tag{57}$$

where E and C are given by (28). We again introduce (15) and proceed as in Section 2, keeping both E and C constant. For stability, the "modified potential"

$$U = -\frac{1}{2}\left[\frac{E^2}{g_{00}} + \frac{C^2}{g_{33}}\right] \tag{58}$$

must be a positive quadratic form in ε and η. For circular motion parallel to (but not in) the plane of symmetry a lengthy calculation shows this to require

$$\sin^2\theta > \frac{a + b - c}{d}, \tag{59$_o$}$$

where the quantities

$$\begin{aligned}
a &\equiv 2g^2(g + 3h)^3(r^2 - 2mr)(r - m)^2,\\
b &\equiv qg^2(m/q + 2h)^2(g + 3h)\{g(r^2 - 2mr)[3(r - m) + 2qg]\\
&\quad + 3(g + 6h)(r - m)^3\},\\
c &\equiv 2q^2g^3(m/q + 2h)^3[g(r^2 - 2mr) + (g + 6h)(r - m)^2],\\
d &\equiv (r - m)(g + 3h)^3\{g^2(r^2 - 2mr)[2(r - m) + 3qg]\\
&\quad + (g^2 + 6hg - 9h^2)(r - m)^3\},
\end{aligned} \tag{60$_o$}$$

are positive for all $r > 2m$. This expression approaches the Newtonian value $8/15$ as $m/r \to 0$ and vanishes as $r \to 2m$. It may be shown that, for a given m and q, (59$_o$) determines a unique minimum value of r_o within which *stable* circular motion is not possible, and whose value increases with q. In addition, numerical evaluation of the equations shows that the limiting values of q and r for stable circles are given by $q = 24.2333$ and $r = 4.9061\, m$. In all cases, the stability condition imposed by (59$_o$) and (60$_o$) is everywhere more restrictive than the existence condition following from Eq. (44$_o$). For extremely large q, the minimum radius for stable orbits and the corresponding azimuth are given by

$$(r - m)^2 = \frac{2qm^2}{15}\left[1 + 16\left(\frac{15}{2q}\right)^{1/2}\right] \tag{61$_o$}$$

and

$$\sin^2 \theta = \frac{8}{15}\left[1 + 4\left(\frac{15}{2q}\right)^{1/2}\right]. \quad (62_o)$$

The stability within the plane of symmetry can be investigated similarly, and leads to the two conditions

$$2qh\,[9mqh + 8mr - 13m^2 - (r + qh)(r + 2qh)] + 2m(r - 3m)(r - 2m) + (r^2 - 2mr)\,[q(g + h) - m] > 0, \quad (63_s)$$

and

$$m - q(g + h) > 0. \quad (64_s)$$

For $q = 0$, we find that $r > 6m$, which is the stability condition obtained for motion in the Schwarzschild field[11]. For $q \ll 1$, Eqs. (63_s), (37), and (38) give

$$r > 6m\left(1 - \frac{7q}{300}\right), \quad (65_s)$$

and (64_s) is always satisfied. For $q \gg 1$, on the other hand, (63_s) is always satisfied, and the limit determined by Eq. (64_s) precisely equals that following from Eq. (47_o) for the maximum radius of orbits parallel to the plane of symmetry; this limit is given by Eq. (49_o). In fact, these two limits coincide for all $q > 2.2544$ for which circular non-coplanar motion exists. Therefore the minimum radius for a stable orbit within the plane of symmetry is larger than the maximum radius for such orbits outside it.

4 DISCUSSION

In the preceding sections we have shown that there exist exact solutions for circular motion of a particle in a monopole-prolate quadrupole field in planes parallel to the plane of symmetry of the quadrupole. These solutions exist for all q and arbitrarily small r in Newtonian mechanics and for values of r down to $2.4481\,m$ [the lower limit being determined by Eq. (44_o)], and of $q > 2.2544$, in general relativity[12]. Both relativistic limitations arise from the requirement that the orbital velocity cannot exceed the local velocity of light.

In both cases the azimuth θ of the circular motion must exceed a minimum value. For the Newtonian case this is given by $\sin^{-1}(2/5)^{1/2}$ from

Eq. (9_o). In the general relativistic case the equations lead to an r-dependent condition on the azimuth which is everywhere more restrictive than the Newtonian one.

In both cases r increases with θ, and reaches its maximum value at the plane of symmetry. Within this plane circular orbits are possible for all $r > (q/5)^{1/2}\, m$ in Newtonian mechanics and for all r greater than a value depending on q [determined by Eqs. (52_s) and (53_s)] in general relativity; for $1 < q < 2$, r can reach the minimum value $2m$.

Within the plane of symmetry the Newtonian orbits are stable for all $r > (3q/5)^{1/2}\, m$, and the relativistic ones for all r greater than a value depending on q [determined from (63_s) and (64_s)], but in any case not smaller than $r = 2.4481\, m$. For $q = 0$, both the existence and the stability limits, $r = 3m$ and $r = 6m$, respectively, agree with those of the orbits in a Schwarzschild field[11].

Outside the plane of symmetry, the Newtonian orbits are stable for all possible r down to $r = (2q/15)^{1/2}\, m$, and the relativistic ones for all possible r down to $2.4481\, m$, as discussed above. Thus in both cases stable orbits of smaller radius are possible outside the plane of symmetry than within. Also, in both cases, the orbits outside the plane are stable up to a maximum value of the radius reached at the plane of symmetry, which coincides exactly with the minimum value of the radius for stable circular orbits within that plane. In the relativistic case, circular orbits are possible only in this plane for $q < 2.2544$, as noted earlier.

The analytical expressions obtained in Section 3 for the limits imposed on r and θ are of such complexity that they allow ready evaluation only for very small or very large q; the results for these cases were summarized above. For intermediate q, it appears that one must resort to numerical methods[13].

It was irrelevant for our calculations whether the fields used were those of point particles, or fields exterior to a body of finite extension. However, for a discussion of the range of validity of our results we must distinguish between these two situations. For point particles our results are valid up to $r = 0$ in the Newtonian case, and up to $r = 2m$ in the general relativistic one. For an extended body of mass M, on the other hand, the quadrupole moment is generally much smaller than MR^2, where R is some average radius of the body. Thus even the maximum radius of the orbit, given by Eq. (18_o) or (49_o), generally would place it in the interior of the body; one could only expect it to be exterior to the body for mass distributions which deviate very much from sphericity. If this is the case, however, higher multipole moments would have to be taken into account.

The Newtonian solutions were obtained by elementary methods. Inspection of Eqs. (2)–(7_0) shows that in addition to an attractive radial force, they require only the presence of a θ-component of the force directed away from the plane $\theta = \pi/2$. Nothing specific to gravitation, nor to the form of the quadrupole potential, is essential. Thus the generalization to an electric monopole-quadrupole field is trivial, as is the treatment of 2^n-pole fields for reasonably small n. It appears that similarly one can find exact solutions for circular motion outside the plane of symmetry for the metrics given by Erez and Rosen for higher multipoles with reasonable ease.

In the Newtonian case, the condition for stability in the monopole-quadrupole field could be obtained easily. However, the extension to motions in higher multipole fields rapidly leads to algebraic equations in r and $\sin^2 \theta$ of too high an order to permit easy conclusions. In the general relativistic case, even the quadrupole calculations are quite lengthy, and the stability conditions for motions in fields of higher multipoles can be expected to lead to unmanageable expressions.

It appears that the solutions given here are the only ones possible which are plane, apart from motions in the plane of symmetry or in planes containing the axis of symmetry, which will be considered elsewhere.

All numerical calculations for this paper were done by computer. The authors are indebted to Professor Leonard Auerbach for his advice and help in writing the programs.

REFERENCES

1. J. DROSTE, *Versl. K. Akad. Wet. Amsterdam* **25,** 163 (1916); [English translation: *Proc. K. Akad. Wet. Amsterdam* **19,** 197 (1916)]; C. DE JANS, *Acad. Roy. Belg. Mém.,* 8° (2) **7**, fasc. 5 (1923); Y. HAGIHARA, *Jap. J. Astr. Geophys.,* **8**, 67 (1931); B. MIELNIK and J. PLEBAŃSKI, *Acta Phys. Polon.,* **21**, 239 (1962); A. W. K. METZNER, *J. Math. Phys.,* **4**, 1194 (1963). For a very detailed review of the problem and an extensive bibliography see H. ARZELIÈS, *Relativité Généralisée. Gravitation,* Vol. II, Gauthier-Villars, Paris, 1963, Chapter VII.
2. P. SCONZO and J. BENEDETTO, I.B.M. Publication, *Task* #0232-G (1965) (unpublished). It is remarkable that the solutions were found by making use of the mathematical equivalence of the Newtonian equations of motion in the plane of symmetry with the general relativistic equations in the Schwarzschild field. Special combinations of multipole fields, for which exact solutions can be found also outside the plane of

symmetry, have been investigated by J. P. Vinti, *J. Res. N.B.S.* **62 B,** 105 (1959). These can be shown to correspond to the field of two fixed monopoles, which was studied earlier by J. Weinacht, *Math. Ann.* **91,** 279 (1923).

3. G. Erez and N. Rosen, *Bull. Res. Council Israel,* **8F** , 47 (1959).
4. H. Weyl, *Ann. Phys.*, **54**, 117 (1917).
5. D. M. Zipoy, *J. Math. Phys.*, **7**, 1137 (1966).
6. D. Brouwer and G. M. Clemence, *Methods of Celestial Mechanics*, Academic Press, New York and London, 1961 and references given there; W. D. MacMillan, in F. R. Moulton and collaborators, *Periodic Orbits*, Carnegie Inst., Washington, 1920, pp. 99–150; D. Brouwer, *Astron. J.*, **51**, 223 (1946).
7. E. T. Whittaker, *Analytical Dynamics*, 4th ed., Cambridge University Press, Cambridge, 1937, Chapter VII.
8. The expression for γ given in ref. 3 is not quite correct, as already noted by A. G. Doroshkevich, Ya. B. Zel'dovich and I. D. Novikov, *J. Exptl. Theoret. Phys.* (U.S.S.R.), **49**, 170 (1965); [English translation: *Soviet Physics JETP*, **22**, 122 (1966)]. Unfortunately, their corrected expression contains a misprint; we hope that ours won't. The actual form of γ does not enter any of the calculations of our paper, but is given for reference only.
9. See e.g. C. Møller, *The Theory of Relativity*, Oxford University Press, Oxford, 1952, § 86. The constant $-\frac{1}{2}$ was inserted in the definition (56) to reduce Eq. (57) to (12) in the Newtonian limit.
10. For the details of this and some of the subsequent calculations see A. Armenti, jr., Temple University Thesis, Philadelphia, Pa., 1970 (unpublished). It should be noted that there the opposite signature was used for the metric, as well as a different definition for the relativistic Lagrangian.
11. J. Droste, ref. 1; K. Schwarzschild, *Sitzber. preuss. Akad. Wiss., Physik-math. Kl.* 189 (1916); C. de Jans, ref. 1, § 24–28; P. Goldhammer, *Nuovo Cimento* **20**, 1205 (1961).
12. To avoid misunderstandings, it should be recalled that we are using spherical coordinates throughout; thus r is the distance to the origin and (except in the plane of symmetry) *not* the radius of the circle. Furthermore, in this section all results obtained in Section 2 are converted to the units used in Section 3.
13. It might be worth noting that in the units used here the quadrupole moment of a celestial body would be very large indeed; e.g. on the basis of a homogeneous nonrotating model, the oblateness reported for the sun [R. H. Dicke and H. M. Goldenberg, *Phys. Rev. Lett.*, **18**, 313 (1967)] corresponds to $|q| = 2.67 \times 10^7$.

PAPER 2

Does a gravitational field influence chemical equilibria?

NANDOR L. BALAZS

Department of Physics, State University of New York, N.Y., Stony Brook U.S.A.

ABSTRACT

We show here that (a) the usual thermodynamical equilibrium conditions for chemically reacting mixtures in the presence of a gravitational field are incorrectly given if the rest masses change; (b) the correct conditions coincide (up to small terms) with the conditions derived from relativistic thermodynamics; (c) these conditions imply that the chemical equilibrium constant obtained for ideal gas reactions depends on the gravitational potential.

I

The classical argument runs as follows. Consider the thermal equilibrium conditions derived from an extremal principle. Let $\tilde{u}\,(s, n_1, n_1, \ldots, n_i, \ldots)$ be the internal energy density, excluding the rest energy density, where s is the entropy density, and n_i the number density of species i; ϕ is the gravitational potential. Then, according to Gibbs that state realizes itself, for which the variation of the total energy is an extremum, keeping the total entropy fixed. The variation of n_i is subject to different conditions, depending whether chemical transformations can occur or not. Let us leave these restrictions open for the moment. Then we find that in equilibrium the following conditions must be satisfied:

$$\delta\,[\int \tilde{u}\,\mathrm{d}\tau + \Sigma \int n_i m_i \phi\,\mathrm{d}\tau - \lambda \int s\,\mathrm{d}\tau] = 0; \qquad (1)$$

2 Kuper/P

i.e.

$$(\partial \tilde{u}/\partial s)_{n_i} = \lambda, \tag{2}$$

$$\Sigma \int (\partial \tilde{u}/\partial n_i + m_i \phi)\, \delta n_i \, d\tau = 0, \tag{3}$$

(the integration extends over the system). Eq. (2) gives the condition that the absolute temperature $(\partial \tilde{u}/\partial s)_{n_i}$ must be independent of the position within the system (λ is a constant Lagrange multiplier). If *no* chemical reaction is possible $\delta \int n_i \, d\tau = 0$ for all i. Adjoining these conditions with the Lagrange multipliers σ_i we find from (3)

$$(\partial \tilde{u}_i/\partial n_i)|_s + m_i \phi + \sigma_i = 0; \tag{4}$$

$\partial \tilde{u}_i/\partial n_i$ is the usual chemical potential $\tilde{u}_i$ of species i. For an ideal gas (4) gives the usual barometric formula. If chemical reactions are possible the number of particles of each constituent is not conserved any more; the change in the number of particles of the different specimens are not independent, since these changes must follow the reaction equation

$$\Sigma \nu_i C_i = 0, \tag{5}$$

where C_i is the name of the i-th species, and ν_i is the stochiometric coefficient of the i-th species. From this it follows that

$$\delta n_1/\nu_1 = \delta n_2/\nu_2 = \cdots = \delta n_i/\nu_i = \delta \xi,$$

or

$$\delta n_i = \nu_i \delta \xi$$

for all i where $\delta \xi$ is independent of i. With this assignment of the variations, we get from (3)

$$\Sigma \nu_i (\tilde{u}_i + m_i \phi) = 0. \tag{6}$$

From this one usually concludes that if there is no rest mass variation in the reaction $\Sigma \nu_i m_i = 0$, the gravitational field does not influence the reaction; thus under these conditions in a reaction among ideal gases the equilibrium constant is independent of height. If on the other hand there is a rest mass variation we seem to get the result that the gravitational field will influence the reaction. This result however is in conflict with the result obtained by Tolman[1] according to which the chemical equilibria "between reacting substances will be characterized by the same conditions—measured by a local observer—as would be calculated on a classical basis", i.e. we ought to get a gravitational-potential-independent answer. Can we resolve this conflict?

II

Before we consider the general case let us work out a specific example to find out where the truth lies. Consider a thermal-equilibrium system of two-state atoms in a radiation field under the influence of a gravitational field with the potential ϕ. (All our calculations are for weak fields to bring out the physical content; they can be easily generalized for strong fields as well.) In thermal equilibrium the temperature will not be constant. If the temperature at sea level is given by T_0, (where $\phi = 0$) at another level one finds

$$T = T_0 (1 - \phi/c^2). \tag{7}$$

Now consider the two-state atoms present, both at sea level and at the level ϕ. They will be in thermal equilibrium with the radiation field. Let n_e and n_g be the densities of the excited and unexcited atoms, respectively. The frequency of the light quantum absorbed in an excitation should be denoted by ω' at sea level. Will the ratio $n_e/n_g = f$ vary with height? Let us look first at the radiation field. At sea level the radiation intensity will be distributed according to the Planck curve with respect to the temperature T_0. At a higher level the Planck curve will shift toward smaller frequencies, each frequency with its corresponding red shift. (In addition the area will be smaller since the energy density is also less.) Now at sea level the operating frequency is ω', and for definiteness let it be above the maximum. What will be the operating frequency at the higher level? At first sight it would seem as if this frequency would also be red shifted. This is however wrong. The operating frequency is the same, and precisely by virtue of this fact does one notice a red shift. A quantum emitted at sea level will arrive with a shifted frequency and by comparing this with the unaltered operating frequency at the higher level one notices a red shift. This being the case, the energy density of the radiation field at the operating frequency will be different at the two heights. In our case, because one picked the frequency ω' above the maximum at sea level, the energy density at the operating frequency will be less at the higher level. For this reason the ratio of excited to unexcited atoms will also shift. This can be seen as follows. At any level the rate of emission and the rate of absorption will be given by $w_e = Q(n_{\omega'} + 1)$, $w_a = Qn_{\omega'}$ where $n_{\omega'}$ is the number of photons at the operating frequency emitted in a given solid angle; Q is a factor which may be ϕ dependent as far as *this* argument is concerned. Balancing the rate of emission $n_e w_e$ with the rate of absorption $n_g w_a$ we immediately find that Q cancels, and n_e/n_g is given by $n_{\omega'}/n_{\omega'} + 1$. Since $n_{\omega'}$

varies with height so will $n_e/n_g = f$. Since $n_{\omega'}/n_{\omega'} + 1 = \exp(-\hbar\omega'/kT)$ with $T = T_0(1 - \phi/c^2)$, we get

$$f(\phi) = f(0)\,\mathrm{e}^{-\hbar\omega'\phi/kT_0c^2}. \tag{8}$$

Thus the equilibrium constant for *this* reaction varies with height. We expect that this will be true in general.

III

We will show now the following: (a) The general theory gives an equilibrium condition which has the same form as in the absence of a gravitational field, bearing out Tolman's contention; (b) notwithstanding this the chemical equilibrium is influenced by the presence of the gravitational field; (c) the classical result (6) is incorrect if the rest mass can change in the reaction. Thus there is no conflict between the correct classical expression and Tolman's condition. The general conditions can be immediately derived. Let $u = \tilde{u} + \Sigma n_i m_i c^2$ be the energy density including the rest-mass density. Then the general equilibrium conditions for weak fields can be written as

$$\delta\left[\int u(1 + \phi/c^2)\,\mathrm{d}\tau - \lambda \int s\,\mathrm{d}\tau\right] = 0, \tag{9}$$

with suitable restrictions on the variations of the quantities n_i. Performing the variations we get

$$(\partial u/\partial s)(1 + \phi/c^2) = \lambda, \tag{10}$$

$$\Sigma \int (\partial u/\partial n_i)(1 + \phi/c^2)\,\delta n_i\,\mathrm{d}\tau = 0. \tag{11}$$

Eq. (10) gives the answer that the temperature varies according to the law described in II. If there are no chemical reactions (11) gives

$$\mu_i(1 + \phi/c^2) + \sigma_i = 0. \tag{12}$$

If chemical reactions occur we get

$$\Sigma\nu_i\mu_i(1 + \phi/c^2) = 0,$$

or

$$\Sigma\nu_i\mu_i = 0. \tag{13}$$

Thus any explicit ϕ-dependence cancels. However, the temperature is still ϕ dependent, which will make the equilibrium concentrations ϕ-dependent. This settles point (a) and (b).

The classical result (6) is incorrect for the following reason. If the rest masses can vary we must include in (1) the extra term $\int \Sigma n_i m_i c^2 \, d\tau$ under the variation, to measure the energy of each constituent from a common zero. In place of (6) we now get

$$\Sigma \nu_i (\tilde{\mu}_i + m_i \phi + m_i c^2) = \Sigma \nu_i [\mu_i (1 + \phi/c^2) - \tilde{\mu}_i \phi/c^2] = 0, \qquad (14)$$

which coincides with (13) up to the small term $\tilde{\mu}_i \phi/c^2$. Thus the classical result (6) is incorrect, while of course the classical result (4) giving rise to the barometric formula is unaffected.

We can now verify (8). The chemical potential of the two-state atom will be given for the ground state and the excited state as follows:

$$\mu_g = RT \log n_g + f(T) + m_0 c^2; \qquad (15)$$

$$\mu_e = RT \log n_e + f(T) + \hbar\omega' + m_0 c^2. \qquad (16)$$

From the condition $\Sigma \nu_i \mu_i = 0$ it immediately follows that

$$n_e/n_g = e^{-\hbar\omega'/RT} = (n_e/n_g)_0 \, e^{-\hbar\omega'\phi/RT_0 c^2},$$

which is the same as (8).

I express my gratitude to the Organizing Committee for their kind invitation; my thanks are also due to the State University of New York and The National Science Foundation for their partial financial support.

REFERENCE

1. R. C. Tolman, *Relativity Thermodynamics and Cosmology*, Clarendon Press, Oxford, 1935.

PAPER 3

Status of canonical quantization*

PETER G. BERGMANN
Syracuse University, N.Y., U.S.A.

ABSTRACT

Various approaches to canonical (local) quantization are related to each other, and their current status evaluated. Beginning with a review of Hamiltonian theory with first-class constraints this review discusses Schrödinger quantization, Hamilton–Jacobi theory (with and without superspace), Feynman integrals, and the sandwich conjecture, and concludes with some speculations of a general nature.

INTRODUCTION

A number of groups throughout the world have now persisted for almost two full decades in their attempts to quantize the gravitational field that is described classically by Einstein's field equations. Every now and then one group or the other has claimed definitive success in this endeavor. If I have remained a skeptic, it is, perhaps, that I expect certain things of the quantum theory of gravitation, and that the original expectations of others have been different. My original purpose was to attempt a truly intimate fusion of the quantum concept with the invariance group that is usually called the principle of general covariance. The resulting theory would give us answers to such questions as the nature of a fully quantized geometry of space-time, the role of world points in this geometry, the "softening-up the light cone", which was postulated occasionally by Pauli, and the effect of this "softening" not only on the divergences associated with the gravitational field but with all

* This work has been partially supported by the Aerospace Research Laboratories and Office of Scientific Research, both of the United States Air Force.

other fields as well. All this was to happen in a theory that was fully elaborated, not a speculative extrapolation of the weak-field or any other approximation running counter to the spirit of general covariance. Alas, I am afraid that we are a long way from this goal, though I am not quite certain what metric to employ within the topological space of all quantum theories of gravitation, with which to measure the distance yet to be traversed by weary theorists.

In this talk I shall confine myself to a review of attempts at quantization that are based on Dirac's classical canonical (Hamiltonian) formulation of Einstein's theory of the gravitational field. This formulation begins with the choice of six variables g_{mn} and six canonically conjugate momentum densities p^{mn} as the primary variables in terms of which the field is to be described. Even on one space-like three-surface the values of these variables cannot be prescribed freely; they are constrained by four relationships at each point of the three-surface, the so-called Hamiltonian constraints, H_s and H_L. The generator of an infinitesimal propagation to a neighboring three-surface, that is to say the Hamiltonian of the theory, is an integral over the three-surface of a linear combination of the Hamiltonian constraints. The coefficient of H_L is the measure for the normal distance between the initial and the follow-up surfaces, whereas the coefficients of the three H_s relate the sideways displacements of points with identical coordinate values on the two surfaces.

Dirac's formulation, in this pristine version, is free of coordinate conditions. The full covariance of the theory is expressed by the lack of restrictions on the coefficients of the four constraints in the Hamiltonian, permitting arbitrary normal and sideways displacements in the progress from one three-surface to the next. Covariance is assured further by the first-class character of the constraints themselves. According to Dirac a first-class variable is one whose Poisson brackets with all constraints are themselves constraints (i.e. vanish if the constraints are satisfied). Poisson brackets between first-class variables are themselves first-class variables; hence the Poisson brackets between first-class constraints are also first-class constraints. Thus any theory with first-class constraints incorporates two distinct Lie algebras; one generated by all the first-class variables, the other, a normal sub-algebra of the first, generated by the first-class constraints. And finally, there exists a factor algebra, which is the quotient between the former and the latter Lie algebras. All these play roles in the quantization programs that are based on the canonical formalism.

In any program of canonical quantization, the resulting quantum theory is as "local" as is the preceding classical formalism. The quantization procedure is based on the premise that the Lie algebra of quantum commutators possesses certain analogies to the Lie algebra of Poisson brackets, though one cannot be a straightforward isomorphism, or even homomorphism, of the other. Rather, one attempts to recover in the quantum theory certain aspects of the classical Lie algebra that one considers of particular relevance for the physical interpretation. Further, one requires that quantum commutators should differ from corresponding Poisson brackets by expressions that contain a positive power of Planck's constant $\hbar$, so as to assure correspondence between classical and quantum theories in the so-called shortwave limit.

In what follows I shall comment on Hamilton–Jacobi theories, Schrödinger theories, the role of superspace, and on Feynman integrals using the notion of superspace. I believe that in some respects the different approaches "agree more than you think".

HAMILTONIAN THEORY WITH FIRST-CLASS CONSTRAINTS

In classical mechanics Hamiltonian theory is most appropriate for the formulation of the Cauchy problem: In phase space the trajectories form a congruence of curves, so that through each point passes exactly one of them. The time coordinate t is entirely separate from the configuration coordinates q_k, and it is not transformed in the course of canonical transformations, though it may enter into the transformation equations of the q's.

An apparent symmetry between the q's and t may be produced by a trick. If t be replaced in its role as the independent variable by an *ad hoc* parameter θ along each trajectory, one can construct a new phase space, larger by two dimensions than the original phase space, whose canonical coordinates are the original q_k, p_k $(k = 1, \ldots, n)$, plus q_{n+1} $(= t)$ and p_{n+1} $(= -H)$. These $2(n + 1)$ coordinates are restricted to a *constraint hypersurface* within the *extended phase space*, by the *Hamiltonian constraint*,

$$\bar{H} \equiv H(q_1 \cdots q_n, p_1 \cdots p_n, t) + p_{n+1} = 0. \quad (1)$$

Multiplied by an arbitrary coefficient $a(q, p, t)$, the constraint $\bar{H}$ will serve as the Hamiltonian as well. If a trajectory is begun on the constraint hypersurface, it will lie wholly on that hypersurface by virtue of the equations of motion. Choice of the coefficient $a(q, p, t)$ determines the parametrization of the trajectories, up to a constant of integration, as it equals the rate of change of t, $(dt/d\theta)$.

In the presence of the single Hamiltonian constraint (1), Dirac's first-class variables are simply what one usually refers to as constants of the motion. Canonical mappings generated by first-class variables are the only ones that map the constraint hypersurface onto itself and, incidentally, trajectories onto trajectories. As mentioned before, these mappings form a Lie algebra, and hence an (infinitesimal) group. As the first-class variables may also be characterized as those variables that are constant along any one trajectory, they are all functions defined on a *reduced phase space*, whose points are those trajectories lying on the constraint hypersurface of the extended phase space. The Poisson brackets of the reduced phase space are identical with those formed between first-class variables with the help of all the $2(n + 1)$ coordinates of the extended phase space. As will be discussed in a later section, the introduction of canonical coordinates appropriate to the reduced phase space is accomplished by means of Hamilton's principal function.

The transition from the ordinary phase space (or rather, from its product by the time axis) to the extended phase space, and thence via the constraint hypersurface to the reduced phase space, converts the Hamiltonian formalism from one in which the time coordinate is singled out to one in which the time plays a role analogous to the configuration coordinates. That is why this reformulation is useful for relativistic theories, in which the symmetry between space and time coordinates is to be stressed.

Within the reduced phase space the whole history of a dynamical system is reduced to a single point; to this extent the sense of unfolding of the trajectory in the course of time is lost. Alternative coordinate systems used to coordinatize the reduced phase space correspond to alternative methods of identifying a total dynamical system by means of sufficient but not redundant data, such as Cauchy data at different times, or a complete set of constants of the motion. Description within the reduced phase space has been both characterized and criticized as a "frozen formalism". With exactly the same right one might criticize the description of a mechanical system by means of a single representative point in gamma-space as failing to exhibit the rich diversity of a many-body system. To my mind it is not a matter of principle but of convenience whether one goes from mu-space into gamma-space, and whether one proceeds from the extended towards the reduced phase space. In both instances, no valid information is discarded; it can always be recovered if desired.

In general relativity there is not one Hamiltonian constraint but infinitely many, four per space point (in Dirac's formalism). Whereas in classical mechanics one Hamiltonian constraint suffices to propagate the system from one instant in time to the next one, in general relativity there are infinitely many different ways to proceed from one coordinatized three-surface in space-time to a neighboring one, and all the constraints of the theory are required to do justice to this richness. As the constraints are all first-class, they form a Lie algebra, which corresponds to the commutator algebra of infinitesimal coordinate transformations.

Propagation from one three-surface to another, infinitesimally close to the first, with the two surfaces coordinatized so that the points on the two surfaces with equal coordinate values $(x^1 \cdots x^3)$ are in infinitesimal proximity to each other, is, of course, to be interpreted as a species of infinitesimal coordinate transformation. And this is why one cannot distinguish between the Hamiltonian (or, perhaps, better: Hamiltonians) that generates propagation from three-surface to three-surface and the generator(s) of infinitesimal coordinate transformations. First-class variables in general relativity, that is to say dynamical variables commuting with all constraints, are both invariants under infinitesimal coordinate transformations and constants of the motion. Whereas the extended phase space in Dirac's formulation is the function space of $g_{mn}(x^1 \cdots x^3)$, $p^{mn}(x^1 \cdots x^3)$, the constraint hypersurface consists of those fields satisfying the four constraints $H_s = 0$, $H_L = 0$ at every space point. Regardless of the particular choice of Hamiltonian, that is to say, of the coefficients ξ^s, ξ^L,

$$H = \int d^3x\,(\xi^s H_s + \xi^L H_L) \tag{2}$$

the constraints are preserved along any trajectory that is begun on the constraint hypersurface. Intuitively, a trajectory on the constraint hypersurface consists of a one-parametric succession of three-surfaces, which together form a congruence of space-like hypersurfaces in a Ricci-flat space-time manifold. Given a particular Ricci-flat space-time, there is, of course, an infinity of such congruences. Furthermore, given a segment of such a congruence, the succession may be continued in an infinity of ways. This is because the coefficients of the constraints appearing in the Hamiltonian (2) may be chosen at will as functions over the space-time manifold. And one may come to the conclusion that the one-dimensional trajectory on the constraint hypersurface does not represent the most natural representation of the underlying dynamics, but that the appropriate infinite-dimensional structure consists of all points on the constraint hypersurface that are accessible to each other by way of some trajectory generated by a Hamiltonian of the form (2). This structure is usually referred to as the *equivalence class* of all those points on the constraint hypersurface that belong to the same Ricci-flat manifold, to the same physical situation. The distinct points that belong to the same equivalence class correspond to all possible sets of Cauchy data that identify that manifold.

The reduced phase space is a symplectic manifold whose points each represent a whole equivalence class on the constraint hypersurface. It is that manifold on which all first-class variables (= invariants = constants of the motion) are defined. In other words, first-class variables are those variables that are constant within each equivalence class. Again, the Poisson bracket of two variables in the reduced phase space is the Poisson bracket between the corresponding first-class variables in the extended phase space. In the reduced phase space the constraints are empty: Every point in the reduced phase space corresponds to a distinct Ricci-flat space-time manifold.

SCHRÖDINGER THEORY

Ordinarily quantum mechanics is not concerned with the incorporation of constraints in the theory. As there is no *a priori* method for dealing with them, one might consider various approaches, such as setting them zero outright, or merely requiring their expectation values to vanish. A good deal depends on the physical significance of the constraint. If a top is constrained by bearings, these bearings might in turn be thought of as subject to quantum rules, and hence to uncertainty relations. But if a classical theory is put forward in a form in which Cauchy data in violation of the constraints are to be considered unphysical, one might prefer to exclude "unphysical" states from the quantum theory as well.

Guidance is provided by the parametrized formulation of classical mechanics that results in the Hamiltonian constraint (1). The configuration space in this formulation is $(n+1)$-dimensional, with the coordinate q_k, t. If we compare the classical Eq. (1) with the standard form of the Schrödinger equation, we find that the latter may be cast in the form

$$\bar{H}|\rangle = 0. \tag{3}$$

Only quantum states satisfying this stringent constraint condition represent acceptable histories.

Adopting a wave-mechanical representation one may interpret the Schrödinger equations as a condition on wave functions defined on the $(n+1)$-dimensional configuration space. These wave functions form a linear vector space, and those among them satisfying Eq. (3) a linear subspace. The operator p_{n+1}, for one, and its wave-mechanical representation, $(\hbar/i)(\partial/\partial t)$, are well-defined operators in the enlarged linear vector space. The $2(n+1)$ canonical variables also satisfy standard commutation relations. But this function space is not a Hilbert space, in that it is not possible to define on it a useful metric, and the operators defined on it do not represent observables. Wave functions obeying Eq. (3) obviously cannot be square-integrable over the $(n+1)$-dimensional configuration space. And at least some of the operators, such as t, cannot have physically sensible expectation values. Moreover, the space of all functions of q_k, t contains almost all functions not corresponding to physically permissible states, and these functions should not be endowable with non-zero probabilities of existence.

For this matter, even a variable such as q_k does not have an expectation value, but only the variable $q_k(t)$ at a particular stated time t. The normal procedures of quantum mechanics are applicable, not to the space of functions of q_k, t, but to the subspace of functions obeying Eq. (3). In this subspace we can define, in the usual manner, a norm, expectation values of observables, etc. The transition from the space of functions of q_k, t to the subspace of functions obeying the Schrödinger equation (3) is analogous to the transition in classical mechanics from the extended to the reduced phase space. The reduced phase space may, for instance, be coordinatized by the coordinates of system trajectories at a fixed time t_0, and these will be canonical coordinates. Likewise, a Schrödinger wave function is completely determined by its form at the fixed time t_0. Any observable at an arbitrary time t may be expressed in terms of the observables $q_k(t_0)$, $p_k(t_0)$, and its expectation value be formed accordingly. This procedure is, of course, nothing but the transition to the Heisenberg picture.

If the original function space is discarded altogether and replaced by the Hilbert space of states which, in the large vector space, obey Eq. (3), then we are replacing, in effect, the customary formulation of quantum mechanics by one based on the classical foundation of the reduced phase space, and free of any direct reference to time development. In such a formulation the Schrödinger and Heisenberg pictures coalesce into one, Eq. (3) is empty, the Hamiltonian vanishes identically, and all observables are automatically constants of the motion. This is the extreme form of a "frozen formalism" in quantum theory. A return to the more customary versions of the theory is possible if the Hamiltonian, and hence the unitary matrix $U(t)$,

$$U(t) = \exp\left(-\frac{i}{\hbar} Ht\right), \tag{4}$$

is known.

A similar program of quantization may be applied to general relativity. Starting with the linear vector space of functionals of the field $g_{mn}(x^1 \cdots x^3)$, one may introduce the operators of multiplication by $g_{mn}(x_0)$ and $p^{mn} = (\hbar/i)(\delta/\delta g_{mn})$, which satisfy standard commutation relations for the components of a field. These operators are defined on a functional space without metric, and they have no expectation values. Physically possible states are represented by functionals that satisfy the $4 \times \infty^3$ constraints

$$H_s(x^1 \cdots x^3)|\rangle = 0, \quad H_L(x^1 \cdots x^3)|\rangle = 0. \tag{5}$$

In view of the fact that the classical expressions for the constraints are, respectively, linear, and inhomogeneous-quadratic in the momentum densities, and highly non-linear in the components of the three-metric, the ordering of factors in the operators appearing in Eq. (5) is not straightforward. In fact, attempts to arrange the factors so as to reproduce the Lie algebra of the infinitesimal coordinate transformations generated by these constraints have not been successful, and there is some indication that no such ordering exists. On the other hand, g_{mn} and p^{mn} are not observables, and it is not clear that it is ever necessary to construct the constraint operators (5) in the functional space in which alone they are non-trivial.

Leaving the resolution of these questions to the future, we may remark that the constraint equations (5) imply that permissible functionals, those in the reduced linear vector space, are those that are invariant under infinitesimal coordinate transformations. That is to say, they are functionals of three-dimensional metric fields that remain unchanged both in form and in value under coordinate transformations within the three-surface and under transitions to neighboring three-surfaces imbedded in the same Ricci-flat manifold.

Again, if the original functional space is discarded and only the linear subspace defined by Eq. (5) retained, Hermitian operators mapping that subspace on itself may represent observables, which, just as in the classical theory, are necessarily constants of the motion as well as invariant under coordinate transformations. Only the linear subspace (5) is metrizable in a physically meaningful manner, though it is undoubtedly not separable.

Much about this scheme of quantization is persuasive, such as that a metric should be defined only among state vectors that satisfy the constraints; that the only observables are constants of the motion, and that they are also represented by operators that map the L_2-space of permissible state vectors on itself; that the notions of Hermitian and of unitary operators, confined as they are to metric spaces, apply only to operators defined on the state space (5); and finally that the construction of the enveloping linear functional space contains elements that depend on largely arbitrary aspects of the chosen formalism of approach, whereas the structure of the reduced space of state vectors (5) and of the operators defined on it exhaust the structural characteristics of the quantum theory of the gravitational field. But all these assets do not compensate for its principal defect: so far it has remained largely an empty scheme. No one has succeeded in constructing a complete set of commuting observables that could serve for indexing the state vectors. And as a result there is no way of constructing the matrix elements of those observables (constants of the motion) which have in fact been obtained nearly ten years ago. In the absence of a developed theory, which would indicate the relationships between conventional field variables and the postulated observables, we must look for alternative approaches. The next section will deal with another classical approach to quantization, Hamilton–Jacobi theory.

HAMILTON–JACOBI THEORY

I shall take the somewhat unconventional point of view that Hamilton's principal function generates the mapping of the constraint hypersurface on the reduced phase space. This assertion is not much more than a change in terminology from the more usual statement that the principal function is the generator of a canonical transformation that leads to a canonical coordinate system consisting entirely of constants of the motion, and in which the Hamiltonian vanishes identically.

Let the principal function S depend on the arguments q_k, t, and P_i, where the indices k and i range from 1 to n. The Hamilton–Jacobi equation,

$$H\left(q_{k'} \frac{\partial S}{\partial q_k}, t\right) + \frac{\partial S}{\partial t} = 0, \quad k = 1 \cdots n, \tag{6}$$

may equally well be written in the form

$$\bar{H}\left(q_r, \frac{\partial S}{\partial q_r}\right) = 0, \quad r = 1, \ldots, n+1, \tag{7}$$

where $\bar{H}$ is the Hamiltonian constraint (1). If the Q_i, P_i $(i = 1 \cdots n)$ are to be the new canonical coordinates (which, of course, are all constants of the motion), then the transformation equations

$$p_r = \frac{\partial S}{\partial q_r}, \quad r = 1, \ldots, n+1, \qquad Q_i = \frac{\partial S}{\partial P_i}, \quad i = 1 \cdots n, \tag{8}$$

fix the Q_i, P_i as functions of the arguments q_r, p_r, provided the latter satisfy the constraint condition $\bar{H} = 0$. The function S is defined on the constraint hypersurface, which is conveniently coordinatized by means of the $(2n+1)$ coordinates q_r, P_i.

Consider now a $(2n)$-dimensional phase space, with j first-class constraints, $j < n$. Denote these constraints by the symbol $C_a\,(q_k, p_k)$ $(a = 1 \cdots j)$. The variational principle with subsidiary conditions,

$$\delta S = 0, \quad S = \int p_k \, \mathrm{d}q_k, \quad C_a = 0, \tag{9}$$

has as its solutions the trajectories that obey the Hamiltonian equations of motion,

$$\frac{\mathrm{d}q_k}{\mathrm{d}\theta} = \frac{\partial H}{\partial p_k}, \qquad \frac{\mathrm{d}p_k}{\mathrm{d}\theta} = -\frac{\partial H}{\partial q_k}, \qquad H = \xi^a C_a, \qquad C_a = 0, \tag{10}$$

where θ is an arbitrary parameter along the trajectories and where the coefficients ξ^a may be chosen at will. All the points on the constraint hypersurface that may be connected with each other by trajectories obeying Eq. (10) form one equivalence class, which is a j-dimensional subspace of the $(2n-j)$-dimensional constraint hypersurface. If the constraints are considered the generators of transformations that lead to equivalent descriptions of one-and-the-same physical situation, then each equivalence class represents one distinct physical situation, and the points forming the equivalence class correspond to the diverse equivalent descriptions of that situation. Again, the reduced phase space is a symplectic manifold whose points correspond to whole equivalence classes; all observables (properties of a physical situation that are independent of the choice of description) are constant within an equivalence class, and hence functions on the reduced phase space.

Being the first-class variables of the constraint hypersurface and the extended phase space, they are the generators of those infinitesimal canonical mappings that map the constraint hypersurface on itself, and they provide canonical mappings of the reduced phase space on itself.

Hamilton's principal function now obeys not one partial differential equation but j such equations, which are of the form

$$C_a\left(q_k, \frac{\partial S}{\partial q_k}\right) = 0. \tag{11}$$

If S is to be considered a function of the q_k and a set of new variables P_i ($i = 1 \cdots n - j$), Eqs. (8) generate a canonical transformation to the coordinates Q_i, P_i of the reduced phase space, with all the same properties that were established for the case $j = 1$. The Hamilton–Jacobi theory contains no reference to the coefficients ξ^a of Eq. (10).

The general theory of relativity has the form just contemplated, except that the indices k, i, and a are to be taken from three-dimensional label spaces. The constraints

$$H_s = \left(\frac{\delta S}{\delta g_{sn}}\right)_{,n} = 0 \tag{12}$$

imply that the value of the functional S is to be independent of the choice of coordinate system on a given three-surface, though not of the intrinsic geometry of that surface. "The same geometry" is meant to denote the set of all metric fields g_{mn} that can be transformed into each other by curvilinear three-dimensional coordinate transformations. Thus, if we substitute the three-geometry ${}^3\mathcal{G}$ (metric structure) of a three-dimensional manifold for the explicit metric field $g_{mn}(x^1 \cdots x^3)$ as part of the arguments on which the functional S depends, then the first three Hamiltonian constraints of general relativity (12) are automatically satisfied, and S must now be found as a solution of the $1 \times \infty^3$ remaining constraints $H_L = 0$.

This, basically, is the idea of "superspace". Its substitution for the function space of metric fields relieves us of three-quarters of the Hamilton–Jacobi equations to be satisfied by the principal functional S. The constraint hypersurface is reduced, but only partially. If we designate a complete commuting set of constants of the motion by Q_i, the index i to be taken from an appropriate label space having, presumably, $2 \times \infty^3$ points, and their canonical conjugates by P_i, then the partially reduced constraint hypersurface is coordinatized by ${}^3\mathcal{G}$, P_i.

Superspace is the set of all positive-definite three-geometries. It can serve as the configuration space of Hamilton–Jacobi theory. If it is to fulfill that purpose, we must endow it with topological properties. There remains the set of constraints H_L, which involve functional derivatives of S with respect to the configuration variables, the ${}^3\mathscr{G}$. The same functional derivatives are required to form the canonically conjugate variables, so that Hamilton's principal functional may indeed mediate the mapping of the partially reduced onto the fully reduced phase space. Thus it would appear that in order to be useable superspace must have enough of a topology for the equivalent of functional derivatives and of the corresponding integrals to be formed, such as

$$\delta A\,({}^3\mathscr{G}) = \int \frac{\delta A}{\delta^3 \mathscr{G}}\, \delta^3 \mathscr{G}\, \mathrm{d}^3\, \Sigma. \tag{13}$$

$\mathrm{d}^3\Sigma$ is to symbolize the notion that a three-geometry has ∞^3 degrees of freedom, a notion sufficiently hazy that it is probably not wholly in error.

In a sense superspace may be thought of as a projection of the function space whose coordinates are the fields $g_{mn}\,(x^1 \cdots x^3)$. This function space has a natural topology, in that, for instance, a neighborhood of one particular field may include all those positive-definite fields which differ from the given field nowhere by more than an amount ε in any one component. The elements of superspace correspond to whole classes of metric fields, those that can be transformed into each other by a coordinate transformation. One might, therefore, think of endowing superspace with a topology that is induced by the projection from function space. This topology, however, has certain shortcomings. Those three-geometries that contain isometry groups correspond to smaller classes of fields of the function space than others. This blemish might be avoided if most of the elements of superspace were replicated a sufficient number of times so that the three-geometries could be coordinatized, as it were, by some invariant coordinate system that is unique up to a remaining group of coordinate transformations corresponding to the maximal isometry group in a three-dimensional manifold (which is itself six-dimensional). It is in this sense that I understand Peres' construction.

Let us assume, for the sake of discussion, that all difficulties have been overcome and that superspace has been topologized, and even coordinatized, in a satisfactory manner. There remains then the infinity of constraints having roughly the form $\int \xi^L H_L\, \mathrm{d}^3\Sigma$, with all possible weighting functions ξ^L. The variability of this weighting function represents the variability in the spacing of successive three-surfaces covering the four-dimensional space-time manifold. Even in superspace the set of all elements belonging to one physical situation is not a one-parametric set but represents an infinite-dimensional structure, whose elements, to be sure, may be connected with each other by one-parametric sets, trajectories as it were.

Some years ago I had thought that in view of the fact that the invariants P_i and the superspace variables ${}^3\mathscr{G}$ together determine a point on the constraint hypersurface and hence, *a fortiori*, an equivalence class, a point in the reduced phase space, and because such a point in the reduced phase space may also, alternatively, be described uniquely by the set of observables Q_i, P_i, the Q_i by themselves must be determined by the three-geometry ${}^3\mathscr{G}$. This is, as it turns out, a fallacious argument. It is true that some invariant information is provided by the knowledge that a particular three-geometry may be imbedded in the (otherwise as yet unknown) Ricci-flat space-time manifold, but this information cannot be reduced to the numerical value of any observable.

FEYNMAN INTEGRATION: SANDWICH CONJECTURE

In attempting to guess at a proper quantum theory that is the counterpart of a given classical theory, one can begin with Hamilton–Jacobi theory and proceed to a species of Schrödinger theory, thereby inverting the WKB procedure. This procedure is ambiguous, in that in a highly non-linear Hamiltonian the sequence of configuration and momentum operators is not well-defined. If there are many constraints, which for physical reasons must obey a definite Lie algebra, factor-ordering may present a well-nigh insurmountable problem. There are also questions of principle. As mentioned before, the configuration and momentum operators are defined on a functional space that is not the Hilbert space of physically admissible states, and which is not metrizable. Hence it is meaningless to require these operators to be Hermitian. In fact, the commutation relations between such operators, though well defined mathematically, bear no relation to uncertainty relations between physical variables, as these variables are in principle not observable. For all these reasons, alternative quantization procedures are well worth looking for.

One such approach is through Feynman's integrals over classical paths. To be sure, "paths" in general relativity are not something as unambiguous, or even as natural as in classical mechanics, as we have seen that the concept of equivalence class, which all deterministic physical theories have in common, coincides in classical mechanics with an ordinary one-dimensional trajectory but is an infinite-dimensional structure of great complexity in general relativity. But there are some paths in general relativity, too, even if they may have to be brought about by coordinate conditions or other restrictions.

In a Feynman integral that is to equal the matrix element between two points in configuration space, all possible paths, in configuration space, are permitted to contribute, each path being represented by a complex number of magnitude 1, whose phase equals the classical action, divided by $\hbar$. For the integral to exist it is necessary that the paths between infinitesimally close points in configuration space are well-defined. In classical mechanics, and with suitably defined converging procedures, this condition is of course satisfied, or this whole approach would never have been taken seriously. That its generalization to Hamiltonians with different structures is by no means straightforward was pointed out by W. Pauli.

What about general relativity? Rather than go straight at the Feynman integral, one might ask the more modest question whether, with two three-geometries given, the "connecting path", that is to say a common Ricci-flat four-geometry, exists and is unique. Both Komar and I have, independently at first, and later together, worried about this question recently. Our result is that this is not so, that in general there is an infinity of four-geometries even in the event that the two given three-geometries differ from each other only infinitesimally. Our reasons are that the differential equations to be solved in the "thin sandwich" case, surely the mathematically least complex, are not always elliptic, and further, even if they were, that no set of boundary conditions appears physically particularly attractive. The whole "sandwich conjecture" is, however, so difficult, and there are so many important subsidiary considerations, that our results should not be considered definitive at this time. Perhaps a slightly different point of view will lead to a more attractive answer.

GENERAL REMARKS: CONCLUSION

Perhaps one or the other of the approaches which I have touched lightly will lead to a complete generally covariant quantum theory of the gravitational field, which will satisfy us in every respect. Though I believe that we are still far from this objective, there are some aspects of the quantum-gravitational field on which it is possible to comment or, more properly speaking, to speculate today.

First, it appears to me clear, even from a classical point of view, that the observables of any consistently covariant theory will not be local in the conventional sense. This is because one requires a rigid, non-dynamic metric structure of space-time to identify a point within a frame of reference that is determined by data lying outside the quantum dynamics of the field. Such a rigid framework is simply not available in a general-relativistic theory. Though I consider it reasonable that eventually the symmetry group of general relativity may be broken, I doubt that such breaking will signal a return to pre-general-relativistic physics. Conceivably world points will be identifiable by means of "coincidences". That is to say, a world point will be determined by means of the numerical values that certain scalar fields assume there. Such a determination will not be unique from a global point of view, and the same method of identification will not work for manifolds that are not sufficiently generic, but, most important, such identification will depend on the choice of properties. For instance, it is entirely reasonable to speak of the set of all points that form the light cone centered on a particular predetermined point (which in turn might be identified by some particular coincidence), but it will not be possible to assign these points other simple properties, which would lead to an ordinary pseudo-Riemannian manifold.

In the "weak-field" or linear approximation to the field equations, which amounts to the treatment of a spin-two, zero-rest-mass field *à la* Pauli–Fierz, one cannot measure field strengths, i.e. the analog of an affine connection, by any conceivable procedure, but one can measure components of the curvature tensor. Perhaps this feature will in some sense be transferable to the full theory, but this is highly uncertain, as the weak-field theory is basically Poincaré-invariant, with a gauge group attached, and thus possesses a rigid space-time structure.

Wherever in physics a fully elaborated quantum theory has been successful, this theory has added to the classical theory more than just the uncertainty relations, born of the classical Poisson brackets, important as the uncertainty principle is. The relativistic theory of the electron adds the notion of spin, to give an early example. Second quantization adds the concepts of quantum statistics, with the important phenomena of Bose condensation and of the exclusion principle, respectively, as well as the possibility of creation and annihilation of particles. Though some of these phenomena, which are usually regarded as specifically quantum-physical effects, can be incorporated in classical models, this classical reincarnation probably cannot be brought about in every instance. A more important remark, in this connection, is that, historically, classical physics has not led to the introduction of the new concepts, or to their universal acceptance; even in retrospect the heuristic motivation within the classical context appears weak.

It is, of course, extremely difficult to predict at this stage of the theoretical development whether the quantum theory of gravitation will make similar novel contributions to our conceptual inventory. Wheeler has suggested that in quantum physics there may be possibilities of change in topology that are not compatible with the classical theory. Classically, the emergence of a multiple connection in the course of time, and of its associated cleft, necessitates the presence of a singular region. Whether the quantum theory of the gravitational field can incorporate this or other possible processes is a fascinating question. A negative outcome might be predicted for any process that would violate a superselection rule, such as the conservation of charge in electrodynamics.

General relativity also possesses a species of superselection rules, which are based on the underlying invariance group and on Noether's theorem. But I do not believe that the gravitational superselection rules bear on changes in topology; hence I see no reason for ruling out Wheeler's suggestions.

Another interesting question is the role of singularities in classical and in quantum relativity. Through the work of Komar, Penrose, Hawking, Geroch, and others we have become aware of the likelihood that under certain circumstances the field equations of general relativity bring about the appearance of singularities in the course of time in manifolds that initially are everywhere regular. This statement may not be entirely accurate, in that I have used such terms as "initially" and "everywhere", which may seem innocuous, but actually require careful technical delineation, as the new theorems deal with manifolds that do not in all respects resemble Minkowski manifolds. Thus it is not yet clearly understood just how great an impact the new theorems have on classical Einstein manifolds. Whether the quantum theory leads to even stronger incursions of singularities or whether on the contrary quantization tends to "soften" the inevitability of their appearance, these questions are, as far as I am aware, largely unexplored, and most likely cannot be explored in any convincing manner until the formal theory has progressed far beyond its present stage.

In conclusion, I should like to repeat once more what to me remain the principal motivations for wanting to construct a quantum theory of gravitation. They are (1) the great likelihood that if most physical fields are quantum fields, the gravitational field will be no exception; (2) the expectation that a fully elaborated quantum theory of gravitation will be far more than a routine replica of other presently known quantum field theories; and (3) finally that any experience that we gain in striving to unify the principles of general covariance and of universal complementarity may teach us something worthwhile about the limitations of either.

PAPER 4

The impossibility of free tachyons

A. BERS*, R. FOX, C. G. KUPER and S. G. LIPSON
Technion—Israel Institute of Technology, Haifa, Israel

ABSTRACT

Despite the possibility of superluminal group velocities appearing in a Lorentz-invariant theory, it is shown that no Lorentz-invariant wave equation can be used for superluminal signal propagation. Causality is therefore not violated. However the Green function for imaginary-mass Klein–Gordon waves exhibits an absolute instability. Tachyons are excluded not by demanding causality, but by demanding that the tachyon vacuum be stable against fluctuations.

1 INTRODUCTION

There has recently been much interest in the properties of hypothetical systems[1,2] in which group velocities exceed the speed of light. The principal issue has been the question whether such systems would violate the axioms of special relativity. In particular they appear at first sight to violate causality.

Feinberg[2], discussing imaginary-mass Klein–Gordon particles ("tachyons"[3]), has an argument purporting to show that causality violation by tachyons is not observable. Feinberg considers information transfer by emission of a tachyon by one observer and its absorption by a second observer. But emission and absorption of tachyons are not Lorentz-invariant concepts, and there exist Lorentz frames in which the second observer's "event" is tachyon *emission*. We then have the problem of distinguishing between spontaneous

* Guggenheim Memorial Foundation Fellow: on sabbatical leave from the *Department of Electrical Engineering and Research Laboratory of Electronics, M.I.T., Cambridge, Massachusetts, U.S.A.*

and stimulated emission, and we cannot definitely identify causality-violating processes. Feinberg's argument is incomplete: we can envisage a slightly more sophisticated Gedanken experiment in which some well-defined pattern of correlated tachyons serves as a signal[4]. This might give a well-defined pattern of stimulated-emission processes which could be distinguished from the spontaneous-emission background, thus exhibiting causality violation.

The postulate that tachyons have superluminal group velocities calls to mind the well-known fact that the group velocity of electromagnetic waves in a dispersive medium can exceed unity for frequencies close to a resonance—and hence causality is apparently violated. However, this violation is obviously spurious, because the macroscopic refractive index is a consequence of scattering of electromagnetic waves by electrons and nuclei, while between scattering processes the propagation velocity is equal to unity[5]. The paradox was resolved by Sommerfeld and Brillouin[6], who showed that one must distinguish between the signal velocity and the group velocity, and that for electromagnetic waves the signal velocity always satisfies $v_s \leqslant 1$.

In the recent past the method of Sommerfeld and Brillouin has been extended to dispersive media that may be unstable[7]. In such cases it is found that even in the absence of resonances the group velocity may fail to give the proper signal-propagation characteristics.

Using these newly developed techniques for studying signal propagation in a general dispersive medium[7], we have recently shown[8]

(a) that the signal propagates at a velocity $v_s < v_\infty$ where v_∞ is the limiting phase velocity at infinite frequency, and thus

(b) that in particular no Lorentz-invariant wave equation can be used for superluminal signal propagation—i.e. causality is not violated—but

(c) that the solutions of the imaginary-mass Klein–Gordon equation are absolutely unstable to arbitrary fluctuations and in any Lorentz frame.

Hence free tachyons have to be rejected as impossible *not* on causal grounds but on stability grounds.

2 NON-VIOLATION OF CAUSALITY

Consider a general wave equation[9]

$$\mathscr{H}\left(\frac{\partial}{\partial x}, \frac{\partial}{\partial t}\right)\phi = 0 \tag{2}$$

where $\mathscr{H}$ is a polynomial. As usual we define the Green function to be the solution of the inhomogeneous equation

$$\mathscr{H}\phi_G = \delta(x)\,\delta(t).$$

Using conventional Fourier–Laplace techniques, the solution of (3) is

$$\phi_G(x, t) = \int_L \frac{d\omega}{2\pi} \int_F \frac{dk}{2\pi} \mathscr{H}^{-1}(ik, -i\omega)\, e^{i(kx-\omega t)}. \tag{3}$$

Here the "Fourier" contour F is the real axis and the "Laplace" contour L passes above all singularities of $\mathscr{H}^{-1}$ for real k (see Figure 1). The Fourier integral F is easily seen to be

$$\int_F \frac{dk}{2\pi} \mathscr{H}^{-1}(ik, -i\omega) \exp ikx = \pm\, i \sum_n \exp ik_n(\omega)\, x\, R_n(\mathscr{H}^{-1}), \tag{5}$$

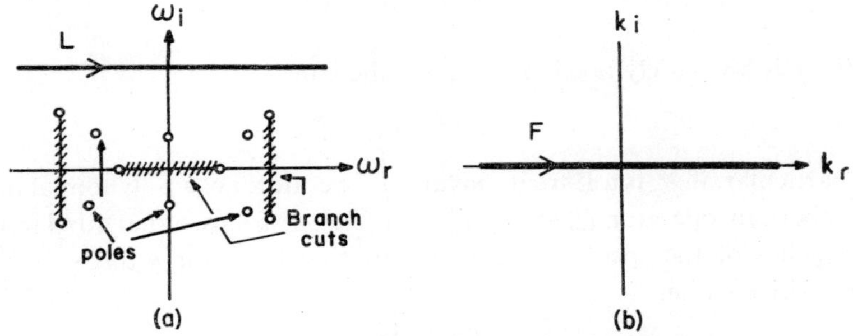

Figure 1. Contours of integration L and F respectively in the complex-ω and complex-k planes. L passes above all the singularities (poles and branch points) of $\mathscr{H}^{-1}(ik_n(\omega), -i\omega)$

where R_n is the residue at the nth pole of $\mathscr{H}^{-1}(ik, -i\omega)$. The meaning of the poles of $\mathscr{H}^{-1}$ (i.e. the zeros of $\mathscr{H}$) is that they give the dispersion law for waves

$$\mathscr{H}(ik_n(\omega), -i\omega) = 0. \tag{6}$$

Using (5), we can write

$$\phi_G^{(n)} = \int_L \frac{d\omega}{2\pi} R_n(\mathscr{H}^{-1}) \exp(W_n x) \tag{7}$$

where we have defined:

$$\left.\begin{aligned} W_n &\equiv i\,(k_n(\omega) - \omega\theta), \\ \theta &\equiv t/x. \end{aligned}\right\} \tag{8}$$

Let us close the path of integration by means of an infinite semicircle. According to whether $\lim_{|\omega|\to\infty} \operatorname{Im}(W_n x) > 0$ or < 0, the appropriate closure is in the upper or lower half-plane respectively. For $x > 0$, the condition for $\lim_{|\omega|\to\infty} \operatorname{Im}(W_n x) > 0$ is that

$$\lim_{\omega\to\infty} \{k_n(\omega) - \omega\theta\} > 0 \tag{9}$$

or

$$x > t \lim_{\omega\to\infty} \omega/k_n(\omega). \tag{10}$$

When Eq. (10) is satisfied, we close the contour of integration in the upper half-plane, and since all the singularities of $\mathscr{H}$ exp $W_n x$ lie below C,

$$\phi_{\mathrm{G}}^{(n)}(x, t) = 0. \tag{11}$$

Thus the nth branch of the dispersion law gives no signal outside the cone

$$x = v_\infty^{(n)} t \equiv t \lim_{\omega\to\infty} \omega/k_n(\omega), \tag{12}$$

and there is absolutely no signal outside the cone

$$x = v_\infty t \equiv t \max v_\infty^{(n)}. \tag{13}$$

In particular, if $\mathscr{H}$ is a Lorentz-invariant operator, i.e. a polynomial in the d'Alembertian operator $\Box \equiv \partial^2/\partial t^2 - \partial^2/\partial x^2$, it follows that for large ω, *all* branches of the spectrum have the limiting behavior $k_n(\omega) \sim \pm\omega$ for large positive ω, i.e.

$$v_\infty = 1.$$

In particular the Klein–Gordon equation preserves causality whatever the sign of the mass term.

Bludman and Ruderman[10] have studied the propagation of sound in ultra-dense matter, and have found a wave equation which is *not* Lorentz-invariant. The spectrum has an "acoustic" and an "optical" branch. Despite the lack of Lorentz-invariance, the acoustic branch of the Bludman–Ruderman spectrum satisfies $v_\infty = 1$. Thus the signal velocity is <1, although $v_g > 1$ for small k.

3 TACHYON INSTABILITY

For the particular case of the Klein–Gordon equation with real or imaginary mass[2]

$$\left(\frac{\partial^2}{\partial t^2} - \frac{\partial^2}{\partial x^2} \mp \mu^2\right)\phi = 0, \tag{15}$$

it is straightforward to calculate ϕ_G explicitly:

$$\phi_G = \tfrac{1}{2}J_0\,[\mu\sqrt{(t^2 - x^2)}]\,u\,(t - x)\,u\,(t + x), \tag{16a}$$

$$\phi_G = \tfrac{1}{2}I_0\,[\mu\sqrt{(t^2 - x^2)}]\,u\,(t - x)\,u\,(t + x), \tag{16b}$$

for ittyons and tachyons respectively. Here J_0 and I_0 are the Bessel and modified Bessel functions of order zero, and u is the unit step function. Equation (16b) is sketched in Figure 2. We note that in agreement with the result

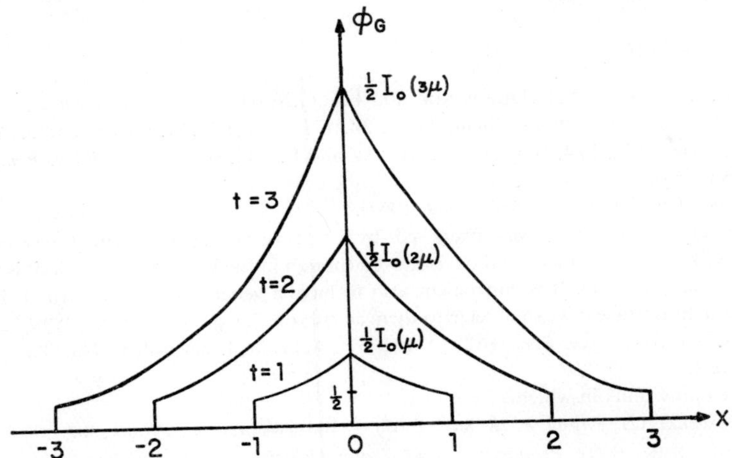

Figure 2. Green's function response of tachyon fields (Eq. (16b))

of Section 2, the front of the disturbance travels out in $\pm x$ with the velocity of light despite the fact that the group velocity is superluminal. For any fixed x once a tachyon disturbance arrives it grows indefinitely in time—this is the character of an absolute instability[11]. In fact from Eq. (16b) we find the asymptotic behavior for constant x, $t \to \infty$:

$$\phi_G\,(x, t) \sim \tfrac{1}{2}\,(2\pi\mu t)^{-1/2}\,\exp \mu t,$$

which can be deduced directly from the dispersion relation for tachyon fields $D\,(\omega, k) = -k^2 + \omega^2 + \mu^2$, together with well-known stability criteria[7]. Furthermore since the dispersion relation for tachyon fields is Lorentz invariant[12], the above results show that free tachyon fields are absolutely unstable in any Lorentz reference frame.

We conclude that the construction of stable free-tachyon wave-packets is not possible. Hence the transmission of information by tachyon fields is ruled out, since the information would be destroyed by the inherent instability of these fields. The question of causal or non-causal communication by means of tachyon fields thus becomes secondary. The instability of all tachyon wave packets, which we have exhibited here, implies that even the tachyon vacuum is unstable against fluctuations. Speculations on the possible existence of tachyons are thus inconsistent, since the exponential growth of amplitude would violate energy conservation.

REFERENCES

1. O. M. P. Bilanuik, V. K. Deshpande and E. C. G. Sudarshan, *Am. J. Phys.*, **30**, 718 (1962); S. Tanaka, *Progr. Theor. Phys.*, **24**, 171 (1960); M. M. Broido and J. G. Taylor, *Phys. Rev.*, **174**, 1606 (1968); J. Dhar and E. C. G. Sudarshan, *Phys. Rev.*, **174**, 1808 (1968).
2. G. Feinberg, *Phys. Rev.*, **159**, 1089 (1967).
3. The name "tachyon" was proposed by Feinberg (1967), from the Greek *ταχυς* ('swift'). We shall adhere to the name even though in the light of our results it is somewhat inappropriate. It is convenient also to have a generic name for normal slower-than-light particles; we shall refer to them as "ittyons", from the Hebrew איטי ('slow').
4. R. G. Newton, *Phys. Rev.*, **162**, 1274 (1967), A. Peres, Festschrift to Narlikar (1971), in press.
5. We employ units in which $c = \hbar = 1$.
6. A. Sommerfeld, *Physik Z.*, **8**, 841 (1907), L. Brillouin, *Ann. Physik*, **44**, 203 (1914); L. Brillouin, *Wave Propagation and Group Velocity;* Academic Press, New York, 1960.
7. A. Bers and R. J. Briggs, *M.I.T. Research Laboratory of Electronics Quarterly Progress Report* No. **71**, 15 October 1963, pp. 122–131; *Bull. Am. Phys. Soc.*, **9**, 304 (1964); R. J. Briggs, *Electron Stream Interaction with Plasmas*, M.I.T. Press, Cambridge, Mass., 1964.
8. R. Fox, C. G. Kuper and S. G. Lipson, *Nature* (1969), in press; A. Bers and C. G. Kuper, unpublished; R. Fox, C. G. Kuper and S. G. Lipson, *Proc. Roy Soc. A* **316**, 512 (1970).
9. We restrict ourselves to a world of one spatial dimension for simplicity.
10. S. A. Bludman and M. A. Ruderman, *Phys. Rev.*, **170**, 1176 (1968).
11. L. D. Landau and E. M. Lifshitz, *Fluid Mechanics*, Pergamon Press, Oxford, 1959, p. 113; P. A. Sturrock, *Phys. Rev.*, **112**, 1488 (1958).
12. This is easily shown by using the transformations of ω and k: $\omega' = (\omega - kv_0)\gamma$, $k' = (k - \omega v_0)\gamma$ where v_0 is the velocity of the frame and $\gamma = (1 - v_0^2)^{-1/2}$.

PAPER 5

Selected topics in the problem of energy and radiation

J. BIČÁK

Charles University, Prague, Czechoslovakia

In connection with the detailed study of gravitational radiation considerable attention has been paid to the concept of energy of the gravitational field.

Needless to say, the analogies with more simple fields, in particular with the electromagnetic field, have proved to be among the most useful guides in developing the classical theory of gravitational radiation and in attempts to understand gravitational energy; indeed, does not the very association of radiation and energy come from electromagnetism?

In this paper, I wish to discuss only a few problems selected from the wide energy—radiation branch of relativity theory. When solving these problems, the analogies mentioned will often be used and emphasized, but at the same time some important differences between the electromagnetic and gravitational fields will be indicated.

1 THE UNIQUENESS OF ENERGY-MOMENTUM TENSORS IN SPECIAL RELATIVITY

In relativistic field theories the most convenient method of finding the conserved quantities is based on Noether theorems. If we start out from the Lorentz invariant Lagrange function, we can form a generally asymmetric canonical energy-momentum tensor $t^{\mu\nu}$, or in rewriting the Lagrangian of a non-gravitational field in a manifestly covariant manner, obtain the symmetric tensor $T^{\mu\nu}$. Both tensors give the same total energy and momentum since they differ by the divergence of the superpotential: $T^{\mu\nu} = t^{\mu\nu} + U^{\mu\nu\lambda}_{,\lambda}$;

$U^{\mu\nu\lambda} = -U^{\mu\lambda\nu}$. (Strictly speaking, also the total quantities may be affected by the divergence of the superpotential in the case of spatially isolated but *radiating* systems for which, in general, $U \sim r^{-2}$ at infinity.) Admissible changes of Lagrangian lead to the known ambiguities of energy-momentum tensors, the non-unique part being always formed by the superpotential[1].

It is not, however, obvious that *all* energy-momentum tensors can be obtained by these techniques. Moreover, there are theories in which the field equations are not derivable from variational principle, or at least, for which we do not know how to derive them (e.g. the "already unified" theory of Rainich).

At the Warsaw conference[2], Fock summarized the work of his group on the uniqueness of the energy-momentum tensors of the electromagnetic field, of incoherent matter, and of a perfect fluid. The uniqueness has been proved without using of Lagrangian formalism, requiring only that the energy-momentum tensor be a symmetric tensor of the second order, formed with the field variables, and conserved as a consequence of field equations.

Here we shall analyze the important cases of wave fields and the gravitational field in this context. Fock's procedure cannot be directly applied for the fields in question, since these are described by equations of second order. There is, however, no great difficulty in modifying Fock's technique and requirements for these cases; rather, the differences appear in the final results. As a matter of fact the resulting expressions, which are now functions of the field variables and also of their first derivatives, are mostly non-unique. Supplementary conditions, as for example positive definiteness of the total energy or gauge-invariance, must be imposed in order to avoid at least some of the arbitrariness.

The simplest wave field described by a second-order equation is the neutral meson field satisfying the Klein–Gordon equation[3]

$$\eta^{\mu\nu}\varphi_{,\mu\nu} + k^2\varphi = 0.$$

We shall demonstrate our procedure on this simple case. The energy-momentum tensor $T^{\mu\nu}$, taken as an arbitrary function of the variable φ and of the first derivatives $\varphi_{,\sigma}$, must satisfy the conservation law $T^{\mu\nu}_{,\nu} = 0$ as an algebraic consequence of the field equation. If the Klein–Gordon equation multiplied by Lagrange multipliers λ^μ is added to the conservation law,

$$\frac{\partial T^{\mu\nu}}{\partial \varphi_{,\sigma}}\varphi_{,\sigma\nu} + \frac{\partial T^{\mu\nu}}{\partial \varphi}\varphi_{,\nu} + \lambda^\mu (\varphi_{;\sigma}^{;\sigma} + k^2\varphi) = 0,$$

the equations obtained must be fulfilled identically in the independent second derivatives. Since the quantities $\partial T^{\mu\nu}/\partial\varphi_{,\sigma}$, $\partial T^{\mu\nu}/\partial\varphi$ and λ^μ are not functions of $\varphi_{,\sigma\nu}$, $T^{\mu\nu}$ must satisfy the equations

$$\left.\begin{aligned} \frac{\partial T^{\mu\nu}}{\partial\varphi_{,\sigma}} + \frac{\partial T^{\mu\sigma}}{\partial\varphi_{,\nu}} + 2\lambda^\mu\eta^{\sigma\nu} &= 0, \\ \frac{\partial T^{\mu\nu}}{\partial\varphi}\varphi_{,\nu} + \lambda^\mu k^2\varphi &= 0; \end{aligned}\right\} \tag{1}$$

the multipliers λ^μ are easily expressed by contracting the first equation in σ and ν. Now, the most general tensor of second order which is constructed out of the quantities φ and $\varphi_{,\sigma}$ is obviously the symmetric tensor

$$T_{\mu\nu} = \mathscr{A}\varphi_{,\mu}\varphi_{,\nu} + \mathscr{B}\eta_{\mu\nu},$$

where $\mathscr{A}$ and $\mathscr{B}$ are arbitrary real functions of φ and $\varphi_{,\sigma}\varphi^{,\sigma}$. Substituting this tensor into (1), and using the independence of the first derivatives $\varphi_{,\sigma}$, we find out that the conditions (1) determine the energy-momentum tensor uniquely up to two arbitrary constants A and B which are defined by the choice of units and the condition at infinity:

$$T_{\mu\nu} = A\,(\varphi_{,\mu}\varphi_{,\nu} - \tfrac{1}{2}\varphi_{,\sigma}\varphi^{,\sigma}\eta_{\mu\nu} + \tfrac{1}{2}k^2\varphi^2\eta_{\mu\nu}) + B\eta_{\mu\nu}.$$

(Hereafter, an energy-momentum tensor containing only this sort of ambiguity will be referred to as unique; trivial additive constants will always be omitted.)

We may proceed similarly in the case of more complex fields. In contrast to the scalar field and to the systems described by equations of the first order, it now appears unmanageable to start out from quite general functions of field variables and their derivatives. Our tensors are as a rule constructed as arbitrary functions of the field variables and quadratic functions of their first derivatives; the analogy with the electromagnetic field (rewritten in terms of the potential) makes this assumption plausible. The detailed calculations, which are rather lengthy, can be found in [4]; here, we shall just outline the final results and then turn to some new related questions.

In the case of the vector (Proca) field, when the field equations are

$$\varphi_{\iota,\sigma}^{;\sigma} + k^2\varphi_\iota = 0 \quad \text{and} \quad \varphi^{\sigma}_{,\sigma} = 0,$$

the resulting symmetric energy-momentum tensor has the form

$$\begin{aligned} T_{\mu\nu} = {} & A_1\,(\varphi_{\mu,\sigma}\varphi_\nu^{;\sigma} - \varphi_{\sigma,\mu}\varphi_\nu^{;\sigma} - \varphi_{\sigma,\nu}\varphi_\mu^{;\sigma} + \tfrac{1}{2}\varphi_{\sigma,\tau}\varphi^{\tau,\sigma}\eta_{\mu\nu} - k^2\varphi_\mu\varphi_\nu) \\ & + A_2\,(\varphi_{\sigma,\mu}\varphi^{\sigma}_{,\nu} - \tfrac{1}{2}\varphi_{\sigma,\tau}\varphi^{\sigma,\tau}\eta_{\mu\nu} + \tfrac{1}{2}k^2\varphi_\sigma\varphi^\sigma\eta_{\mu\nu}), \end{aligned} \tag{2}$$

where A_1, A_2 are arbitrary constants. This tensor is, of course, non-unique. If we go over, by putting $k = 0$, to the Maxwell field described in terms of the potential φ_μ, the term $A_3\,(\varphi_{\mu,\nu} + \varphi_{\nu,\mu})$ appears in addition. The ambiguities may easily be eliminated by demanding the gauge invariance of the energy-momentum tensor of the electromagnetic field. For the Proca field, however, the dynamical characteristics need not be gauge invariant. Nor does the requirement of positive definiteness of the total energy exclude the arbitrariness because the terms in the coefficient A_1 do not contribute to the total energy, expressed by decomposition into plane waves. Indeed, the tensor (2) can be cast into the form $T_{\mu\nu} = A_2\,(\cdots) + A_1 U^{\sigma}_{\mu\nu,\sigma}$, where the superpotential $U^{\nu\sigma}_{\mu}$ is antisymmetric in upper indices.

It is interesting to note, however, that the arbitrariness can be avoided by treating the equations of Proca field as a limiting case of the non-linear equations

$$\varphi^{;\sigma}_{\iota,\sigma} - \varphi^{\sigma}_{,\iota\sigma} = f(\eta)\,\varphi_\iota, \tag{3}$$

in which $f(\eta)$ is an arbitrary real function of the variable $\eta \equiv \varphi_\sigma\varphi^\sigma$ (the generalized Lorentz condition follows from the equations above). If $f(\eta) = k^2 +$ (small parameter) $\times$ (arbitrary function of η), Eq. (3) becomes approximately the Proca field equation and becaue $\mathrm{d}f/\mathrm{d}\eta \neq 0$, it can be shown that the resultant energy momentum tensor is unique ($A_1 = A_2$). We shall see another possibility of avoiding the non-uniqueness in the following.

2 ENERGY-MOMENTUM COMPLEXES OF THE GRAVITATIONAL FIELD

It is well known that in general relativity a number of complexes t^{ν}_{μ}, constructed from the metric tensor $g_{\mu\nu}$ and derivatives up to the second order have been proposed for the gravitational field. For the weak field, when the metric tensor can, in a suitable coordinate system, be written in the form $g_{\mu\nu} = \eta_{\mu\nu} + h_{\mu\nu}$, the Einstein equations in vacuo become

$$R_{\mu\nu} \equiv \tfrac{1}{2}\eta^{\varrho\sigma}\,(h_{\varrho\sigma,\mu\nu} + h_{\mu\nu,\varrho\sigma} - h_{\varrho\mu,\nu\sigma} - h_{\varrho\nu,\mu\sigma}) = 0. \tag{4}$$

These equations can formally be regarded as the Lorentz covariant equations of the tensor field $h_{\mu\nu}$ in flat space, invariant with respect to the gauge group

$$h_{\mu\nu} \to h'_{\mu\nu} = h_{\mu\nu} + \xi_{\mu,\nu} + \xi_{\nu,\mu}.$$

From this point of view it is natural to ask what is the most general energy-momentum complex which is exactly conserved modulo the field Eq. (4).

We confine ourselves to complexes which are quadratic forms of $h_{\mu\nu,\lambda}$; however, bearing in mind the non-linear case, we do not require the symmetry of $t_{\mu\nu}$. Our requirement $R_{\mu\nu} = 0 \Rightarrow t^{\mu\nu}_{,\nu} = 0$ leads to the equations

$$t^{\mu\nu}_{,\nu} + \lambda^{\mu\varrho\sigma} R_{\varrho\sigma} = 0, \tag{5}$$

which must be satisfied identically in $h_{\mu\nu,\lambda}$ and $h_{\mu\nu,\lambda\varkappa}$. These equations restrict the initial complex (which appears to be a linear combination with arbitrary constant coefficients of twenty independent quantities quadratic in $h_{\mu\nu,\lambda}$) to a four-parameter system as follows:

$$\left.\begin{aligned} t_{\mu\nu} = {} & A_1 (h_{,\tau} h^{\tau\sigma}_{,\sigma} \eta_{\mu\nu} - h_{,\tau} h^{,\tau} \eta_{\mu\nu} + h_{,\mu} h_{,\nu} + h^{,\tau} h_{\mu\nu,\tau} - h_{,\tau} h^{\tau}_{\mu,\nu} - h_{,\mu} h^{,\sigma}_{\nu\sigma}) \\ & + A_2 (h^{,\tau} h_{\mu\nu,\tau} - h_{,\nu} h^{,\sigma}_{\mu\sigma} - h_{\mu\nu,\tau} h^{\tau\sigma}_{,\sigma} + h^{\varrho}_{\mu,\varrho} h^{\sigma}_{\nu,\sigma} - h^{\varrho,\sigma}_{\mu} h_{\varrho\nu,\sigma} + h^{\varrho\sigma}_{,\nu} h_{\varrho\mu,\sigma}) \\ & + A_3 (2h^{\varrho\sigma,\tau} h_{\varrho\tau,\sigma} \eta_{\mu\nu} - h^{\varrho\sigma,\tau} h_{\varrho\sigma,\tau} \eta_{\mu\nu} - h_{,\tau} h^{,\tau} \eta_{\mu\nu} + 2h^{\varrho\sigma}_{,\mu} h_{\varrho\sigma,\nu} \\ & + 2h^{,\tau} h_{\mu\nu,\tau} - 2h^{,\tau} h_{\tau\mu,\nu} + 2h^{,\tau} h_{\tau\nu,\mu} - 4h^{\varrho\sigma}_{,\mu} h_{\varrho\nu,\sigma}) \\ & + A_4 (h^{\varrho\sigma}_{,\varrho} h^{\tau}_{\sigma,\tau} \eta_{\mu\nu} - h^{\varrho\sigma,\tau} h_{\varrho\tau,\sigma} \eta_{\mu\nu} - 2h^{\varrho\sigma}_{,\varrho} h_{\sigma\nu,\mu} + 2h^{\varrho\sigma}_{,\mu} h_{\varrho\nu,\sigma}). \end{aligned}\right\} \tag{6}$$

Here, A_1, A_2, A_3 and A_4 are arbitrary constants, and $h \equiv h^{\sigma}_{\sigma}$. Except for the choice $A_3 = 0$, A_1, A_2, $A_4 \neq 0$ when $t_{\mu\nu}$ is the divergence of a superpotential, every other choice of parameters leads to a complex which may be regarded as the energy-momentum complex of the gravitational field in the linear theory. In particular, for $-2A_1 = A_3$, $A_2 = A_4 = 0$ we get the linearized Einstein's complex (also referred to as Einstein's pseudotensor). In contrast to the non-linear theory, this complex cannot be derived from a superpotential, but every other complex can be obtained by adding a superpotential to the Einstein complex. Of course, recalling the known argument of Weyl[5], we could not hope to find a gauge invariant expression (in contrast with electrodynamics). Fortunately, the situation is not too awkward, since all energy-momentum tensors (6) can be shown to change only by a divergence under gauge transformations. Let us remark, however, that there exists a uniquely determined symmetric complex ($A_1 = A_2 = -4A_3$, $A_4 = 0$) which might be preferred owing to the possibility of formulating the angular momentum conservation law.

In the case of full non-linear theory in vacuo we can look analogously for complexes quadratic in $g_{\iota\varkappa,\lambda}$ with transformation properties like a tensor density of weight *one* under linear transformations (note that e.g. the Landau–Lifshitz complex is thereby excluded from our considerations)[30]. Replacing $\eta_{\mu\nu} \to g_{\mu\nu}$, $h_{\mu\nu,\lambda} \to g_{\mu\nu,\lambda}$, we obtain, in addition to the equations formally analogical to (5), also the equations $(\partial t^{\mu\nu}/\partial g_{\varrho\sigma})\, g_{\varrho\sigma,\nu} + \lambda^{\mu\varrho\sigma} \underset{1}{R}_{\varrho\sigma} = 0$, where $\underset{1}{R}_{\varrho\sigma}$ includes the terms of Ricci tensor quadratic in $g_{\varrho\sigma,\nu}$. It can be seen that these conditions limit the complex uniquely to the Einstein complex. This fact does seem to support one of the present opinions on conservation laws in general relativity expressed explicitly by Trautman[6], for example. Accepting the view that the energy concept can be meaningful only in special situations in general relativity, the Einstein complex is quite satisfactory.

3 ON THE ENERGY OF PERTURBATIONS OF THE GRAVITATIONAL FIELD AND SOME RELATED PROBLEMS

After reviewing the above results, we would probably now be inclined to believe more deeply in the effectiveness and, surely, in the comfort of working with a variational principle. Although this seems to be quite justifiable, our procedure appears to be the only rigorous way of obtaining some negative results. More positively, we note that the symmetrical energy-momentum tensor, which is included in the family (6), cannot be obtained in any straightforward way from a variational principle. Nevertheless it is a perfectly meaningful expression.

In the last few years, a great deal of attention has been paid to the study of perturbations of a given (background) metric corresponding to an arbitrarily curved space-time. Let us mention the approach to quantization started by Lichnerowicz[7], who considers the background vacuum field as *c*-numbers and the microscopic variations of the metric tensor as subject to more or less "classical" quantization, a number of investigations on the stability of solutions, a quite recent study of non-radial pulsations of stellar models by Thorne *et al.*[8], and the thorough analysis of gravitational radiation in the high frequency limit by Isaacson[9]. What can we say about the energy of perturbations[10]?

Suppose that the background metric $\gamma_{\mu\nu}$ satisfies the vacuum Einstein equations $R_{\mu\nu}(\gamma_{\alpha\beta}) = 0$. We may perturb the metric and to the first order write $g_{\mu\nu} = \gamma_{\mu\nu} + \varepsilon h_{\mu\nu}$, where $g_{\mu\nu}$ represents the total metric, and ε is a small parameter. The variation of the metric tensor leads via the variation of the Christoffel symbols, to the variation of the Ricci tensor. Assuming the total metric again to represent the solution of the Einstein equations in vacuo, we find to lowest order the linear equation for perturbations:

$$\delta R_{\mu\nu} \equiv \tfrac{1}{2}\gamma^{\varrho\sigma}(h_{\varrho\sigma;\mu\nu} + h_{\mu\nu;\varrho\sigma} - h_{\varrho\mu;\nu\sigma} - h_{\varrho\nu;\mu\sigma}) = 0. \tag{7}$$

Here, the covariant derivatives are taken with respect to the background metric which is also used to raise or lower all indices. We see that Eq. (4) of conventional linear theory is the same as the equation just obtained, provided that ordinary derivatives are replaced by covariant ones and the Minkowski metric $\eta_{\mu\nu}$ by the background metric $\gamma_{\mu\nu}$. Let us then ask what is the most general tensor of the second order constructed as a quadratic form of $h_{\mu\nu;\lambda}$ which is covariantly conserved[11] modulo Eq. (7). (By analogy with the electromagnetic field in general relativity, and in the spirit of minimal gravitational coupling, we allow the second derivatives of neither $h_{\mu\nu}$ nor $\gamma_{\mu\nu}$ to appear in our tensors.) Now, if we replace commas by semicolons and $\eta_{\mu\nu}$ by $\gamma_{\mu\nu}$ in the system of energy-momentum tensors (6), we find out that there is *no* choice of parameters A_1, A_2, A_3, A_4 for which the resulting tensor would be covariantly conserved. The technical reason is quite simple: the second derivatives do not commute.

This negative result appears to be only a particular case of a more general problem—namely the possible forms of energy-momentum tensors of higher-spin fields in general relativity. To see this, we notice firstly that Eq. (7) is invariant under the gauge transformation $x'^{\mu} = x^{\mu} + \xi^{\mu}$ (but note that this invariance holds only for Einstein spaces satisfying $R_{\mu\nu}(\gamma_{\alpha\beta}) = \lambda\gamma_{\mu\nu}$); if we define

$$\psi_{\mu\nu} \equiv h_{\mu\nu} - \tfrac{1}{2}\gamma_{\mu\nu}h, \quad \psi \equiv \gamma^{\alpha\beta}\psi_{\alpha\beta}, \quad h \equiv \gamma^{\alpha\beta}h_{\alpha\beta},$$

(7) takes the form

$$\psi_{\mu\nu;\beta}^{\;;\beta} - \tfrac{1}{2}\gamma_{\mu\nu}\psi_{;\beta}^{;\beta} - \psi_{\mu\beta;\nu}^{\;;\beta} - \psi_{\nu\beta;\mu}^{\;;\beta} + 2R_{\sigma\nu\mu\beta}\psi^{\beta\sigma} = 0.$$

We may impose as a choice of gauge the conditions

$$\psi_{;\nu}^{\mu\nu} = 0, \quad \psi = 0, \tag{8}$$

which simplify the equation above to

$$\psi_{\mu\nu;\beta}^{\;;\beta} + 2R_{\sigma\nu\mu\beta}\psi^{\beta\sigma} = 0. \tag{9}$$

It can be seen that (8) and (9) are consistent in an Einstein space. However, these equations may also be regarded as a covariant form of massless spin—2 field equations, the field being treated as a physical (nongravitational) test field. If we now apply our procedure to find all energy-momentum tensors which are quadratic in $\psi_{\mu\nu;\lambda}$ and covariantly conserved as a consequence of (8) and (9), we again get a negative result. A similar result can be found for spin-2 fields of non-zero rest mass, the equations of which may be obtained by adding the term $k^2\psi_{\mu\nu}$ to the left-hand side of (9). Indeed, the gravitational interaction strongly restricts possible forms of energy momentum tensors. We arrived at a unique tensor for the electromagnetic field by demanding gauge invariance; in the case of Proca field, the non-uniquiness was, avoided only by introducing non-linearities into field equations. It can easily be seen, however, that in both cases we get unique expressions when taking the gravitational interaction into account. (The consistent field equations written in terms of the potential φ_μ in a general, not necessarily Einstein space-time are $\varphi_{\mu;\nu}^{;\nu} + R_{\sigma\mu}\varphi^\sigma + k^2\varphi_\mu = 0$, $\varphi^\mu_{;\mu} = 0$; $k = 0$ for the electromagnetic field.)

The difficulties in formulating consistent equations for fields with spin $s > 1$ in a general space-time are familiar. (Some of them were first pointed out by Buchdahl[13]; for zero-rest mass fields, see ref. 14.) Of course, the most straightforward way suggests itself, replace partial derivatives by covariant derivatives in the *first-order* flat-space equations. Starting from the tensor formulation of spin-2 field equations[12], the resulting (consistent) equations have a simple form

$$_\mu H_{\nu\lambda} = \psi_{\mu\lambda;\nu} - \psi_{\mu\nu;\lambda}, \quad _\mu H^{;\nu}_{\nu\lambda} + k^2\psi_{\mu\lambda} = 0, \quad \psi^\sigma_\sigma = 0.$$

It is immediately seen, however, that, in a curved space, the "Lorentz condition" $\psi^{\mu\nu}_{,\nu} = 0$ not only does not follow from these equations but is not compatible with them even in an Einstein space (compare with (8) and (9), which are compatible, at least in an Einstein space). Disregarding these difficulties, we want to emphasize again that no energy-momentum tensor involving only $\psi_{\mu\nu}$ and $\psi_{\mu\nu;\lambda}$ exists ($_\mu H_{\nu\lambda}$ being expressed by means of $\psi_{\mu\nu;\lambda}$) which would be covariantly conserved modulo the equations given above. In particular, the symmetric tensor constructed in the classical paper of Fierz (see [12], Eq. (28.6b)) is excluded by the gravitational intraction, if the latter is introduced in the above straightforward way. On the other hand, it is well known that the first-order equations under consideration can be derived from a variational principle and therefore, there exists a metric (Rosenfeld–Belinfante) symmetric energy-momentum tensor[1] which is covariantly conserved. When expressed explicitly, this tensor can be shown to contain the *second* derivatives of $\psi_{\mu\nu}$, in contrast to lower spin fields, whose energy momentum tensors require only the field variables and their first derivatives. To indicate the influence of the gravitational interaction on energy-momentum tensors of higher-spin fields we have confined ourselves to the particular example of a spin-2 field; one may expect that similar conclusions can be drawn also for the other higher-spin fields.

Now, let us give the explicit form of the metric energy-momentum tensor for the perturbations of a vacuum (background) gravitational field. Eq. (7) is derivable from the Lagrange function

$$L = \tfrac{1}{2} h^{\alpha\beta;\gamma} h_{\alpha\beta;\gamma} - \tfrac{1}{2} h_{;\gamma} h^{;\gamma} + h_{;\alpha} h^{\alpha\beta}_{;\beta} - h_{\gamma\beta;\alpha} h^{\alpha\beta;\gamma}.$$

Using (7), we can cast the metric energy-momentum tensor, which except for a multiplicative constant involving ε^2, is given by the relation $T^{\mu\nu} = -2/\sqrt{-\gamma}\, (\delta\, (L \sqrt{-\gamma}))/(\delta\gamma_{\mu\nu})$, into the following form:

$$\begin{aligned} T_{\mu\nu} = {} & \gamma_{\mu\nu} [\tfrac{1}{2} h^{\varrho\tau;\sigma} h_{\varrho\tau;\sigma} - \tfrac{1}{2} h_{;\varrho} h^{;\varrho} - h^{\varrho\tau;\sigma} h_{\sigma\tau;\varrho} - h_{;\varrho;\sigma} h^{\varrho\sigma}] - h_{\varrho\sigma;\mu} h^{\varrho\sigma}_{;\nu} \\ & + h_{;\nu} h_{;\mu} - h_{;(\mu} h^{\varrho}_{\nu);\varrho} + 2h_{\varrho\sigma;(\mu} h^{\varrho;\sigma}_{\nu)} - 2h_{\mu\varrho;\sigma} h^{\varrho;\sigma}_{\nu} - 2h_{\mu\sigma;\varrho} h^{\varrho;\sigma}_{\nu} \\ & + 2h_{\mu\nu;\varrho} h^{\varrho\sigma}_{;\sigma} + h_{\mu\nu;\varrho} h^{;\varrho} + [h_{;\varrho;(\nu} - 2h^{\sigma}_{(\nu;\varrho;\sigma}]\, h^{\varrho}_{\mu)} \\ & + 2h_{\mu\nu;\varrho;\sigma} h^{\sigma\varrho} + h^{;\sigma}_{;\sigma} h_{\mu\nu}, \end{aligned}$$

where the indices in parantheses are to be symmetrized, $A_{(\alpha\beta)} = A_{\alpha\beta} + A_{\beta\alpha}$. It may be verified by straightforward, though rather lengthy calculation that $T^{\nu}_{\mu;\nu} = 0$ modulo Eq. (7) indeed holds. This tensor may be applied to various situations. For example, it describes the energy content of gravitational waves radiated out from a massive star due to small non-radial pulsations and propagating on the curved static background field of the star. It should be emphasized that this tensor contains the second derivatives of $h_{\mu\nu}$. If the background space is Minkowskian, the metric energy-momentum tensor in the conventional linear theory is retrieved. As indicated in the introduction of this section, the symmetric tensor included in the system (6)—the unique tensor quadratic in the first derivatives—cannot be derived by any straightforward way from a variational principle since it represents neither the canonical nor the metric energy-momentum tensor. Nevertheless, it might be preferred in the usual linear theory owing to its closer analogy with the corresponding expressions in the theories of lower-spin fields.

Let us now turn to the problem of perturbations of the gravitational field which have a character of high-frequency waves. In this physically important case one may well speak of the energy and momentum of perturbations, in close analogy with electromagnetic field, and in fact, take almost any of the tensors (6), in which $h_{\varrho\sigma,\tau}$ is replaced by $h_{\varrho\sigma;\tau}$ and $\eta_{\varrho\sigma}$ by $\gamma_{\varrho\sigma}$, as an energy-momentum tensor. Suppose that the vacuum perturbations have a character of high-frequency waves, the wavelength λ being short compared to the background curvature of space-time:

$$g_{\mu\nu} = \gamma_{\mu\nu} + \varepsilon h_{\mu\nu}(\lambda),$$

whereby

$$\gamma_{\mu\nu}, \gamma_{\mu\nu,\varrho}, \gamma_{\mu\nu,\varrho\sigma} \sim \mathrm{O}(1),$$

$$h_{\mu\nu} \sim \mathrm{O}(1), \quad h_{\mu\nu,\varrho} \sim \mathrm{O}(\lambda^{-1}), \quad h_{\mu\nu,\varrho\sigma} \sim \mathrm{O}(\lambda^{-2})$$

(λ plays the role of another small parameter). Then, all the tensor (6) are, with good accuracy, covariantly conserved modulo (7), since the covariant derivatives commute as far as λ is small. To be more specific, while $h_{\mu\nu;\varrho\sigma} \sim \mathrm{O}(\lambda^{-2})$, the commutator has the form

$$h_{\mu\nu;[\varrho\sigma]} = R_{\mu\lambda\varrho\sigma}(\gamma)\, h^{\lambda}_{\nu} + R_{\nu\lambda\varrho\sigma}(\gamma)\, h^{\lambda}_{\mu} \sim \mathrm{O}(1),$$

so that $t^{\nu}_{\mu;\nu} \sim \varepsilon^2 \times$(first derivative)$\times$(commutator of the second derivatives)

$$\sim \mathrm{O}(\varepsilon^2\lambda^{-1}) \ll t^{\nu}_{\mu} \sim \mathrm{O}(\varepsilon^2\lambda^{-2}) \quad \text{for} \quad \lambda \to 0.$$

If perturbations do not contribute significantly to the background curvature of space-time, both parameters ε, λ are independent, only $\varepsilon \ll \lambda$ must hold. The case $\varepsilon \sim \lambda$ has been carefully studied by Isaacson[9] who was inspired by the Brill–Hartle self-consistent field approximation method. The total background curvature, then, is entirely due to the microscopic waves represented by the perturbations satisfying Eq. (7). From a mathematical viewpoint, by shifting the terms which arise from the approximation method adopted from the left to the right-hand side of the vacuum field equations, Isaacson finds the effective energy-momentum tensor for high-frequency gravitational waves which acts as a source of the background curvature. This tensor is covariantly conserved and involves no second derivatives of $h_{\mu\nu}$, provided that the Brill–Hartle (BH) integral averaging and the commutativity of the second derivatives are used. Again, however, any tensor (6), in which the replacements $h_{\varrho\sigma,\tau} \to h_{\varrho\sigma;\tau}$ and $\eta_{\varrho\sigma} \to \gamma_{\varrho\sigma}$ are performed, and in which $A_3 \neq 0$ (see below), may be taken as the energy-momentum tensor of perturbations (in particular, the Einstein or the symmetric one may be preferred). These tensors do not contain second derivatives from the very beginning (without the BH averaging being necessary) and are, with good accuracy, conserved as a consequence of (7). When performing the BH averaging, and restricting the gauge as in (8), we get the expression

$$t_{\mu\nu}^{\mathrm{BH}} = 2\varepsilon^2 A_3 \langle h^{\varrho\sigma}_{;\mu} h_{\varrho\sigma;\nu} \rangle$$

which, for $A_3 = 1/64\pi$, coincides with that of Isaacson[15]. (The same result follows from the metric energy-momentum tensor for perturbations.)

The high-frequency waves propagating on a Schwarzschild background, after being radiated out due to nonradial pulsations of the spherically symmetric source, have been analyzed by Thorne *et al.*[8]; in their most interesting work, the damping effects have been exhibited for the first time in a physically realistic situation.

4 EXACT RADIATIVE SOLUTIONS OF EINSTEIN EQUATIONS

These, and other recent significant achievments in the theory of gravitational radiation are based on approximation methods confined usually to the first, or the second order[31]. Needless to say (in particular in a non-linear theory such as general relativity) one may always doubt whether exact solutions exist which correspond to the approximate solutions obtained. Then what about the exact radiative solutions? There are Einstein–Rosen cylindrical waves, various types of plane waves and Robinson–Trautman spherical waves which represent the only (known) exact radiative solutions. Also these

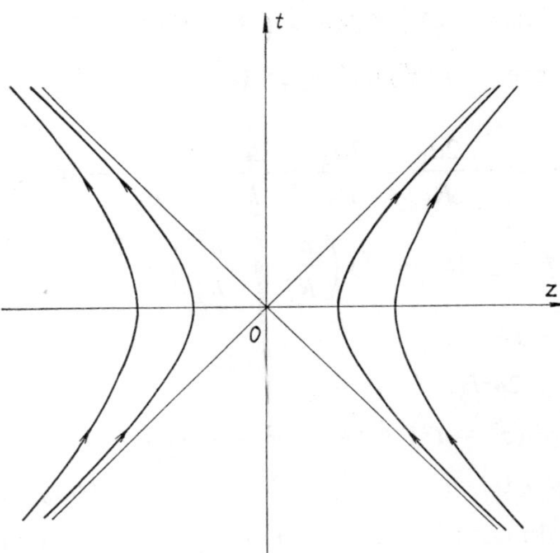

Figure 1.

solutions, however, are known to evoke doubts as to whether, by investigating their properties, physically significant conclusions about the nature of gravitational waves may be drawn. Indeed, the cylindrical waves correspond to a source extending to infinity. Except for general cases of cylindrical waves, none of the solutions in question is of the Petrov type I, while the linear approximation indicates that a realistic solution representing a *finite* radiating source should be just of this most general type. Moreover, the energy density and the momentum carried by the waves usually vanish when calculated by means of various energy-momentum complexes (for a review, see [16]; the calculations based on the Moller's tetrad complex are given in [17]).

In the final section we shall analyze the solution of Bonnor and Swaminarayan[18] (hereafter, BS solution), which is radiative and does not suffer from any of the drawbacks mentioned above. It also indicates an answer to the puzzling question whether the field of freely gravitating particles is radiative.

5 EXACT RADIATIVE SOLUTION REPRESENTING A BOUNDED SYSTEM[19]

As Oscar Wilde observed, the truth is rarely pure and never simple. Although the BS solution

$$\begin{aligned} \mathrm{d}s^2 = -e^{\lambda}\,\mathrm{d}\varrho^2 - \varrho^2 \mathrm{e}^{-\mu}\,\mathrm{d}\varphi^2 + (z^2 - t^2)^{-1} \{(z^2 \mathrm{e}^{\mu} - t^2 \mathrm{e}^{\lambda})\,\mathrm{d}t^2 \\ - (z^2 \mathrm{e}^{\lambda} - t^2 \mathrm{e}^{\mu})\,\mathrm{d}z^2 + 2zt\,(\mathrm{e}^{\lambda} - \mathrm{e}^{\mu})\,\mathrm{d}z\,\mathrm{d}t\}, \end{aligned} \tag{10}$$

where

$$\mu = -\frac{2a_1}{R_1} - \frac{2a_2}{R_2} + \frac{2a_1}{h_1} + \frac{2a_2}{h_2} + \ln k,$$

$$\lambda = \frac{a_1 a_2}{(h_1 - h_2)^2} f - \varrho^2 (z^2 - t^2) \left(\frac{a_1^2}{R_1^4} + \frac{a_2^2}{R_2^4} \right) + \frac{2a_1 R}{h_1 R_1} + \frac{2a_2 R}{h_2 R_2} + \ln k,$$

$$R = \tfrac{1}{2} (\varrho^2 + z^2 - t^2),$$

$$R_i = \{(R - h_i)^2 + 2\varrho^2 h_i\}^{1/2} \quad (i = 1, 2),$$

$$f = 4R_1^{-1} R_2^{-1} \{\varrho^2 (z^2 - t^2) + (R - \varrho^2 - h_1)(R - \varrho^2 - h_2) - R_1 R_2\},$$

$$a_1, a_2, h_1 > 0, \quad h_2 > 0, \quad k > 0 \quad \text{being arbitrary constants,}$$

is not simple, we do not propose to convince the reader that it is, physically, *quite* true. Nevertheless, at present it is the most realistic radiative solution of the Einstein equations, indeed the only known exact solution to represent moving particles, and a great deal of information can be read out of it. Let us first summarize the basic properties of the solution and then turn, to its radiative characteristics.

Provided $a_1, a_2 \neq 0$ the solution describes two pairs of mass points which are represented by the world-lines $\varrho = 0, z_i = \pm(t^2 + 2h_i)^{1/2}$ (see Figure 1), and are thus uniformly accelerated with acceleration $\pm(2h_i)^{-1/2}$ in "background" Minkowskian space-time. From the physical point of view the presence of two pairs of particles has only a formal character, for both pairs have absolutely equal properties, and each of them moves independently of the other (the hypersurfaces $z = \pm t$ are null). The metric has no physical singularities other than the points where particles occur, provided the condition $e^{\lambda} = e^{-\mu}$ for $\varrho = 0$ is fulfilled (which ensures the existence of a local inertial system on the z-axis). This condition can be satisfied by the choice of constants:

$$a_1 = (h_1 - h_2)^2 (2h_2)^{-1}, \quad a_2 = -(h_1 - h_2)^2 (2h_1)^{-1}, \quad k = 1.$$

Since the masses of particles are, except for positive factors, determined by constants a_1, a_2, the BS solution describes in this case two pairs of particles each of which contains one particle with positive and one with negative mass. The particles accelerate each other owing only to their mutual gravitational interaction. In the other cases the motion is caused by stresses on the z-axis. The choice

$$a_1 = (h_1 - h_2)^2 (2h_1)^{-1}, \quad k = 1, \quad a_2 > 0$$

corresponds to four particles with positive masses; one particle of a given pair moves freely, whereas the second is attached to a stress extending to infinity. If we take

$$a_2 = -(h_1 - h_2)^2 (2h_1)^{-1}, \quad k = \exp\{(h_1 - h_2)^2 (h_1 h_2)^{-1} - 2a_1 h_1^{-1}\}$$

and leave a_1 arbitrary, then in each pair the particle with negative mass moves under a stress of finite length and the other (with unrestricted mass) freely. Finally, if $a_1 > 0$, $a_2 = 0$, or $a_2 > 0$, $a_1 = 0$, the solution describes only two particles with the same positive mass connected by a stress of finite length; the source is thus bounded and contains only positive masses in this case. By means of coordinate transformations Bonnor and Swaminarayan demonstrated that, except for those listed above, all other singularities in the metric, in particular at $z = \pm t$ are only apparent. To support this, we calculated the invariant $R_{\varkappa\lambda\mu\nu}R^{\varkappa\lambda\mu\nu}$ which, indeed, remains finite at $z = \pm t$, while it diverges at the places where the particles are.

As to the group of motions, there are two independent Killing vector fields corresponding to the axial symmetry of the solution and to the invariance under the transformations

$$z^* = (z - Vt)(1 - V^2)^{-1/2}, \quad t^* = (t - Vz)(1 - V^2)^{-1/2},$$

where $V = \text{const.}$ (Lorentz transformations parallel to the axis $\varrho = 0$.) It can be seen that by going over to the asymptotically *non*-inertial, uniformly accelerated system in which the particles are at rest, the solution takes the form of the Weyl metric. But only the region $z^2 > t^2$ is mapped by this transformation! On the other hand, the region $t^2 > z^2$ is closely related to the non-singular part of the Einstein–Rosen solution[20]. Similarly as it is done for the Einstein–Rosen metric e.g. in [21] one can prove that no independent Killing vectors exist other than those given above. Of course, both Killing vectors are space-like in the region $t^2 > z^2$.

It is necessary to stress here that the Petrov type of the metric is I at all points of space-time, as was, after all, to be expected.

Now, let us turn to the radiative properties of the solution. These can best be investigated by means of the Bondi method[22]. In our paper[19] the radiative character of the solution has also been proved with the help of the Newman–Penrose technique; since, however, the Bondi method provides a more direct physical insight into the solution, we shall confine ourselves to the analysis in terms of the Bondi news function.

If we pass from coordinates $\{\varrho, z, \varphi\}$ to spherical coordinates $\{r, \theta, \varphi\}$ by means of the transformation $\varrho = r \sin\theta$, $z = r\cos\theta$, and introduce the retarded time $u = t - r$, we obtain the line element in a definite, rather involved form. The expansions of the metric tensor in r^{-1} with u, θ, φ fixed indicate that this metric is not asymptotically flat. In order that our metric be of Bondi's form, we must find a coordinate system $\bar{u}, \bar{r}, \bar{\theta}, \varphi$ such that $\bar{g}_{00}$, $\bar{g}_{01}, \bar{g}_{02}, \bar{g}_{22}, \bar{g}_{33}$ have the asymptotic form required by the Bondi method, and $\bar{g}_{11} = \bar{g}_{12} = 0$, $\bar{g}_{22} \cdot \bar{g}_{33} = \bar{r}^4 \sin^2\bar{\theta}$; then, the lines $\bar{u}, \bar{\theta}, \varphi = \text{const.}$ will be null geodesics, and $\bar{r}$ the luminosity distance. If we suppose that the transformation leading to this form may be expanded in powers of $\bar{r}^{-1}$, the requirements on $\bar{g}_{\mu\nu}$ restrict the undetermined functions of coordinates $\bar{u}$ and $\bar{\theta}$ which stand as the coefficients in the expansions in $\bar{r}^{-1}$. It may be proved that all the coefficients are uniquely determined up to the arbitrariness corresponding to the transformations from the Bondi–Metzner group which, of course, preserve the character of metric. Restricting ourselves, then, to the system in which $r = \bar{r} + O(1)$ and $\theta = \bar{\theta} + O(\bar{r}^{-1})$, we can derive the news function by comparing the expansions of our $\bar{g}_{\mu\nu}$ with the general asymptotic form of Bondi's metric. The news function, expressed as a function of u and θ, is as follows:

$$\mathcal{N} = \tfrac{1}{2}(e^{\beta}\sin^2\theta)^{-1}(1 - e^{\beta} - \alpha_u \sin^2\theta). \tag{11}$$

Here,

$$\alpha = -\frac{2a_1}{U_1} - \frac{2a_2}{U_2}, \quad \alpha_u \equiv \frac{\partial\alpha}{\partial u},$$

$$U_i = (u^2 + 2h_i \sin^2\theta)^{1/2},$$

$$\beta = \frac{4a_1 a_2}{(h_1 - h_2)^2}\left\{\frac{u^2 + (h_1 + h_2)\sin^2\theta}{U_1 U_2} - 1\right\}$$

$$+ \sin^4\theta\left\{\frac{a_1^2}{U_1^4} + \frac{a_2^2}{U_2^4}\right\} - \left\{\frac{2a_1 u}{h_1 U_1} + \frac{2a_2 u}{h_2 U_2}\right\}.$$

As is well-known, the character of the asymptotic expansion of the Riemann tensor tetrad components answers the question whether a given solution is to be classified as radiative. Now, the investigation of the news function enables us to prove that this expansion begins by $\bar{r}^{-1}$ irrespective of the choice of the constants a_i, h_i, k. Therefore, *the solution has radiative character for the particles constrained as well as for the particles moving owing only to their mutual gravitational interaction.* In the classification of Bondi[22], the BS solution belongs to the first, radiative class, and it thus represents a counter-example to the conjecture of Bondi (and of those supporting the view of Infeld) that the fields of freely moving particles are non-radiative, characterized by a vanishing news function.

Now let us ask how far the use of Bondi's method is justified. Doubts of two kinds may possibly arise: (1) Bondi *et al.* are interested only in outgoing radiation, and incoming radiation is supposed to be excluded by boundary conditions, whereas, owing to the time symmetry of the solution, we can investigate the metric and the Riemann tensor along null geodesics pointing into the past null infinity with the same results as when we investigated them along null geodesics pointing into the future null infinity. (2) Bondi method is suitable for the study of isolated systems: can use it also for the system of uniformly accelerated particles, which is not isolated permanently?

The example of the BS solution distinctly illustrates that the outgoing radiation condition used by Bondi *et al.* as a boundary condition on the metric does *not* exclude incoming radiation (contrary to the belief which seems to have once been held)[23]. In the Bondi method it is essential that the outgoing radiation be satisfied for fixed retarded time $\bar{u}$ and angle $\bar{\theta}$. The components of the metric tensor of the BS solution behave, roughly speaking, as the function $f = [(r^2 - t^2)^2 + a^2]^{-1/2}$ (a = const.), which corresponds to a time-symmetric pulse flowing from infinity inwards to the origin and then proceeding outwards again to infinity. Introducing the retarded time $u = t - r$ and the advanced time $v = t + r$ we can make sure that both for fixed retarded time u and for fixed advanced time v the function f is uniformly and radially smooth at infinity so that the outgoing radiation condition is satisfied for fixed u at large r, and the *incoming* radiation condition is satisfied for fixed v at large r. The radiation condition formulated in this way guarantees that the field corresponds either to outgoing or to incoming radiation at great distances, and that it contains no mixture of incoming and outgoing radiation[24].

For the time being we leave aside the question how the incoming radiation might be excluded in the BS solution. If, however, incoming radiation is present, the non-vanishing news function determined by the behaviour of field at the future null infinity need not mean that there is only mass loss due to the outgoing radiation. To investigate the mass *gain* due to *incoming* radiation, we should have to turn to the past null infinity. It should perhaps be remarked here that only the particles represent the source of the radiative field in the BS solution: the gravitational field described by the solution is not a combination of a field due to the particles and of some independent time-symmetric pulse of radiation, as might be assumed at first sight. That this is really so can be demonstrated by analyzing the solution in terms of retarded and advanced potentials[18], but we shall corroborate it in the following also by comparing the angular distribution of the radiated energy and the total rate of radiation from the freely moving particles, with the analogy of Maxwell's theory.

In order to discuss the question whether the Bondi method can be applied to a system which is not isolated permanently, we may refer to the proof of the possibility of introducing the Bondi metric for the BS solution. The retarded time u (or $\bar{u}$) is not, of course, the retarded time of particles performing a hyperbolic motion, but rather it corresponds to the retarded time of a system which is distributed permanently around the origin. When looking for a relation between our retarded time and the actual retarded time of particles, as well as when calculating the radiated energy, a comparison with the analogous case of uniformly accelerated charged particles within special relativity proves to be very valuable. Of course, our analogy with electrodynamics will have a particularly clear meaning if the masses of particles are small. We ought to bear in mind, however, that all limiting procedures are based on an *exact* solution of Einstein equations. The reader should refer to [19] for a detailed discussion (in particular in connection with radiation) of the analogy of the BS solution with Born's solution representing the field of uniformly accelerated charges in electrodynamics[25]. Here we wish to give only a few final results which appear to be remarkably lucid regarding the involved form of the BS metric from which are they derived.

It is well-known that, in an asymptotically flat space-time, the news function enables one to express the angular distribution of radiated energy (radiation pattern) at large distances r in the form

$$I = \frac{1}{4\pi} \mathcal{N}^2 (\bar{u}, \theta) \, r^{-2}, \tag{12}$$

which plays the role of the magnitude of Poynting's vector in electrodynamics. (This expression can be verified by using a number of energy-momentum complexes, especially the Einstein complex and the Landau–Lifshitz complex provided that coordinates are Minkowskian at infinity[26]. An observer at infinity can measure the energy flux I, in which $\mathcal{N}$ is given by (11), in each particular case of the BS solution. Since, however, the retarded time $\bar{u}$ does not represent the retarded time of the particles performing the hyperbolic motion and in addition the different proper retarded times correspond to different particles, there is no point in asking what *total* energy has been radiated at a given $\bar{u}$. We can, however, find the total radiated energy in the case of *freely* moving particles with sufficiently *small* masses. It is necessary to stress in advance that we now consider the whole field in the region $z + t > 0$ (see Figure 1) as the retarded field from the pair moving on $z > 0$, and we do not take into account the other pair, the field of which is considered as the advanced field[27] in $z + t < 0$.

At first, put $h_2 \equiv h$, $h_1 = h + \varepsilon$, where $\varepsilon > 0$ is a small quantity characterizing the mutual distance of the particles. The masses, then, are given by the expressions[28]:

$$m^{(1)} = \frac{\varepsilon^2}{2h\,(2h + 2\varepsilon)^{1/2}}, \quad m^{(2)} = -\frac{\varepsilon^2}{(2h + 2\varepsilon)\,(2h)^{1/2}}.$$

Next, consider two charged particles with the charges $e^{(1)} = m^{(1)}$, $e^{(2)} = m^{(2)}$ represented by the world lines $z^{(1)} = (t^2 + 2h + 2\varepsilon)^{1/2}$, $z^{(2)} = (t^2 + 2h)^{1/2}$ in flat space-time. We speak of $v^{(i)} = (\mathrm{d}z^{(i)}/\mathrm{d}t)$, $a^{(i)} = (\mathrm{d}^2 z^{(i)}/\mathrm{d}t^2)$ as the velocities and accelerations of the charges *as well as* of the gravitating particles (the velocities and accelarations with respect to the metric (10) are, with a good accuracy, equal to $v^{(i)}$ and $a^{(i)}$).

Finally let us compare the radiation patterns, which are calculated by means of the Poynting vector in electrodynamics, and, with the help of the news function (11) in the gravitational case. The news function now takes a simple form owing to the special situation under consideration. It should however be noted that the retarded time u occuring in (11) is a function of the actual retarded time u^* of the pair and of the angle θ at large r and it must be replaced by these quantities if the true radiation pattern is to be obtained. At the moment $u^* = 0$, i.e. when the charges and the gravitating particles are at rest at the turning point, the radiation pattern is given by

$$I_{\mathrm{elm.}} = \frac{\varepsilon^6}{4\pi} \frac{9\cos^4\theta \sin^2\theta}{(2h)^6} \frac{1}{r^2}$$

in the electromagnetic case, and by

$$I_{\mathrm{gr.}} = \frac{\varepsilon^6}{4\pi} \frac{9\cos^2\theta \sin^4\theta}{(2h)^6} \frac{1}{r^2}$$

in the case of freely gravitating masses. These expressions, found on the basis of the exact Born and BS solutions, can also be obtained by means of the standard multiple expansion technique (in the electromagnetic case) and by means of the Bonnor and Rotenberg double series approximation method[29] (in the gravitational case). (This corroboration is possible only at the turning point, since we are not dealing with permanently isolated systems.) Denoting the fourth time derivative of the octupole moment by $\ddddot{O}$, we can cast $I_{\mathrm{elm.}}$ and $I_{\mathrm{gr.}}$ into the forms

$$I_{\mathrm{elm.}} = \frac{1}{4\pi} \frac{\ddddot{O}^2}{144} \cos^4\theta \sin^2\theta \frac{1}{r^2},$$

$$I_{\mathrm{gr.}} = \frac{1}{4\pi} \frac{\ddddot{O}^2}{144} \cos^2\theta \sin^4\theta \frac{1}{r^2}.$$

Both radiation patterns thus depend on the octupole moment of the corresponding pair in the same way; a difference between tensor—and vector—

radiation is exhibited only in a slightly different dependence on the azimuth angle. If we want to calculate the total radiated energies at any time u^*, we first multiply the respective radiation patterns by the factor r^2 $(1 - v \cos \theta)$ (where $v = u^* (u^{*2} + 2h)^{-1/2}$) and then by integration obtain

$$\mathscr{R}_{\text{elm.}} = \frac{24\varepsilon^6}{35\,(2h)^6}, \quad \mathscr{R}_{\text{gr.}} = \frac{18\varepsilon^6}{35\,(2h)^6}.$$

As in the case of one uniformly accelerated charge[25], the pair of masses as well as the pair of charges emit radiation at a constant rate independent of u^*, equal to the rate at the moment $u^* = 0$ when the particles are at rest.

By means of the Poynting vector and the news function one can express also the total *momentum* radiated per unit time u^*. Lengthy integrations lead, in both cases, to the simple results: $\mathscr{P}_z = \mathscr{R}v$, where $\mathscr{R}$ denotes here the respective rates of radiation of energy; the same relation is valid also for one uniformly accelerated charge. $\mathscr{P}_z$ determines the change of the total momentum component in the direction of the symmetry axis, whereas the components in a direction perpendicular to the symmetry axis vanish at all times owing to the axial symmetry. As expected the particles do not radiate momentum at the turning point, since equal energy is emitted in mutually opposite directions.

Let us remark, finally, that the Born solution is the only known exact solution of Maxwell equations, given in closed form, which describes the radiative field of moving charged particles; in other cases the equation for retarded time has not been explicitly solved. It is not easy to believe that another exact solution of Einstein equations of similar type to the BS solution will be found soon.

REFERENCES AND NOTES

1. F. J. Belinfante, *Physica*, **7**, 449 (1940).
2. V. Fock, *Relativistic Theories of Gravitation* (Proceedings of a conference held in Warsaw and Jablonna) Ed. L. Infeld, Warsaw: Pergamon Press, Oxford, 1964.
3. Here and henceforth, $\eta^{\mu\nu}$ is the flat Minkowski metric with the signature $(+ - - -)$; the partial derivatives are denoted by a comma. In subsequent sections, the Riemann metric will be indicated by $g_{\mu\nu}$, and the covariant derivatives by a semicolon.
4. J. Bičák, *Czech. J. Phys.*, **B15**, 81 (1966).
5. H. Weyl, *Amer. J. Math.*, **66**, 591 (1944).
6. A. Trautman, *Lectures in General Relativity*, Brandeis Summer Institute in Theoretical Physics, 1964, Vol. I, Ed. S. Deser and K. W. Ford, Prentice-Hall Inc., New York, N.Y., 1965.

7. A. LICHNEROWICZ, *Relativity, Groups and Topology*. Ed. C. and B. DE WITT, Gordon and Breach Science Publishers Inc., New York, N.Y., 1964.
8. K. S. THORNE and A. CAMPOLATTARO, *Astrophys. J.*, **149**, 591 (1967); R. PRICE and K. S. THORNE, *Astrophys. J.*, **155**, 163 (1969); K. S. THORNE, preprint, May 1969.
9. R. A. ISAACSON, *Phys. Rev.*, **166**, 1263, 1272 (1968).
10. I thank Prof. A. TRAUTMAN for asking me this question.
11. Of course the local covariant conservation law $t^{\mu\nu}_{;\nu} = 0$ does not lead to an integral conservation law in general. However, if a space-time admits a Killing vector field, an integral conservation law follows for a symmetric $t^{\mu\nu}$. Moreover, there are important situations in which a preferred tetrad field exists, and the global conservation law can also be formulated (see [9], for example).
12. E. M. CORSON, *Introduction to Tensors, Spinors and Relativistic Wave Equations*, Blackie and Son Ltd., Glasgow, 1954, p. 120. This book gives a complete list of classical papers on higher-spin fields in special relativity. The equations are based on those of Fierz's (1939) original paper.
13. H. A. BUCHDAHL, *Nuovo Cim.*, **10**, 96 (1958); *ibid* **25**, 486 (1962).
14. R. PENROSE, *An Analysis of the Structure of Space-Time*, Adams Prize Essay, Princeton, 1967.
15. For the rules of BH integral averaging and for the behaviour of $t^{\mathrm{BH}}_{\mu\nu}$ under gauge transformations, the reader should refer to the paper of Isaacson[9], where the integral conservation laws are also formulated.
16. J. WEBER, *General Relativity and Gravitational Waves*, Interscience Publishers Ltd., New York, 1961.
17. K. KUCHAŘ and J. LANGER, *Czech. J. Phys.*, **B13**, 233 (1963).
18. W. B. BONNOR and N. S. SWAMINARAYAN, *Z. Phys.*, **177**, 240 (1964); W. B. BONNOR, *Wiss. Z. der F. Schiller Univ. Jena*, **15**, 71 (1966).
19. The detailed analysis of this solution, in particular from the aspect of radiation, is given in J. BIČÁK, *Proc. Roy. Soc.*, **A302**, 201 (1968). In the present article this work is briefly summarized and a few new points are discussed.
20. N. S. SWAMINARAYAN, *Commun. Math. Phys.*, **2**, 59 (1966).
21. A. Z. PETROV, *Einstein Spaces*, Pergamon Press, Oxford, 1964.
22. H. BONDI, M. G. J. VAN DER BURG and A. W. K. METZNER, *Proc. Roy. Soc.*, **A269**, 21 (1962).
23. For the Bondi method modified to systems of (unbounded) sources with cylindrical symmetry it has been shown by J. STACHEL [*J. Math. Phys.*, **7**, 1321 (1966)] that asymptotic conditions analogous to those used in the asymptotically spherical case do not exclude certain infinite incoming radiation trains.
24. R. K. SACHS, *Phys. Rev.*, **128**, 2851 (1962).
25. As to the puzzling problem of uniformly accelerated charges in electrodynamics, we adhere to the views of T. FULTON and F. ROHRLICH, *Annals Phys.*, **9**, 499 (1960). In any case, regardless of the problematic absence of a damping force, the field of a uniformly accelerated charge possesses a wave zone in which $\mathbf{E} \to \mathbf{H} \to r^{-1}$, $|\mathbf{E}| = |\mathbf{H}|$, and the angular distribution of radiated energy takes the expected form. Moreover, admitting advanced effects, no problems arise.
26. F. H. J. CORNISH, *Proc. Roy. Soc.*, **A282**, 358 (1964).
27. For a justification of this interpretation, see ref. 19. These considerations also suggest

how incoming radiation may be excluded. It seems natural to exclude it as in electrodynamics, and so to restrict the BS solution to a half of space-time $z + t > 0$ and to add a flat Minkowskian metric in $z + t < 0$. Of course, the question arises how to interpret the discontinuity on the null hypersurface $z + t = 0$. (We meet similar problems also in electrodynamics.[25])

28. The expressions defining the masses of the particles in the BS solution coincide with those of H. Bondi, *Rev. Mod. Phys.*, **29**, 423 (1957). A negative mass in general relativity was first considered in that paper. Bondi's sources are represented by extended matter and not by singularities. It can be seen that the Newtonian expression for the acceleration is approximately valid at the turning point of the particle trajectories, for small masses. These facts, and the close analogy with monopole charges, strongly suggest that the particles occurring in the BS solution are gravitational *monopoles*.
29. W. B. Bonnor and M. A. Rotenberg, *Proc. Roy. Soc.*, **A289**, 247 (1966).
30. Some drawbacks connected with the complexes which are *not* tensor densities of weight one (e.g. the Landau-Lifshitz complex) are indicated in C. Møller, *Max-Planck Festschrift,* Deutscher Verlag der Wissenschaft, Berlin, 1959, and in E. Schmutzer, *Relativistische Physik*, B. G. Teubner, Leipzig, 1968, p. 547.
31. Quite recently, very significant progress in the slow-motion approximation method (using matched asymptotic expansions) has been made by W. L. Burke (preprint, California Institute of Technology, February 1970); see also S. Chandrasekhar, F. P. Esposito, *Astrophys J.*, **160,** 153 (1970).

PAPER 6

Applications of SU (2) technique in general relativity

MOSHE CARMELI

Aerospace Research Laboratories, Wright-Patterson Air Force Base, Ohio, U.S.A.

ABSTRACT

Spin weighted field functions employed in discussing gravitational radiation problems in general relativity are considered as functions over the group SU(2). As a result the Newman–Penrose formalism for obtaining exact gravitationally-conserved quantities is given a group-theoretic interpretation.

1 INTRODUCTION

Recently[1] the Newman–Penrose formalism[2] for obtaining exact gravitationally-conserved quantities[3] was discussed and a group-theoretic interpretation was given to it. This was done by relating each triad of the orthogonal vectors on the sphere to an element $g \in O_3$, where O_3 is the three-dimensional rotation group.

As a result, the spin-weighted quantities employed by Newman and Penrose become functions of $g \in O_3$. Using the familiar relation between the special unitary group of order two, SU(2), and O_3, the field functions were written also as functions of $u \in \mathrm{SU}(2)$.

By working with functions defined over the group SU(2) one is able to apply some powerful mathematical methods known from the theory of representations[4].

In this paper we extend this discussion to spinor fields.

In Section 2 we introduce a spin frame[5] at each space-time point and emphasize its dependence on the groups SU(2) and SL(2, c) under change of frame. Here SL(2, c) is the unimodular group of order two.

In Section 3 we find the transformation property of the dyad components under certain change of the spin frame represented by a unitary matrix $\gamma \in \mathrm{SU}(2)$. In Section 4 we fix our coordinate system and spin frame in a very specific way. This enables us to identify two of the coordinates with variables of u. The last Section is devoted to discussion of expansion of spin weighted functions in terms of matrix elements of representations of SU(2); it is pointed out that this expansion is a natural generalization of the Fourier expansion.

2 PRELIMINARIES AND NOTATION

Following Newman and Penrose[5,6], we introduce at each point of (curved) space-time two basis spinors l^A and n^A which satisfy the normalization condition

$$l_A n^A = \varepsilon_{AB} l^A n^B = -l^A n_A = 1, \tag{2.1}$$

or the equivalent condition

$$l_A n_B - n_A l_B = \varepsilon_{AB}. \tag{2.2}$$

The spinors l_A and n_A provide a spin frame. An arbitrary spinor can then be written in terms of them.

In Eq. (2.1) and (2.2), and in the following, spin indices are denoted by capital letters A, B, ..., taking the values 1 and 2. The ε is the Levi–Civita symbol defined by

$$\varepsilon^{AB} = \varepsilon_{AB} = \begin{pmatrix} 0 & 1 \\ -1 & 0 \end{pmatrix}. \tag{2.3}$$

Spin indices are raised and lowered by means of the ε's.

Introduce a generic symbol for l^A and n^A, i.e. ζ_a^A where $a = 1, 2$, defined by

$$\zeta_1^A = l^A, \quad \zeta_2^A = n^A, \tag{2.4}$$

and

$$\bar{\zeta}_{\dot{1}}^{\dot{A}} = \bar{l}^{\dot{A}}, \quad \bar{\zeta}_{\dot{2}}^{\dot{A}} = \bar{n}^{\dot{A}}. \tag{2.5}$$

Latin indices $a, b, c, \ldots$ range over 1, 2; $\dot{a}, \dot{b}, \ldots$ over $\dot{1}, \dot{2}$.

With the new notation Eq. (2.2) then has the form

$$\varepsilon^{AB} = \varepsilon^{ab} \zeta_a^A \zeta_b^B, \tag{2.6}$$

and conversely,

$$\varepsilon_{ab} = \zeta_a^A \zeta_b^B \varepsilon_{AB}, \tag{2.7}$$

where ε^{ab} denotes an array of scalars which have the same numerical values as the components of ε_{AB}. Also define

$$\zeta^{aA} = \varepsilon^{ab} \zeta_a^A, \quad \zeta_b^A = \zeta^{aA} \varepsilon_{ab}. \tag{2.8}$$

The two basis spinors ζ_a^A are called a *dyad.* Field functions will be described in terms of *dyad components* which are scalars under spin frame transformations. The dyad components of a spinor $T_{AB\dot{X}\dot{Y}}$, for example, is

$$T_{ab\dot{x}\dot{y}} = \zeta_a^A \zeta_b^B \bar{\zeta}_{\dot{x}}^{\dot{X}} \bar{\zeta}_{\dot{y}}^{\dot{Y}} T_{AB\dot{X}\dot{Y}}. \tag{2.9}$$

The spin frame is chosen quite arbitrarily. A new spin frame $\zeta_a'^A$ might as well be introduced. The new spin frame is related to the "original" one by

$$\begin{pmatrix} \zeta_1'^A \\ \zeta_2'^A \end{pmatrix} = \begin{pmatrix} \alpha & \beta \\ \gamma & \delta \end{pmatrix} \begin{pmatrix} \zeta_1^A \\ \zeta_2^A \end{pmatrix}. \tag{2.10}$$

Here α, β, γ, and δ are arbitrary complex numbers satisfying the condition

$$\alpha\delta - \beta\gamma = 1. \tag{2.11}$$

We will denote the matrix in (2.10) by g,

$$g = \begin{pmatrix} \alpha & \beta \\ \gamma & \delta \end{pmatrix}. \tag{2.12}$$

The aggregate of all such matrices provides a group of transformations in the complex linear space of spin frames. This is the complex unimodular group of order two, SL(2, c). A subgroup of SL(2, c) which has a special importance in our discussion is that of all unitary matrices, the group SU(2).

Accordingly, the spin frame introduced at each point of space-time is defined up to a transformation $g \in \mathrm{SL}(2, c)$. This can formally be written as

$$\zeta_a^A = \zeta_a^A(g). \tag{2.13}$$

We also have

$$\zeta_a^A = \zeta_a^A(u), \tag{2.14}$$

where $u \in \mathrm{SU}(2)$, since $\mathrm{SU}(2) \subset \mathrm{SL}(2, c)$.

Suppose now we have a spin frame $\zeta_a^A(u)$. Define the unitary matrix γ by

$$\gamma = \begin{pmatrix} e^{-i\lambda/2} & 0 \\ 0 & e^{i\lambda/2} \end{pmatrix}. \tag{2.15}$$

Then under the transformation (2.15) $\zeta_a^A(u)$ go over into

$$\begin{pmatrix} \zeta_1^A(\gamma u) \\ \zeta_2^A(\gamma u) \end{pmatrix} = \begin{pmatrix} e^{-i\lambda/2} & 0 \\ 0 & e^{i\lambda/2} \end{pmatrix} \begin{pmatrix} \zeta_1^A(u) \\ \zeta_2^A(u) \end{pmatrix}. \tag{2.16}$$

Accordingly, we have

$$\zeta_\mp^A(\gamma u) = e^{\mp i\lambda/2} \zeta_\mp^A(u), \tag{2.17}$$

where $\zeta_-^A = \zeta_1^A$ and $\zeta_+^A = \zeta_2^A$.

3 TRANSFORMATION PROPERTY OF DYAD COMPONENTS

The dyad components of a spinor α_A are given by

$$\alpha_{\mp 1/2} = \alpha_A \zeta_\mp^A. \tag{3.1}$$

Since $\zeta_\pm^A$ are functions of $u \in SU(2)$, the dyad components are also functions of u,

$$\alpha_{\mp 1/2} = \alpha_{\mp 1/2}(u). \tag{3.2}$$

Under the transformation (2.15) $\alpha_{\pm 1/2}(u)$ go over into

$$\begin{aligned} \alpha_{\pm 1/2}(\gamma u) &= \alpha_A \zeta_\pm(\gamma u) \\ &= e^{\pm i\lambda/2} \alpha_{\pm 1/2}(u). \end{aligned} \tag{3.3}$$

A symmetric spinor Φ_{AB} will have three dyad components

$$\begin{aligned} \Phi_{\mp 1} &= \Phi_{AB} \zeta_\mp^A \zeta_\mp^B \\ \Phi_0 &= \Phi_{AB} \zeta_+^A \zeta_-^B. \end{aligned} \tag{3.4}$$

Under the transformation (2.15) they transform into

$$\Phi_m(\gamma u) = e^{im\lambda} \Phi_m(u), \tag{3.5}$$

where $m = -1, 0, 1$.

A symmetric 4-spinor Ψ_{ABCD} will have five components

$$\begin{aligned} \Psi_{\mp 2} &= \Psi_{ABCD} \zeta_\mp^A \zeta_\mp^B \zeta_\mp^C \zeta_\mp^D \\ \Psi_{\mp 1} &= \Psi_{ABCD} \zeta_\mp^A \zeta_\mp^B \zeta_\mp^C \zeta_\pm^D \\ \Psi_0 &= \Psi_{ABCD} \zeta_-^A \zeta_-^B \zeta_+^C \zeta_+^D. \end{aligned} \tag{3.6}$$

Again, under (2.15) they transform according to (3.5), where $m = -2, -1, 0, 1, 2$.

Functions of this kind have been introduced by Newman and Penrose to describe the Maxwell and Weyl tensors[5]. Their group dependence was pointed out by Carmeli[1]. A function which satisfies the condition (3.5) is called a *function of spin weight m*.

As has been shown by Naimark, functions of this type provide a natural Hilbert space for the principal series and the complementary series of (infinite) representations of the group $SL(2, c)$[4].

4 CHOICE OF COORDINATES AND TETRAD

So far we considered the local dependence of field functions on $u \in SU(2)$. For asymptotically flat space there is an essentially global dependence on u, the latter is related now to the coordinates of the system. To see this we fix the tetrad and the coordinate system as follows.

The spin frame introduced in Section 2 induces the normalized null tetrad basis l^μ, n^μ, m^μ, $\bar{m}^\mu$ in the corresponding space-time, where

$$\begin{aligned} l^\mu &\leftrightarrow l_A \bar{l}_{\dot{X}} \\ n^\mu &\leftrightarrow n_A \bar{n}_{\dot{X}} \\ m^\mu &\leftrightarrow n_A \bar{l}_{\dot{X}}, \end{aligned} \tag{4.1}$$

with $l_\mu n^\mu = -m_\mu \bar{m}^\mu = 1$. Other scalar products of these vectors vanish.

We now introduce some restrictions on the choice of coordinates and tetrad vectors[7]. A family of null hypersurfaces, designated by a parameter $\tau =$ const is chosen. The vector l_μ is then taken to be $l_\mu = \tau_{,\mu}$. Also τ is taken to be the coordinate x^0, hence $l_\mu = \delta^0_\mu$. The l^μ are tangent to a family of null geodesics lying within the hypersurfaces, and we choose an affine parameter r along these geodesics to be the coordinate x^1. The remaining two coordinates x^2 and x^3 are taken as angular coordinates θ and φ, respectively, whose choice singles out a particular one of the null geodesics on each of the hypersurfaces $\tau =$ const. We also make the further specification that n^μ and m^μ are to be parallely propagated in the direction of l^μ.

In addition to its dependence on φ and θ, we will choose m^μ to depend on another angle φ_2 which might be considered as the angle between the space-like vector Re (m^μ) and the curve $\varphi =$ constant.

For any value of the variables φ, θ and φ_2 we now associate an element $u \in SU(2)$ whose Euler's angles are $\pi/2 - \varphi$, θ, and φ_2,

$$u = u\left(\frac{\pi}{2} - \varphi, \theta, \varphi_2\right). \tag{4.2}$$

Accordingly, the tetrad and the spin frame (4.1) can be considered as functions of τ, r, and $u \in SU(2)$. As a result, the dyad components become also functions of these variables, and their transformation property given in Section 3 is valid.

5 EXPANSION OF FUNCTIONS OF SPIN WEIGHT s

Let $T^j_{mn}(u)$ be the matrix elements of the irreducible representations of SU(2) in their canonical basis. Here $m, n = -j, -j+1, \ldots, j$, where $j = 0, 1/2, 1, 3/2, \ldots$ is the weight of the representation. They satisfy the orthogonality relation[4]

$$\int T^j_{mn}(u)\, \overline{T^{j'}_{m'n'}(u)}\, du = (2j+1)^{-1}\, \delta_{jj'}\delta_{mm'}\delta_{nn'}. \tag{5.1}$$

The integral in (5.1) is an invariant integral which have the property[8]

$$\begin{aligned} \int f(uu_1)\, du &= \int f(u_1 u)\, du \\ &= \int f(u^{-1})\, du \\ &= \int f(u)\, du, \end{aligned}$$

for any $u_1 \in SU(2)$, and

$$\int du = 1.$$

The functions $T^j_{mn}(u)$ form a complete orthogonal system for all functions whose modulus square is integrable with the measure du [9]. Accordingly,

$$f(u) = \sum_{j=0}^{\infty} \sum_{m,n=-j}^{j} \alpha^j_{mn} T^j_{mn}(u), \tag{5.2}$$

where α^j_{mn} are independent of u, and given by

$$\begin{aligned} \alpha^j_{mn} &= (2j+1)\, A^j_{mn}, \\ A^j_{mn} &= \int f(u)\, \overline{T^j_{mn}(u)}\, du. \end{aligned} \tag{5.3}$$

If f is also a function of τ and r, then A^j_{mn} are also functions of these variables.

The expression given above in terms of T^j_{mn} is analogous to that of the Fourier transform. The well known Plancherel's formula for the Fourier transform will have the form

$$\int |f(u)|^2 \, du = \sum_j (2j + 1) \sum_{m,n} |A^j_{mn}|^2 \tag{5.4}$$

in the present case. Just as the usual Fourier transform realizes a decomposition of the regular representation of the additive group of real numbers into its irreducible representations, the generalized Fourier transform (5.3) realizes an isometric mapping of the regular representation of SU(2) onto the direct sum of irreducible representations $u \to T^j(u)$, where each representation $u \to T^j(u)$ is included in this direct sum $2j + 1$ times. An analogous proposition and a formula similar to Eq. (5.4) hold for any compact topological group. A somewhat more complicated formula holds for the group SL(2, c).[4]

Examination of the behavior of different functions under the transformation $u \to \gamma u$, where γ is given by (2.15), shows that a function $f(u)$ of spin weight s can be expanded as[10]

$$f(u) = \sum_{j=|s|}^{\infty} \sum_{n=-j}^{j} \alpha^j_n \, T^j_{sn}(u) . \tag{5.5}$$

For example the dyad components of a symmetric spinor Ψ_{ABCD} can be expanded as ($s = -2, -1, 0, 1, 2$)

$$\Psi_s = \sum_{j=|s|}^{\infty} \sum_{n=-j}^{j} \Psi^j_{sn}(\tau, r) \, T^j_{sn}(u) . \tag{5.6}$$

If Ψ represent the Weyl spinor, then $\Psi^j_{sn}(\tau, r)$ can be determined by the field equations. Under certain conditions, for example, one can assume that

$$\Psi^j_{sn}(\tau, r) = a^j_{sn}(\tau)/r^{s+3} + \cdots . \tag{5.7}$$

Then the field equations will determine some of the coefficients in (5.7).

In particular, if we write

$$\Psi^2_{2n}(\tau, r) = \frac{a_n(\tau)}{r^5} + \frac{b_n(\tau)}{r^6} + \cdots , \tag{5.8}$$

then Einstein's field equations give

$$\frac{db_n}{d\tau} = 0, \quad (n = -2, -1, 0, 1, 2) . \tag{5.9}$$

These five complex constants are the Newman–Penrose constants[3].

REFERENCES

1. M. Carmeli, *J. Math. Phys.*, **10**, 569 (1969).
2. E. T. Newman and R. Penrose, *J. Math. Phys.*, **7**, 863 (1966).
3. E. T. Newman and R. Penrose, *Phys. Rev. Lett.*, **15**, 231 (1965).
4. M. A. Naimark, *Linear Representations of the Lorentz Group*, Pergamon Press, Inc., New York, 1964.
5. E. T. Newman and R. Penrose, *J. Math. Phys.*, **3**, 566 (1962).
6. F. A. E. Pirani, *Lectures on General Relativity*, Vol. 1, Brandeis Summer Institute in Theoretical Physics, Prentice-Hall, Inc., New Jersey, 1964.
7. A. I. Janis and E. T. Newman, *J. Math. Phys.*, **6**, 902 (1965).
8. The concept of invariant integral is discussed in [4]. See also J. D. Talman, *Special Functions: A Group Theoretic Approach*, based on lectures by E. P. Wigner, W. A. Benjamin, Inc., New York, 1968.
9. The functions $T^j_{mn}(u)$ satisfy the following relations

$$K_{\pm}T^j_{mn} = [(j \pm m + 1)(j \mp m)]^{1/2}\, T^j_{m\pm 1,n}$$
$$K_3 T^j_{mn} = mT^j_{mn}.$$

Here the operators $K_{\pm}$ and K_3 are given by

$$K_{\pm} = e^{\mp i\varphi_2}\left(\pm \operatorname{cotg}\theta \frac{\partial}{\partial\varphi_2} + i\frac{\partial}{\partial\theta} \mp \operatorname{cosec}\theta \frac{\partial}{\partial\varphi_1}\right)$$
$$K_3 = i\frac{\partial}{\partial\varphi_2}.$$

See M. Carmeli, *J. Math. Physics*, **10**, 1699 (1969).

10. It will be noted, using Eq. (5.5), that K_+, K_-, and K_3, given in [9], when operating on a function $f_s(u)$ of spin weight s, will give new functions of spin weights $s+1$, $s-1$, and s, respectively,

$$K_{\pm} f_s(u) = f'_{s\pm 1}(u)$$
$$K_3 f_s(u) = f'_s(u).$$

PAPER 7

On the energy tensor of a perfect fluid

C. CATTANEO
Università di Roma, Italy

The standard energy-momentum tensor of a non-viscous relativistic fluid has the well-known form

$$T^{ik} \equiv c^2\varrho_0 u^i u^k + p u^{ik} \equiv (c^2\varrho_0 + p)\, u^i u^k + p g^{ik} \quad (i, k = 1, 2, 3, 4) \qquad (1)$$

where u^i is the unit vector tangent to the trajectories, $u^{ik} \equiv g^{ik} + u^i u^k$ is the space metric tensor in the local rest frame, p is the proper pressure and ϱ_0 is the proper mass density, which includes pure matter-density energy as well as thermodynamic energy. The equations of motion of the fluid can then be obtained by equating to zero the divergence of the tensor T^{ik}

$$\nabla_k T^{ik} \equiv \nabla_k \{(c^2\varrho_0 + p)\, u^i u^k + p g^{ik}\} = 0. \qquad (2)$$

From Eq. (2) we can deduce, with simple adaptations, most of the principal statements of Newtonian fluid dynamics.

In particular, for a barotropic flow in which a direct relation between ϱ_0 and p is assumed,

$$\varrho_0 = \varrho_0(p), \qquad (3)$$

these equations show that sound waves propagate through the fluid with exactly the same velocity as in Newtonian fluid dynamics

$$V = 1/\sqrt{\mathrm{d}\varrho_0/\mathrm{d}p}. \qquad (4)$$

In relativity V must not exceed c:

$$V \leqslant c; \qquad (5)$$

this implies a *separate axiom* which imposes a lower limit to the differential compressibility of the fluid

$$\frac{d\varrho_0}{dp} \geqslant \frac{1}{c^2}. \tag{6}$$

The question arises whether the inequality (5) could be obtained directly from the form of the energy-momentum tensor, without the need of a further *ad hoc* hypothesis.

For this purpose let me start with the following preliminary considerations. For a general continuous material medium it is possible to separate its energy-momentum tensor into a matter *tensor* $c^2\varrho_0 u^i u^k$, which includes all the standard internal matter energies, and a tensor of pure tension τ^{ik}.

In special relativity, and by extension also in general relativity, it is usual to postulate that in a local inertial frame this tensor has the following form

$$\overset{ik}{\tau_0} \equiv \begin{pmatrix} \overset{11}{\tau_0} & \overset{12}{\tau_0} & \overset{13}{\tau_0} & 0 \\ \overset{21}{\tau_0} & \overset{22}{\tau_0} & \overset{23}{\tau_0} & 0 \\ \overset{31}{\tau_0} & \overset{32}{\tau_0} & \overset{33}{\tau_0} & 0 \\ 0 & 0 & 0 & 0 \end{pmatrix} \tag{7}$$

where the nine $\overset{\alpha\beta}{\tau_0}$ components ($\alpha, \beta = 1, 2, 3$) have the usual meaning of space stress components. From this axiom it follows that, if n^k is an arbitrary unit vector orthogonal to u^k, the contact four-force $\tau^{ik}n_k$ acting across a unit area normal to n^k is a purely mechanical force

$$\tau^{ik}n_k u_i = 0 \quad : \tag{8}$$

it seems to me that this last condition is all that our empirical experience suggests about the tensor τ^{ik}. If we accept only this condition, it is easy to see that it implies no more than

$$\tau^{4\beta} = 0, \quad (\beta = 1, 2, 3) \tag{9}$$

and leaves open the possibility that $\overset{44}{\tau_0}$ be different from zero, in other words the possibility of attributing a proper energy content to a field of pure tension.

I wish to emphasize that, in any case, such energy content should not have to be confused with the standard internal energy *produced* by the mechanical tensions through possible matter deformation, since this thermodynamical energy is already included in $\varrho_0 c^2$. This extra energy term would be associated with the tension field independently of the nature of the material medium and it would be present also in an incompressible fluid.

What choice of τ_0^{44} is possible for a nonviscous fluid? We can get a suggestion from a naive elementary example. In Newtonian mechanics an incompressible fluid (ϱ = const) in stationary motion satisfies Bernoulli's equation

$$\tfrac{1}{2}\varrho v^2 + p + W = \text{const} \tag{10}$$

along each flow-line, where W is the potential energy, per unit volume, of the external field. In this equation p evidently plays a role not different from that of W, namely of a specific potential energy. This suggests that p should be a possible value of τ_0^{44} in the case of a non-viscous fluid.

A qualitative argument in favour of attributing a proper-energy content to a field of pure mechanical tension is provided by the unity of physics, since all other physical fields, e.g. Maxwell field, possess an intrinsic energy. With this choice the energy-momentum tensor of the fluid is slightly modified as follows:

$$\overset{*}{T}{}^{ik} \equiv (c^2\varrho_0 + p)\, u^i u^k + p u^{ik} \equiv (c^2\varrho_0 + 2p)\, u^i u^k + p g^{ik}. \tag{11}$$

Accordingly, the equations of motion are modified thus:

$$\nabla_k \overset{*}{T}{}^{ik} \equiv \nabla_k \{(c^2\varrho_0 + 2p)\, u^i u^k + p g^{ik}\} = 0. \tag{12}$$

The modification so introduced in the theory does not change the essential features of the fluid. The internal forces are not changed at all; all the known properties concerning rotational motions, irrotational motions, steady motions, etc., still hold with obvious adaptations.

Also quantitatively only slight modifications occur, at least as long as the pressure is not too large.

In contrast, the velocity of sound waves is significantly modified, even for small pressures. The study of characteristic hypersurfaces of the equations of motion leads to the following modified formula for the sound velocity

$$V = \frac{c}{\sqrt{1 + c^2\, \mathrm{d}\varrho_0/\mathrm{d}p}}. \tag{13}$$

This formula shows that, if we retain the reasonable hypotesis $\mathrm{d}\varrho_0/\mathrm{d}p \geqslant 0$, V never exceeds the speed of light. The limiting value c is attained only in ideal fluids, definable by the condition ϱ_0 = const, which more justly can be called incompressible.

There is no time for details or critical remarks. I wish to emphasize that I have only suggested a physical conjecture, which perhaps is not devoid of plausibility.

REFERENCES

1. W. RINDLER, *Special Relativity*, Oliver and Boyd, Edinburgh, 1966, Chapter VIII.
2. C. CATTANEO, *Rend. Accad. Naz. dei Lincei* **46**, 698 (1969).

PAPER 8

Approximate radiative solutions of Einstein–Maxwell equations

Y. CHOQUET-BRUHAT

Faculté des Sciences, Paris, France

INTRODUCTION

We will construct, by a general method derived from the W.K.B. method, approximate solutions of the Einstein–Maxwell equations, where the metric is the sum of a background metric and a rapidly oscillating perturbation, whereas the electromagnetic field is the sum of a background field and a rapidly oscillating one, which will be found to be of the pure radiation type.

We will show that, despite the non-linear character of Einstein equations, the significant part of the perturbation in the metric, and the radiative electromagnetic field, propagate along the rays associated with wave fronts of the background metric, and satisfy ordinary linear differential equations along those rays. This is related to the "exceptional" character of wave fronts in the Einstein–Maxwell theory: signals are transmitted without deformation, which is not in general the case for non-linear partial differential equations.

The non-linear character of the theory appears in the complementary conditions, which have to be satisfied in the construction of an approximate solution. They impose restrictions on the background, and express the energy loss due to the gravitational and electromagnetic radiations.

1 HYPOTHESES AND DEFINITIONS

Let $\bar{g}_{\alpha\beta}(x)$ be an arbitrary hyperbolic metric on a space time V_4 (four dimensional differentiable manifold), and

$$F_{\alpha\beta}(x, \omega\varphi) = G_{\alpha\beta}(x) + H_{\alpha\beta}(x, \omega\varphi), \tag{1.1}$$

an electromagnetic field on V_4 (exterior differentiable 2-form), sum of a slowly varying field $G_{\alpha\beta}(x)$ and a rapidly oscillating one $H_{\alpha\beta}(x, \omega\varphi)$, where ω is a large (real) parameter, and φ a scalar function on V_4. For any function $f(x, \omega\varphi)$ on V_4 we have, with obvious notations,

$$\partial f/\partial x^\alpha = \partial_\alpha f + \omega f'\varphi_\alpha, \quad \varphi_\alpha = \partial_\alpha\varphi. \tag{1.2}$$

We will look for a metric on V_4:

$$g_{\alpha\beta}(x, \omega, \varphi) = \bar{g}_{\alpha\beta}(x) + \frac{1}{\omega}\overset{1}{g}_{\alpha\beta}(x, \omega\varphi) + \frac{1}{\omega^2}\overset{2}{g}_{\alpha\beta}(x, \omega\varphi) + \cdots, \tag{1.3}$$

such that the Einstein and Maxwell equations are satisfied to order one in ω, i.e. such that there exists a constant M (independent of ω) with:

$$\sup_{x\in V_4} |R_{\alpha\beta} - \tau_{\alpha\beta}| \leqslant M\omega^{-1}, \quad \forall\omega, \tag{1.4}$$

$$\sup_{x\in V_4} |\nabla_\alpha F^{\alpha\beta}| \leqslant M\omega^{-1},$$

$$\sup_{x\in V_4} |\oint \nabla_\alpha F_{\beta\gamma}| \leqslant M\omega^{-1}, \quad \forall\omega. \tag{1.5}$$

2 EINSTEIN'S EQUATIONS

We obtained in [1] the expansion in powers of ω of $g^{\alpha\beta}$, $\Gamma^\lambda_{\alpha\beta}$, and $R_{\alpha\beta}$ for the metric (1.3)

$$g^{\alpha\beta} = \bar{g}^{\alpha\beta} + \frac{1}{\omega}\overset{1}{g}{}^{\alpha\beta} + \frac{1}{\omega^2}\overset{2}{g}{}^{\alpha\beta} + \cdots, \quad \overset{1}{g}{}^{\alpha\beta} = -\bar{g}^{\alpha\lambda}\bar{g}^{\beta\mu}\overset{1}{g}_{\lambda\mu}, \tag{2.1}$$

$$\Gamma^\lambda_{\alpha\beta} = \overset{0}{\Gamma}{}^\lambda_{\alpha\beta} + \frac{1}{\omega}\overset{1}{\Gamma}{}^\lambda_{\alpha\beta} + \cdots, \quad \overset{0}{\Gamma}{}^\lambda_{\alpha\beta} = \bar{\Gamma}^\lambda_{\alpha\beta} + \frac{1}{2}\bar{g}^{\lambda\mu}(\overset{1}{g}{}'_{\beta\mu}\varphi_\alpha + \overset{1}{g}{}'_{\alpha\mu}\varphi_\beta - \overset{1}{g}{}'_{\alpha\mu}\varphi_\mu), \tag{2.2}$$

$$R_{\alpha\beta} = \omega\overset{-1}{R}_{\alpha\beta} + \overset{0}{R}_{\alpha\beta} + \cdots. \tag{2.3}$$

We thus have, if $\tau_{\alpha\beta}$ is the Maxwell tensor of $F_{\alpha\beta}$ and $\bar{\tau}_{\alpha\beta}$ of $H_{\alpha\beta}$:

$$\tau_{\alpha\beta} = \overset{0}{\tau}_{\alpha\beta} + \frac{1}{\omega}\overset{1}{\tau}_{\alpha\beta} + \cdots, \tag{2.4}$$

with (a bar denotes contravariant components taken with $\bar{g}^{\alpha\beta}$)

$$\overset{0}{\tau}_{\alpha\beta} = \tfrac{1}{4}\, \bar{g}_{\alpha\beta}\, (\bar{F}^{\lambda\mu} F_{\lambda\mu}) - \bar{F}^{\lambda}_{\alpha} F_{\lambda\beta}. \tag{2.5}$$

Einstein's equations,

$$R_{\alpha\beta} = \tau_{\alpha\beta}, \tag{2.6}$$

imply thus, to order zero in ω, as in [1],

$$\overset{-1}{R}_{\alpha\beta} = 0, \tag{2.7}$$

which, if φ is a null surface (condition found necessary for the perturbation to have an intrinsic meaning), reduce to four algebraically independent conditions which read, if we choose as coordinate $x^0 = \varphi$ (radiative coordinates)

$$\bar{g}^{ij} \overset{1}{g}{}'_{ij} = 0, \quad \bar{\varphi}^j \overset{1}{g}{}'_{ij} = 0. \tag{2.8}$$

The null character of φ is:

$$\bar{g}^{\alpha\beta} \varphi_\alpha \varphi_\beta = 0, \tag{2.9}$$

i.e., in radiative coordinates

$$\varphi_\alpha = \delta^0_\alpha, \quad \bar{g}^{00} = 0. \tag{2.10}$$

The verification to order one in ω of (2.6) imposes:

$$\overset{0}{R}_{\alpha\beta} = \overset{0}{\tau}_{\alpha\beta}. \tag{2.11}$$

We have shown in ref. 1 that $\overset{0}{R}_{ij}$ takes, in radiative coordinates, a very simple form of a propagation operator along the rays φ_α. We thus obtain here:

$$-\bar{\varphi}^h\, \widetilde{\nabla}_h \overset{1}{g}{}'_{ij} - \tfrac{1}{2} \overset{1}{g}{}'_{ij}\, \bar{\nabla}_\lambda \bar{\varphi}^\lambda + \bar{R}_{ij} - \overset{0}{\tau}_{ij} = 0. \tag{2.12}$$

($\widetilde{\nabla}_h \overset{1}{g}{}'_{ij}$ is the covariant derivative in the metric $\bar{g}_{\alpha\beta}$ of the tensor with components $\overset{1}{g}{}'_{ij}$ and 0.)

The $\overset{1}{g}_{\alpha 0}$ may be taken as zero (they have no intrinsic meaning, cf. ref. 1), and the equations $\overset{0}{R}_{\alpha 0} = \overset{0}{\tau}_{\alpha 0}$ may always be satisfied, algebraically, by the choice of $\overset{2}{g}_{ij}$.

The conditions (2.8) and (2.12) imply

$$\bar{\varphi}^j\, (\bar{R}_{ij} - \overset{0}{\tau}_{ij}) = 0, \quad \bar{g}^{ij}\, (\bar{R}_{ij} - \overset{0}{\tau}_{ij}) = 0. \tag{2.13}$$

Conversely, if (2.12) and (2.13) are satisfied, (2.8) may be looked upon as an initial condition.

3 MAXWELL EQUATIONS

The study of the Maxwell equations by the W.K.B. method has been done in the flat case by Kline[2] and in a given curved space by Ehlers[3]. The results are then analogous to those found by Lichnerowicz[4] in the study of discontinuities of the electromagnetic field and their propagation. We will include, now, the effect of the perturbation (1.3) of the metric.

Maxwell equations in empty space

$$\nabla_\alpha F^{\alpha\beta} = 0 \quad \text{and} \quad \oint \nabla_\alpha F_{\beta\gamma} = 0, \tag{3.1}$$

imply, to order zero in ω [for (1.1), using (1.2)]:

$$\varphi_\alpha \overline{H}'^{\alpha\beta} = 0 \quad \text{and} \quad \oint \varphi_\alpha H'_{\beta\gamma} = 0. \tag{3.2}$$

Thus $H'_{\alpha\beta}$ has to be a pure radiation field (cf. ref. 4) and φ_α a null vector in the metric $\bar{g}_{\alpha\beta}$ (2.9). Since the term independent of ω in $H_{\alpha\beta}$ may be incorporated in $G_{\alpha\beta}$, we will suppose that (3.2) is also satisfied by $H_{\alpha\beta}$. Then there exists a vector b_α (x, ω, φ), orthogonal to φ_α such that:

$$H_{\alpha\beta} = \varphi_\alpha b_\beta - \varphi_\beta b_\alpha, \quad \bar{b}^\alpha \varphi_\alpha = 0. \tag{3.3}$$

To satisfy Eq. (3.1), to order one in ω, one has to impose:

$$\overline{\nabla}_\alpha (\bar{G}^{\alpha\beta} + \overline{H}^{\alpha\beta}) + (\overset{0}{\Gamma}{}^\alpha_{\alpha\lambda} - \overline{\Gamma}{}^\alpha_{\alpha\lambda}) (\bar{G}^{\lambda\beta} + \overline{H}^{\lambda\beta}) + \varphi_\alpha [(\bar{g}^{\alpha\lambda} \overset{1}{g}{}^{\beta\mu} + \overset{1}{g}{}^{\alpha\lambda} \bar{g}^{\beta\mu}) F_{\lambda\mu}]' = 0. \tag{3.4}$$

It is then straightforward to see, using conditions of order zero on Einstein (2.7) and Maxwell (3.2) equations, that all the complementary terms involving products in $\overset{1}{g}$ and H vanish, and that (3.4) reduces to

$$\overline{\nabla}_\alpha (\bar{G}^{\alpha\beta} + \overline{H}^{\alpha\beta}) + f^\beta = 0, \tag{3.5}$$

where

$$f^\beta \equiv \tfrac{1}{2} \bar{g}^{\alpha\mu} \overset{1}{g}{}'_{\alpha\mu} \varphi_\lambda \bar{G}^{\lambda\beta} + \varphi_\alpha (\bar{g}^{\alpha\lambda} \overset{1}{g}{}'^{\beta\mu} + \overset{1}{g}{}'^{\alpha\lambda} \bar{g}^{\beta\mu}) G_{\lambda\mu} = 0. \tag{3.6}$$

The fact that no quadratic term in the oscillatory parts appears in (3.4) shows the exceptional character (in the sense of Lax–Boillat) of wave fronts in the Einstein–Maxwell theory.

The second Maxwell equation (3.1) imposes, to order one:

$$\oint \overline{\nabla}_\alpha (G_{\beta\gamma} + H_{\beta\gamma}) = 0. \tag{3.7}$$

Eqs. (3.5), (3.7) and (3.2) lead to propagation equations by simple combinations (one has to use the fact that $\overline{\nabla}_\alpha \varphi_\varrho = \overline{\nabla}_\varrho \varphi_\alpha$, which gives $\bar{\varphi}^\varrho \overline{\nabla}_\varrho \varphi_\alpha = 0$, since the rays are null geodesics of $\bar{g}_{\alpha\beta}$):

$$2\bar{\varphi}^{\varrho}\bar{\nabla}_{\varrho}H_{\alpha\beta} + H_{\alpha\beta}\bar{\nabla}_{\varrho}\bar{\varphi}^{\varrho} + \bar{\varphi}^{\varrho} \oint \bar{\nabla}_{\alpha}G_{\beta\varrho}$$

$$- \varphi_{\alpha}(\bar{\nabla}_{\varrho}\bar{G}^{\varrho}_{\beta} + f_{\beta}) - \varphi_{\beta}(\bar{\nabla}_{\varrho}\bar{G}^{\varrho}_{\alpha} + f_{\alpha}) = 0. \tag{3.8}$$

From (3.8) one deduces easily, by (3.3), a propagation equation for b^{α}, which reads, in radiative coordinates,

$$\bar{\varphi}^{j}\bar{\nabla}_{j}b_{i} + \tfrac{1}{2}b_{i}\bar{\nabla}_{\varrho}\bar{\varphi}^{\varrho} - g_{i} = 0, \tag{3.9}$$

$$g_{i} \equiv -\bar{\varphi}^{j} \oint \bar{\nabla}_{0}G_{ij} + \bar{\nabla}_{\varrho}G^{\varrho}_{i} + f_{i} = 0. \tag{3.10}$$

Conversely (3.9) and (3.3) will imply (3.5) and (3.7) if $G_{\alpha\beta}$ satisfies the necessary conditions (in radiative coordinates)

$$\bar{\varphi}^{j} \oint \bar{\nabla}_{i}G_{hj} = 0. \tag{3.11}$$

4 INTEGRATION OF PROPAGATION EQUATIONS

The propagation Eqs. (2.12) and (3.9) are, if $\bar{g}_{\alpha\beta}$ and $G_{\alpha\beta}$ are given, a system of ordinary differential equations along the rays $\bar{\varphi}^{\alpha}$. They may be written, $U(x, \xi)$ being a vector with components $\overset{1}{g}{}'_{ij}$, b_i:

$$\frac{\mathrm{d}}{\mathrm{d}t}(kU) = A(x)\,U + B(x, \xi), \quad \frac{\mathrm{d}}{\mathrm{d}t} \equiv \bar{\varphi}^{\lambda}\lambda_{\lambda}. \tag{4.1}$$

In the above equation, k is the dilatation factor of the rays, i.e.

$$\frac{1}{2}\bar{\nabla}_{\lambda}\varphi^{\lambda} = \frac{\mathrm{d}}{\mathrm{d}t}\log k; \tag{4.2}$$

$A(x)$ is a known matrix, which vanishes with the Christoffel symbols of $\bar{g}_{\alpha\beta}$; and $B(x, \xi)$ is a vector with components $k(\bar{R}_{ij} + \overset{0}{\tau}_{ij})$ and kg_i.

These equations reduce, if $A = B = 0$, to conservation equations, and give kU = constant, along the rays. In the general case their integration is straightforward and gives (the integration being performed along the rays $\bar{\varphi}^{\alpha}$, cf. ref. 1, Section 8)

$$k(x)\,U(x, \xi) = \Phi(x)\,[\Psi(x, \xi) + \Theta(x, \xi)], \tag{4.3}$$

where

$$\Phi = \exp \int_0^t A\,\mathrm{d}\tau, \tag{4.4}$$

and

$$\Psi = \int_0^t B\Phi^{-1}\,\mathrm{d}\tau. \tag{4.5}$$

Θ is determined by the values of U on an initial submanifold S, transversal to the rays:

$$[U(x, \xi)]_{x \in s} = \Theta. \tag{4.6}$$

5 CONCLUSIONS

The constructed perturbation $\overset{1}{g}_{\alpha\beta}$, and the field $H_{\alpha\beta}$ will satisfy the inequalities (1.4) and (1.5) if they are uniformly bounded (in ω and x) together with their second (respectively first) derivatives.

A study analogous to the one given in [1] shows that the background metric has to satisfy equations of the type

$$\bar{R}_{\alpha\beta} - \bar{\tau}_{\alpha\beta} = \tau\varphi_\alpha\varphi_\beta, \tag{5.1}$$

with

$$\tau > 0. \tag{5.2}$$

The electromagnetic energy $|\bar{b}^\alpha b_\alpha|$ of $H_{\alpha\beta}$, and the energy of the gravitational radiation $|\frac{1}{4}\overset{1}{g}'^{ij}\overset{1}{g}'_{ij}|$ due to the perturbation $\overset{1}{g}_{\alpha\beta}$, will sum up and impose a choice of the background, whereby the mass loss due to both these radiations appears explicitly.

REFERENCES

1. Y. Choquet-Bruhat, *Comm. Math. Phys.*, **2**, 16 (1969) and *C.R. Acad. Sc.*, **258**, 1089 (1964).
2. M. Kline and I. W. Kay, *Electromagnetic theory and geometrical optics*, Interscience, New York, N.Y., 1965.
3. J. Ehlers, *Zeitschr. f. Naturforschung*, **22**, 1328 (1967).
4. A. Lichnerowicz, *Ann. di mat. pura ed app.*, **50**, 1 (1960).

PAPER 9

The rotating Einstein–Rosen bridge

JEFFREY M. COHEN

Institute for Space Studies, Goddard Space Flight Center, NASA, U.S.A.

ABSTRACT

The Einstein–Rosen bridge is a model for a non-rotating mass constructed out of nothing but curved empty space. It is shown that the model can be generalized to describe a slowly-rotating mass with a well-defined angular momentum. The gravitational radius of a Einstein–Rosen bridge of mass 2.1 solar masses is about 6 km, while the physical radius of certain recently-found neutron-star models of the same mass is only about twice as large.

---

1 INTRODUCTION

Out of nothing but curved empty space, Einstein and Rosen[1] constructed a model for a non-rotating mass. They accomplished this by analytically continuing the Schwarzschild metric. When expressed in isotropic coordinates, the space-like surface t = const of the Schwarzschild metric

$$\mathrm{d}s^2 = -V^2\,\mathrm{d}t^2 + \psi^4/\mathrm{d}R^2 + R^2\,\mathrm{d}\theta^2 + R^2\sin^2\theta\,\mathrm{d}\varphi^2$$

$$V = (R-\alpha)/(R+\alpha), \quad \psi = 1 + \alpha/R, \quad \alpha = m/2 \tag{1}$$

can be interpreted as two asymptotically-flat spaces connected by a bridge, known as the Einstein–Rosen bridge. The two-dimensional section t = const, $\theta = \pi/2$, is multiply connected and can be imbedded in a three-dimensional Euclidean space. A two-dimensional representation of the surface is shown in Figure 1.

The Einstein–Rosen bridge can be generalized to describe rotating as well as non-rotating bodies. Our main interest will be to obtain an expression for the angular momentum of such a model for mass.

For many years, it has been argued that neutron stars and Einstein–Rosen bridges do not occur in nature because no one had seen any and because they

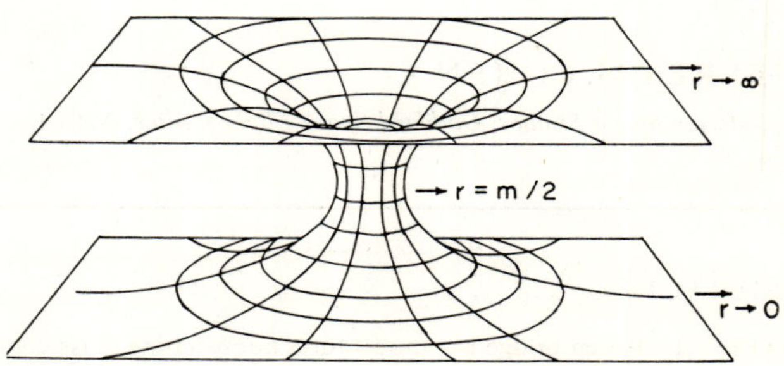

Figure 1. Two dimensional section $\theta = \pi/2$ of Einstein–Rosen bridge. This imbedding in Euclidean space is obtained by introducing cylindrical coordinates r, φ, z with the radial parameter $r = R\psi^2$ having the property that areas of spheres in the Schwarzschild metric $= 4\pi r^2$. The cross-section of constant φ can be obtained by equating the line element $ds^2 = dr^2 + dz^2$ and the line element (1) giving $dr^2 + dz^2 = dr^2/(1 - 2mr^{-1})$. Integration of this equation yields $(z - z_0)^2 = 8m(r - 2m)$. Setting $z_0 = 0$ and revolving this parabola about the z axis yields an imbedding of the $\theta = \pi/2$ section of the Einstein–Rosen bridge which is shown above

are so compact that many people found it difficult to imagine ever seeing them even if they existed. For example, a neutron-star model[2] with mass 2.1 solar masses has a radius of about 13 km while the radius of the throat of an Einstein–Rosen bridge of the same mass is about 6 km.

It is now believed that pulsars are rotating neutron stars[3,4]. If this is true, we have observed objects with radius of the order of their Schwarzschild radius.

To obtain the angular momentum of the Einstein–Rosen bridge, there are conceptual as well as mathematical difficulties. In order to simplify the physical interpretation of the results for the Einstein–Rosen bridge, neutron-star models will be treated first and the results will be compared with those for the Einstein–Rosen bridge.

2 ROTATING EXTERIOR METRIC

The metric exterior to any slowly rotating body (which is spherical when non-rotating), such as a thin shell[5–9] or a neutron star, is given by

$$ds^2 = -A^1\,dt^2 + B^2\,dr^2 + r^2\,d\theta^2 + r^2 \sin^2\theta\,(d\varphi - \Omega\,dt)^2 \tag{2}$$

where

$$A^2 = B^{-2} = 1 - 2mr^{-1} \tag{3}$$

$$\Omega = 2J/r^3. \tag{4}$$

When the integration constant J vanishes, the metric (2) reduces to the standard Schwarzschild metric.

That the mass m can be measured in the asymptotic region far from the body, is well known. In a similar way, the constant J can be measured. It will be shown that J is the angular momentum of the rotating source.

3 ANGULAR MOMENTUM

In spaces with symmetries, the conservation law of general relativity

$$T^{\mu\nu}_{;\nu} = 0 \tag{5}$$

gives rise to conserved quantities. The isometry group associated with such symmetries is generated by a Killing vector ξ^μ satisfying Killing's equation: $\xi_{\mu;\nu} + \xi_{\nu,\mu} = 0$. Contraction of the Killing vector ξ_μ with Eq. (5) yields a quantity which can be transformed into the divergence

$$(\xi_\mu T^{\mu\nu})_{;\nu} = 0. \tag{6}$$

Integration of Eq. (6) over a portion of space-time σ, bounded only in time, and use of the n-dimensional divergence theorem yields an integral over the boundary $\partial\sigma$:

$$0 = \int_\sigma (\xi_\mu T^{\mu\nu})_{;\nu}\,dv_4 = \int_{\partial\sigma} \xi_\mu T^{\mu\nu}\,d\sigma_\nu. \tag{7}$$

If the source is bounded in space or falls off sufficiently rapidly at spatial infinity or if there is no source as with the Einstein–Rosen bridge, the integral vanishes over the time-like portion of the boundary at spatial infinity. Consequently, the second integral (7) reduces to the difference of value of an integral over any two space-like surfaces. Being independent of the space-like surface Σ, the integral is a conserved quantity:

$$J_c = \int_{\Sigma} \xi_\mu T^{\mu\nu} \, d\sigma_\nu . \tag{8}$$

Here Σ denotes a three-dimensional space-like surface and $d\sigma_\nu$ its surface element. In the Newtonian limit, Eq. (8) reduces to the Newtonian expression for the angular momentum of a rotating body. Thus it seems reasonable to define J_c as the general relativistic expression for the angular momentum of a body[10,11,12].

It is rather difficult for an observer to get inside a neutron star, or any star for that matter, in order to measure the angular momentum or the mass. For the Einstein–Rosen bridge, it is impossible to go in through the throat and back out again. These difficulties are circumvented if the mass and angular momentum are measured via the asymptotic form of the metric. The mass of a star is usually defined via the asymptotic form of the metric. This mass can be measured, e.g., by putting a test body in orbit around it. Similarly, the angular momentum can be found via an integral over a two-dimensional surface far from the body.

This can be done when the Killing vector ξ^μ is tangent to the space-like surface Σ, since the Einstein tensor $G^{\mu\nu}$ can be expressed in terms of a covariant divergence in this case[13]. Elimination of $T^{\mu\nu}$ from Eq. (8) via Einstein's equations and use of the divergence theorem yields

$$8\pi \, J_c = \int_{\partial\Sigma} \xi_\mu \left(p^{\mu\nu} - \eta^{\mu\nu} p\right) d\sigma_\nu^2 . \tag{9}$$

Here $\partial\Sigma$ denotes the two-dimensional boundary of Σ, $d\sigma_\nu^2$ an area element of $\partial\Sigma$, $p_{\mu\nu} = n_{\mu;\nu}$ denotes the second fundamental form on the space-like surface with normal n^μ and $p = p^\mu_\mu$. By integrating over the two-dimensional boundary $\partial\Sigma$ one can find the total angular momentum generated by all the sources enclosed within the surface.

4 ROTATING NEUTRON STARS

Before returning to the Einstein–Rosen bridge, neutron stars will be considered since the physical interpretation of the results is simpler and since pulsars are at present believed to be neutron stars. The results for neutron stars are also useful since they can be compared with those for the Einstein–Rosen bridge in order to expedite the physical interpretation of the results for the bridge.

The metric outside a slowly-rotating neutron star[9] is given in Eq. (2). Substitution of this metric into the expression (9) for the angular momentum of a body yields:

$$J_c = J. \tag{10}$$

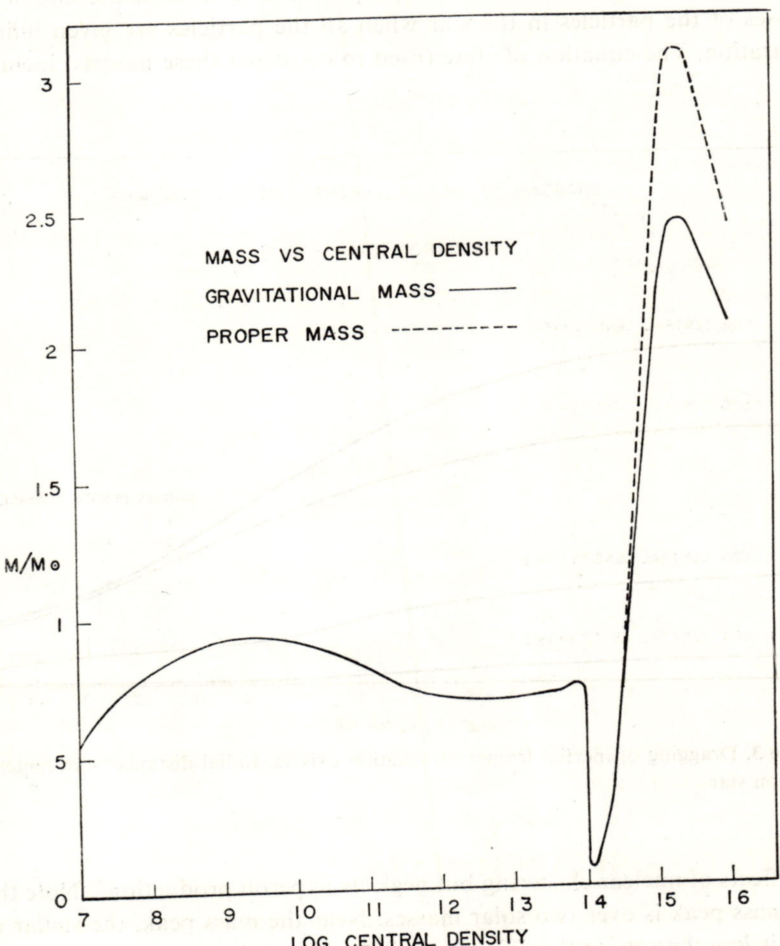

Figure 2. Gravitational mass and proper mass vs. central density

Thus, the integration constant J (which appears in the exterior metric) is equal to the angular momentum of the rotating neutron star.

The metric within a rotating neutron star has the same form as the exterior metric. Unfortunately, it has proved difficult to obtain an analytic solution for the metric coefficients. They were determined via machine integration of Einstein's equations, giving a family of slowly rotating neutron star models[4]. The gravitational mass is plotted as a function of central density in Figure 2. The upper dotted curve denotes the "proper mass" defined as the sum of the masses of the particles in the star when all the particles are given infinite separation. The equation of state (used to construct these models) includes

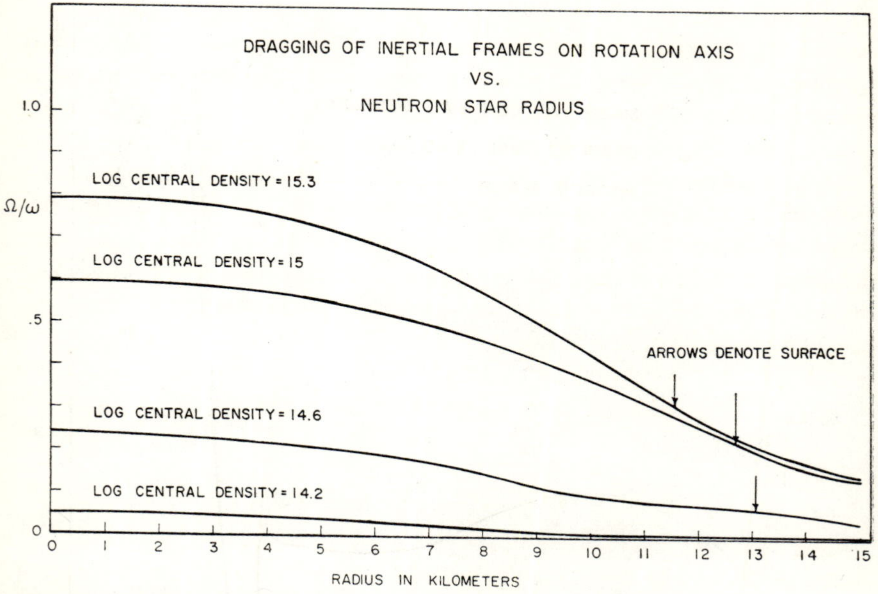

Figure 3. Dragging of inertial frames on rotation axis vs. radial distance from center of neutron star

the effects of nuclear clustering but neglects hyperon production[2]. Note that the mass peak is over two solar masses. Near the mass peak, the stellar radius is less than twice the Schwarzschild radius. (The Schwarzschild radius is the maximum radius of the throat of the Einstein–Rosen bridge.)

A rotating neutron star generates an induced rotation of the inertial frames both inside and outside the star[5–9]. The angular velocity of inertial frames is a complicated vector function of position[8]. However, along the rotation axis, the angular velocity of inertial frames is parallel to the rotation axis and equal in magnitude to Ω given in Eq. (2)[9]. In Figure 3 is given the angular velocity of inertial frames along the rotation axis Ω vs. radius for various neutron star models[14].

According to the suggestion of Finzi and Wolf[4], the energy source of the Crab nebulae is the loss of rotational energy from a rotating neutron star which is slowing down—the Crab pulsar. If neutron-star models are given the rotational period ($T = 0.033\,090\,14$ seconds) and the fractional change of rotational period ($\Delta T/T = 4.22 \times 10^{-13}$) of the Crab pulsar[15,16], it can be shown that the rate of rotational energy loss is larger than 1.5×10^{38} erg/second for models with mass greater than 0.4 solar masses[14]. This energy generation rate is in accordance with various estimates of the total amount of electromagnetic energy emitted by the Crab nebulae[17–19]. This electromagnetic radiation is only a lower limit to the total energy being pumped into the Crab since additional energy may go into the acceleration of protons, etc.[14].

In the calculation of the rotational kinetic energy[9] of stellar models, account was taken of the contributions from the pressure, gravitational potential, doppler shift, and the induced rotation of inertial frames both inside and outside of the stellar model. For completeness, the expression for the rotational kinetic energy is given below[9]:

$$E_{\text{rot}} = (4\pi/3) \int_0^R (\varrho + p)\, r^4 B A^{-1} (\omega - \Omega)^2 \, \mathrm{d}r + \int_0^R r^4 \Omega_r^2 (12AB)^{-1} \, \mathrm{d}r + J^2/R^3. \tag{1}$$

Here ϱ is the energy density, p the pressure, R the radius of the star, ω the angular velocity of the star as seen by an observer at infinity, and J is the angular momentum of the star. The last term in Eq. (11) is the contribution to the rotational kinetic energy from the gravitational field *outside* the star. As a check on the computer code, the angular momentum was calculated both from Eqs. (8) and (10) and the same result was obtained. This is because the two-sphere (over which the integral (9) was evaluated) bounds the three-space containing the neutron star. The situation is more complicated with the Einstein–Rosen bridge[1].

5 ANGULAR MOMENTUM OF ROTATING EINSTEIN-ROSEN BRIDGE

We now have enough information to return to the Einstein–Rosen bridge and to obtain an expression for its angular momentum. The coordinate transformation[6]

$$r = R\psi^2, \quad \varphi = \bar{\varphi} + 2J\,(2m)^{-3} \tag{12}$$

brings the Schwarzschild metric into isotropic form and brings the metric (2) into the form:

$$ds^2 = -V^2\,dt^2 + \psi^4\,(dR^2 + R^2\,d\theta^2 + R^2\sin^2\theta\,(d\varphi - \bar{\Omega}\,dt)^2). \tag{13}$$

Here V and ψ are defined below Eq. (1), and $\bar{\Omega}$ is given by:

$$\bar{\Omega} = 2J\,[(R\psi^2)^{-3} - (2m)^{-3}]. \tag{14}$$

Inspection of Eq. (8) shows that, since the stress-energy tensor $T^{\mu\nu}$ vanishes for the Einstein–Rosen bridge, so does the conserved quantity J_c. The same result is obtained from Eq. (9) in a more subtle way. The integral in Eq. (9) is evaluated over the boundary of the space-like surface Σ. But in this case, the boundary $\partial\Sigma$ consists of two disconnected two-spheres which give equal and opposite contributions to J_c. That this result is consistent with de Rham's second theorem[20] will be shown in the appendix. (This result is a special case of de Rham's theorem which can be obtained in a similar manner.)

If an observer on the upper sheet of the Einstein–Rosen bridge (Figure 1) measures the angular momentum via Eq. (9), he cannot tell from the asymptotic metric whether there is, e.g., a rotating neutron star or a rotating bridge generating the angular momentum. Consequently, he will use only one two-sphere to measure the angular momentum and in each case he will find that the angular momentum equals J from Eq. (4). In other words, an observer on the upper sheet will see the same asymptotic metric as an observer exterior to a neutron star. Consequently, each will assume that the space is bounded by a single two-sphere and calculate the integral (9) accordingly. And each observer will find that the angular momentum equals J. Thus, the rotating Einstein–Rosen bridge is a model for rotating mass with a well defined angular momentum.

In this paper we have seen that the Einstein–Rosen bridge, like neutron stars, has reasonable physical properties. Despite this, many people have claimed that neutron stars do not exist because of their high density and because they had never been observed. It is now believed that the recently discovered pulsars are rotating neutron stars. Possibly, the Einstein–Rosen bridge will also be observed in the not too distant future. Perhaps there is one near the center of the galaxy which emits gravitational waves[21] each time an object falls in.

Acknowledgement

For helpful discussion, I am indebted to K. S. Thorne and E. Lubkin. I should like to thank the staff of the Technion for their hospitality and financial aid. This work was supported in part by an NAS-NRC Resident Research Associateship sponsored by the National Aeronautics and Space Administration.

REFERENCES

1. A. Einstein and N. Rosen, *Phys. Rev.*, **48**, 73 (1935).
2. J. M. Cohen, W. Langar, I. Rosen and A. G. W. Cameron, in press.
3. T. Gold, *Nature*, **218**, 731 (1968).
4. A. Finzi and R. A. Wolf, *Ap. J. Letters*, **155**, L107 (1969).
5. J. M. Cohen, *Fourth Summer Seminar on Applied Mathematics*, Cornel University, 1965, in: *Lectures in Applied Mathematics*, Vol. 8, *Relativity Theory and Astrophysics*, Ed. J. Ehlers, Amer. Math. Soc., Providence, R.I., 1967.
6. D. R. Brill and J. M. Cohen, *Phys. Rev.*, **143**, 1011 (1966).
7. J. M. Cohen and D. R. Brill, *Nuovo Cimento*, **56B**, 209 (1968).
8. J. M. Cohen, *Phys. Rev.*, **173**, 1258 (1968).
9. J. M. Cohen, *Astrophys. and Space Sci.*, in press.
10. J. M. Cohen, *J. Math. Phys.*, **8**, 1477 (1967); **9**, 905 (1968).
11. A. Trautman, *Gravitation*, Ed. J. Witten, Wiley, New York, N.Y., 1963.
12. A. Komar, *Phys. Rev.*, **127**, 1411 (1962).
13. Y. Foures-Bruhat, *Gravitation*, Ed. L. Witten, Wiley, New York, N.Y., 1963.
14. J. M. Cohen and A. G. W. Cameron, *Nature*, in press.
15. H. D. Craft, R. V. E. Lovelace, D. W. Richards and J. E. Sutton, 1969, in press.
16. W. J. Cocke, M. J. Disney and D. J. Taylor, *Nature*, **221**, 525 (1969).
17. I. S. Shklooskii, *Soviet Astr.*, **AJ10**, 6 (1966).
18. R. C. Hanes, D. V. Ellis, G. J. Fishman, J. D. Kurfers and W. H. Tucker, *Ap. J. Letters*, **151**, 19 (1968).
19. D. B. Melrose, private communication, 1969.
20. G. de Rham, *Variétés Différentiables*, Hermann, Paris, 1960.
21. J. Weber, This Volume, p. 309.

APPENDIX

Application of de Rham's theorem to Einstein–Rosen bridge

In this appendix, it will be shown that the results of Section V are consistent with de Rham's second theorem concerning the periods associated with p-cycles:

If Z is a p-cycle and ω is a closed form, the *value of the integrals*

$$\int_z \omega \tag{A1}$$

can be assigned arbitrarily subject only to the consistency relation that whenever

$$\sum_i a_i z_i = \text{boundary}, \tag{A2}$$

then

$$\sum_i a_i \int_{z_i} \omega = 0. \tag{A3}$$

Here the quantities

$$\int_{z_i} \omega \tag{A4}$$

are called the periods of the p-cycle, closed forms η are defined via

$$\mathrm{d}\eta = 0, \tag{A5}$$

and a p-cycle is a p-dimensional surface with no boundary.

For the Einstein–Rosen bridge treated in Section V, the p-cycles consist of two disconnected two-spheres, one on the upper sheet and one on the lower sheet of Figure 1. These spheres are represented by circles of constant r in Figure 1.

The closed form ω is given by

$$\omega = * \, [\xi_\mu \, (p^\mu_\nu - \delta^\mu_\nu p) \, \omega^\nu] \tag{A6}$$

where the ω^ν are orthonormal basis vectors, and *[ ] denotes a differential form dual to [ ]. The integral (A1) with ω given in Eq. (A6) is identical with that of Eq. (9). Exterior differentiation of Eq. (A6) yields

$$\mathrm{d}\omega = * \, [\xi_\mu \, (p^{\mu\nu} - \eta^{\mu\nu} p)]_{;\nu} \tag{A7}$$

which vanishes if the stress-energy tensor $T^{\mu\nu}$ equals zero. Thus, ω is a closed form if $T^{\mu\nu}$ vanishes.

Consequently, from de Rham's theorem we conclude that the value of the integral (9), when integrated over a two-sphere on the upper sheet, is arbitrary and can be determined only from the physics. This is the way things should be. However, no matter what value this integral has, de Rham's theorem guarantees that the integral (8) vanishes.

PAPER 10

Distribution theory and thin shells in general relativity

MARION D. COHEN
Newark College of Engineering, N.J., U.S.A.

AND

JEFFREY M. COHEN
Goddard Institute for Space Studies, N.Y., U.S.A.

ABSTRACT

The problem, previously unsolved, of computing the value of $\int_{-\infty}^{\infty} \theta\delta \, dx$, where θ is the Heaviside function and δ is the Dirac delta function comes up quite frequently in physics, particularly in the general-relativistic theory of thin shells. In this paper, it is shown that the value is undetermined—i.e., it depends on the particular physical situation—but that it must always lie between 0 and 1. Several intuitive arguments are presented which lead to the same conclusion. If the value were $\frac{1}{2}$, as is generally believed, then the results would be inconsistent with Einstein's equations.

The problem, previously unsolved, of computing the value of $\int_{-\infty}^{\infty} \theta\delta \, dx$, where θ is the Heaviside function and δ is the Dirac delta function, comes up quite frequently in physics, particularly in the general relativistic theory of thin shells. As an example, consider the metric for a thin spherical shell[1]:

$$ds^2 = -A^2 \, dt^2 + B^2 \, dr^2 + r^2 (d\theta^2 + \sin^2 \theta \, d\varphi^2), \tag{1}$$

where

$$A^2 = B^{-2} = 1 - \frac{2m}{r}, \quad \text{for} \quad r > r_0 \tag{2}$$

$$A^2 = 1 - \frac{2m}{r_0}, \quad B^2 = 1, \quad \text{for} \quad r < r_0 \tag{3}$$

and r_0 = radius of shell.

Here the non-vanishing components of the stress-energy tensor are:

$$T^{00} = K\delta\,(r - r_0), \quad T^{22} = T^{33} = S\delta\,(r - r_0), \tag{4}$$

(where $4\pi \int \delta r^2 \, \mathrm{d}r = 1$) and the solution of Einstein's equations reduces to the determination of K and S. In order to find these quantities, we consider first the G^{00} equation, which yields:

$$2S\left(4\pi \int_0^\infty B\,\delta r^2 \,\mathrm{d}r\right) = \frac{(1 - A_0)^2\, r_0}{2A_0}, \quad A_0 = \left(1 - \frac{2m}{r_0}\right)^{1/2}. \tag{5}$$

This last equation gives us S, as soon as we know $4\pi \int_0^\infty B\delta r^2 \,\mathrm{d}r$. Now r^2 is continuous, B is continuous except for a jump at $r = r_0$, and thus the problem is analogous to determining $\int_{-\infty}^{\infty} \theta\delta \,\mathrm{d}x$ in one-dimensional cartesian coordinates.

Let us recall that δ has the property that:

$$\int \delta\varphi \,\mathrm{d}x = \varphi(0) \tag{6}$$

for all testing functions φ. This suggests writing:

$$\int \delta\theta \,\mathrm{d}x = \theta(0), \tag{7}$$

even though θ is not a testing function. Thus the problem reduces to that of finding the value of the Heaviside function θ at the point 0, where it jumps from 0 to 1.

Now, what is meant by the value of any given distribution at any given point? This question was answered in part by Lojasiewicz[2], whose theory we now briefly present. For simplication, we take 0 to be the point. The theory is motivated by the observation that for any continuous function f and for any real number x, we have:

$$f(0) = \lim_{\varepsilon \to 0} f(\varepsilon x). \tag{8}$$

Lojasiewicz considered substituting an arbitrary distribution T for f in the above equation and thereby defining:

$$T(0) = \lim_{\varepsilon \to 0} T(\varepsilon x). \tag{9}$$

The trouble with this is two-fold—first, the limit may not exist; second, even if it does exist, it is a *distribution*, not a real number, as desired. However, Lojasiewicz proved that, if the distribution $\lim_{\varepsilon\to 0} T(\varepsilon x)$ exists, then it is a constant distribution. He then went on to define $T(0)$ to be that constant. According to this:

$$T(0) = \frac{\lim_{\varepsilon\to 0} \int T(\varepsilon x)\,\varphi\,\mathrm{d}x}{\int \varphi\,\mathrm{d}x} = \frac{\lim_{\varepsilon\to 0} \int \frac{1}{\varepsilon} T\varphi\left(\frac{x}{\varepsilon}\right)\mathrm{d}x}{\int \varphi\,\mathrm{d}x} \tag{10}$$

for any testing function φ. Equivalently:

$$T(0) = \lim_{\varepsilon\to 0} \int \frac{1}{\varepsilon}\, T\varphi\left(\frac{x}{\varepsilon}\right)\mathrm{d}x, \tag{11}$$

for any testing function φ such that $\int_{-\infty}^{\infty} \varphi = 1$.

The main problem with Lojasiewicz's theory is that there are quite a few T for which $T(0)$ does not exist—and θ is one of them. For if we let φ be any testing function with support to the right of 0, we get:

$$\lim_{\varepsilon\to 0^+} \int \frac{1}{\varepsilon}\, \theta\varphi\left(\frac{x}{\varepsilon}\right)\mathrm{d}x = 1 \tag{12}$$

and

$$\lim_{\varepsilon\to 0^-} \int \frac{1}{\varepsilon}\, \theta\varphi\left(\frac{x}{\varepsilon}\right)\mathrm{d}x = 0, \tag{13}$$

and therefore $\lim_{\varepsilon\to 0} \int (1/\varepsilon)\,\theta\varphi\,(x/\varepsilon)\,\mathrm{d}_x$ cannot exist.

In this paper, we give a theory concerning $T(0)$ which includes Lojasiewicz's theory in those cases where his theory applies, and which itself applies to every distribution T—in particular, θ. Note first that:

$$T(0) = \lim_{\varepsilon\to 0} \int \frac{1}{\varepsilon}\, T\varphi\left(\frac{x}{\varepsilon}\right)\mathrm{d}x = \lim_{\varepsilon\to 0^+} \int \frac{1}{\varepsilon}\, T\varphi\left(\frac{x}{\varepsilon}\right)\mathrm{d}x. \tag{14}$$

Secondly, the most natural testing functions to consider are the non-negative ones. This fact has been used in defining many concepts in the theory of distributions—e.g., positive distributions. With these two facts in mind, we define the *set of values of a given distribution T at the point 0* as the set of all

$$\lim_{\varepsilon \to 0+} \int \frac{1}{\varepsilon} T\varphi\left(\frac{x}{\varepsilon}\right) dx,$$

where φ ranges over all non-negative testing functions with $\int \varphi \, dx = 1$.

According to this definition, the set of values is always connected—i.e., an interval in the real line—excluding the points $-\infty$ and $+\infty$. For suppose a and b are in the set of values. Then we have non-negative testing functions φ_1 and φ_2 such that

$$\int \varphi_1 \, dx = \int \varphi_2 \, dx = 1, \tag{15}$$

$$\lim_{\varepsilon \to 0+} \int \frac{1}{\varepsilon} T\varphi_1\left(\frac{x}{\varepsilon}\right) dx = a, \tag{16}$$

and

$$\lim_{\varepsilon \to 0+} \int \frac{1}{\varepsilon} T\varphi_2\left(\frac{x}{\varepsilon}\right) dx = b. \tag{17}$$

Now, suppose $\alpha + \beta = 1$, $\alpha \geqslant 0$, $\beta \geqslant 0$ are given. Then if we let $\varphi = \alpha\varphi_1 + \beta\varphi_2$, φ has the properties of being non-negative and such that

$$\lim_{\varepsilon \to 0+} \int \frac{1}{\varepsilon} T\varphi\left(\frac{x}{\varepsilon}\right) dx = \alpha a + \beta b, \tag{18}$$

and $\int \varphi \, dx = 1$. Then $\alpha a + \beta b$ is in the set of values.

Now, let us return to our old problem, the Heaviside function θ. If we ask what is the set of values of θ at 0, we find that it is the closed interval, [0, 1]. For if φ is a non-negative testing function, then:

$$\lim_{\varepsilon \to 0+} \int \frac{1}{\varepsilon} \theta\varphi\left(\frac{x}{\varepsilon}\right) dx = \lim_{\varepsilon \to 0+} \int_0^\infty \frac{1}{\varepsilon} \varphi\left(\frac{x}{\varepsilon}\right) dx = \int_0^\infty \varphi \, dx, \tag{19}$$

which is between 0 and 1. On the other hand, if we choose φ with support to the left of 0, we get:

$$\lim_{\varepsilon \to 0+} \int \frac{1}{\varepsilon} \theta\varphi\left(\frac{x}{\varepsilon}\right) dx = 0, \tag{20}$$

and if we choose φ with support to the right of 0:

$$\lim_{\varepsilon \to 0+} \int \frac{1}{\varepsilon} \theta\varphi\left(\frac{x}{\varepsilon}\right) dx = \int \varphi = 1. \tag{21}$$

Therefore the set of values of θ at 0 is the closed interval [0, 1]. The actual value depends on the physical situation one is confronted with.

Another way to deal with the problem is to use the definition of θ, rather than of δ. For all testing functions φ, θ has the property:

$$\int_{-\infty}^{\infty} \varphi\theta \, \mathrm{d}x = \int_{0}^{\infty} \varphi \, \mathrm{d}x, \tag{22}$$

Again, this suggests writing:

$$\int_{-\infty}^{\infty} \theta\delta \, \mathrm{d}x = \int_{0}^{\infty} \delta \, \mathrm{d}x, \tag{23}$$

even though δ is certainly not a testing function. However, what is $\int_0^\infty \delta \, \mathrm{d}x$? Is the point 0 to be integrated over, or not? If so, then $\int_0^\infty \delta \, \mathrm{d}x = 1$; if not, $\int_0^\infty \delta \, \mathrm{d}x = 0$. Perhaps some fraction of the point 0 is to be integrated over? Thus we again have a choice of values of $\int \theta\delta \, \mathrm{d}x$ ranging from 0 to 1.

Since $\theta' = \delta$, θ is positive, and θ must be monotone increasing. This implies that $\theta(0)$ must be between 0 and 1.

If the integral $\int \theta\delta \, \mathrm{d}x$ had some definite value between these two limits, independent of the physical situation, then the results of distribution theory would be inconsistent with general relativity. Returning to our example of the thin spherical shell, the G^{00} equation also yields:

$$4\pi \int r^2 B\delta \, \mathrm{d}r = (1 - A_0)\, r_0/K. \tag{24}$$

Combining this with our previous result, we obtain:

$$S = m\,(A_0^{-1} - 1)/4, \tag{25}$$

which completes the solution.

Now, we have

$$4\pi \int r^2 B\delta \, \mathrm{d}r = (1 - A_0)\, r_0/m; \tag{26}$$

thus $4\pi \int r^2 B\delta \, \mathrm{d}r$ depends on r_0 and m. Since $r_0 \geqslant 2$ m, for static solutions to Einstein's equations, $4\pi \int r^2 B\delta \, \mathrm{d}r$ can range from 1 to 2, depending on the ratio $2m/r_0$. This is consistent with the result of this paper, for:

$$4\pi \int r^2 B\delta \, \mathrm{d}r = \int B\delta \, \mathrm{d}x = B(r_0), \tag{27}$$

which, according to our theory, in the range from $B(r_0^-) = 1$ to $B(r_0^+) = (1 - 2mr_0^{-1})^{-1/2} \to \infty$, if $r_0 \to 2m$.

In this paper, we have found that the value of the integral of a piecewise continuous function times a delta "function" must lie between the upper and lower limits of the piecewise continuous function, but that the actual value can be determined only from the physical situation. In the case of thin spherical shells, the function B jumps from 1 to $B(r_0^+)$. The actual value can be determined from the physical situation—namely, the ratio $2m/r_0$.

Acknowledgement

We should like to thank the staff of the Technion for their hospitality and financial support at the Haifa Conference. One of us (J.M.C.) holds an NAS-NRC Resident Research Associateship at the Goddard Institute for Space Studies.

REFERENCES

1. J. M. Cohen and M. D. Cohen, *Nuovo Cimento*, **60B**, 241 (1969).
2. S. Lojasiewicz, *Bull. Acad. Polon. Sci. Cl. III*, **4**, 239 (1956).

PAPER 11

Twistors, symplectic structure and Lagrange's identity

M. CRAMPIN and F. A. E. PIRANI
King's College, University of London, England

ABSTRACT

This note describes the relation between the natural symplectic structure on the cotangent bundle over a space-time manifold, Lagrange's identity for Jacobi fields, and the symplectic structure developed by Penrose in his theory of twistors.

The content of this note, which is largely expository, arose from a discussion with Roger Penrose about the symplectic structure for twistors described by him in a recent paper[1]. The way in which the symplectic structure appears there is a little mysterious; the following is an attempt to clarify the situation. The main point to be made is that there is a natural symplectic structure on the cotangent bundle (phase space) over space-time, closely related to Lagrange's identity, which is concerned with the behaviour of neighbouring geodesics. It was by considering similar relations between null geodesics that Penrose discovered his twistor, simplectic structure, and it occurred to us that the two structures might be related. It turns out that for null geodesics they are identical.

First we establish the notation and some basic geometrical facts. Let M be a differentiable manifold of dimension n. In terms of local coordinates $\{x^a\}$, a tangent vector u at a point x of M may be written $u = u^a\,(\partial/\partial x^a)$; if u is tangent to the curve γ given by $x^a = \gamma^a(t)$, then $u^a = \mathrm{d}\gamma^a/\mathrm{d}t = \dot{\gamma}^a$. The collection of pairs (x, u), $x \in M$, u a tangent vector at x, is called the tangent bundle of M, written $T(M)$. It is a $2n$-dimensional manifold: one may take $\{x^a, u^a\}$ for coordinates. The projection map $\pi: T(M) \to M$ is given by $\pi(x, u) = x$. In classical mechanics, if M is the configuration space of a system, $T(M)$ is the space of generalized coordinates and velocities.

A curve γ on M lifts naturally to a curve on $T(M)$. To the point $\gamma(t)$ of M is associated the point $(\gamma(t), \dot{\gamma}(t))$ of $T(M)$ projecting onto $\gamma(t)$. For example, given an affine connection on M, the geodesic with tangent vector u at x has equations $\mathrm{d}x^a/\mathrm{d}t = u^a$, $\mathrm{d}u^a/\mathrm{d}t + \Gamma^a_{bc}u^b u^c = 0$, and lifts therefore to a curve through (x, u) whose tangent vector is

$$u^a \frac{\partial}{\partial x^a} - \Gamma^a_{bc}u^b u^c \frac{\partial}{\partial u^a}.$$

The fact that there is a unique geodesic through a given point of M with a given tangent vector implies that there is a unique lifted geodesic through each point of $T(M)$. In other words, the lifted geodesics form a congruence of curves on $T(M)$, the integral curves of the vector field

$$u^a \frac{\partial}{\partial x^a} - \Gamma^a_{bc}u^b u^c \frac{\partial}{\partial u^a}.$$

Here lies the advantage of the tangent bundle machinery—it spreads out the geodesics.

The cotangent bundle $T^*(M)$ is the collection of pairs (x, p), p a cotangent vector at x, that is, a linear functional on the tangent space at x. In terms of coordinates, $p = p_a\, \mathrm{d}x^a$, where $\{\mathrm{d}x^a\}$ is the basis for the cotangent space dual to the basis $\{\partial/\partial x^a\}$ for the tangent space. The cotangent bundle is the phase space of mechanics.

A symplectic structure on a manifold of even dimension is a choice of a two-form ω (skew twice-covariant tensor field) of maximal rank on the manifold, whose exterior derivative $\mathrm{d}\omega$ is zero. The contangent bundle is equipped with a natural symplectic structure, given by the two-form $\omega = \mathrm{d}p_a \wedge \mathrm{d}x^a$. (Clearly $\mathrm{d}\omega = 0$; and the n-fold exterior product $\omega \wedge \omega \wedge \cdots \wedge \omega = n!\, \mathrm{d}p_1 \wedge \cdots \wedge \mathrm{d}p_n \wedge \mathrm{d}x^1 \wedge \cdots \wedge \mathrm{d}x^n \neq 0$, which is a way of saying that ω has maximal rank.) This is the form which is described in classical mechanics as being conserved by Hamilton's equations.

If ϕ is a differentiable map of manifolds, $\phi_{\#}$ denotes the differential of ϕ, that is, the induced map of tangent spaces: in terms of coordinates, $\phi_{\#}$ is the Jacobian matrix of ϕ.

We show that the two-form ω defining the natural symplectic structure on $T^*(M)$ is not dependent on the special coordinates chosen, by expressing it in a coordinate-free way. Define a one-form θ (covariant vector field) on $T^*(M)$ by $\theta(\xi) = p(\pi_{\#}\xi)$, where ξ is a tangent vector (to $T^*(M)$) at (x, p). Then $\pi_{\#}\xi$ is a tangent vector to M at x, and p is a cotangent vector at x, so $p(\pi_{\#}\xi)$ is well defined and depends linearly on ξ. It is easy to check that ω is the exterior derivative of θ.

Now there is in general no natural correspondence between tangent and cotangent bundles unless some additional structure is given. A time-independent Lagrangian—which is just a function L on $T(M)$—provides the necessary additional structure, for it defines the

$$\text{Legendre map} \quad \Lambda : (x, u) \to \left(x, \left. \frac{\partial L}{\partial u^a} \right|_{(x,u)} \mathrm{d}x^a \right).$$

A Lagrangian L is said to be regular if

$$g_{ab} = \frac{\partial^2 L}{\partial u^a \, \partial u^b}$$

is nonsingular everywhere. This is the condition that $\Lambda_{\#}$ be an isomorphism of tangent spaces at each point. Let L be a regular Lagrangian, and define g^{ab} by $g^{ab}g_{bc} = \delta^a_c$, as usual. Then the Euler–Lagrange equations

$$\frac{\mathrm{d}}{\mathrm{d}t}\left(\frac{\partial L}{\partial u^a}\right) - \frac{\partial L}{\partial x^a} = 0$$

may be rewritten

$$\frac{\mathrm{d}x^a}{\mathrm{d}t} = u^a, \quad \frac{\mathrm{d}u^a}{\mathrm{d}t} = g^{ab}\left(\frac{\partial L}{\partial x^b} - u^c \frac{\partial^2 L}{\partial x^c \, \partial u^b}\right).$$

Thus the solution curves of the Euler–Lagrange equations are the projections onto M of the integral curves of the vector field

$$\Gamma = u^a \frac{\partial}{\partial x^a} + g^{ab}\left(\frac{\partial L}{\partial x^b} - u^c \frac{\partial^2 L}{\partial x^c \, \partial u^b}\right)\frac{\partial}{\partial u^a}.$$

Our previous example of the geodesics is a special case, at least where they arise from a metric; of course, L is then $\frac{1}{2}g_{ab}u^a u^b$.

If one assumes not merely that L is regular but that Λ has a differentiable inverse, then one may construct the vector field $\Lambda_{\#}\Gamma$ on $T^*(M)$. It turns out that

$$\Lambda_{\#}\Gamma = \frac{\partial H}{\partial p_a}\frac{\partial}{\partial x^a} - \frac{\partial H}{\partial x^a}\frac{\partial}{\partial p_a},$$

where H (a function on $T^*(M)$) is the Hamiltonian corresponding to L. This is not entirely unexpected, for it says that the integral curves of $\Lambda_{\#}\Gamma$ satisfy Hamilton's equations.

It is now possible to state in what sense precisely the two-form ω is conserved by Hamilton's equations. The Lie derivative $£_X\Omega$ of a p-form Ω with respect to a vector field X is given conveniently by the formula $£_X\Omega = \mathrm{d}(\iota(X)\Omega) + \iota(X)\,\mathrm{d}\Omega$, where $\iota(X)\,\Omega$ is the $(p-1)$-form $\Omega\,(X, \cdot, \ldots, \cdot)$. The two-form ω satisfies $\mathrm{d}\omega = 0$, and it is easy to check that $\iota(\Lambda_{\#}\Gamma)\,\omega = -\mathrm{d}H$. Thus $£_{\Lambda_{\#}\Gamma}\omega = \mathrm{d}\,(-\mathrm{d}H) = 0$: ω is dragged along the integral curves of $\Lambda_{\#}\Gamma$.[2]

Let us transfer this information back to the tangent bundle. It follows from the equation that $£_\Gamma\,(\Lambda^{\#}\omega) = 0$, where $\Lambda^{\#}\omega$ is the pull-back of ω to $T(M)$: for any vectors ξ, η at a point (x, u) of $T(M)$, $(\Lambda^{\#}\omega)\,(\xi, \eta) = \omega\,(\Lambda_{\#}\xi, \Lambda_{\#}\eta)$. In coordinates, if

$$\xi = \xi^a\frac{\partial}{\partial x^a} + \Xi^a\frac{\partial}{\partial u^a}, \quad \eta = \eta^a\frac{\partial}{\partial x^a} + H^a\frac{\partial}{\partial u^a},$$

then

$$(\Lambda^{\#}\omega)\,(\xi, \eta) = \frac{\partial^2 L}{\partial u^a\,\partial u^b}(\eta^a\Xi^b - \xi^a H^b) + \frac{\partial^2 L}{\partial x^a\,\partial u^b}(\xi^a\eta^b - \eta^a\xi^b).$$

Specializing to the metric case, where $L = \tfrac{1}{2}g_{ab}u^a u^b$, we find that

$$(\Lambda^{\#}\omega)\,(\xi, \eta) = g_{ab}\,[\eta^a\,(\Xi^b + \Gamma^b_{cd}\xi^c u^d) - \xi^a\,(H^b + \Gamma^b_{cd}\eta^c u^d)].$$

Now extend the vectors ξ, η along the integral curve $(\gamma, \dot{\gamma})$ of Γ through (x, u) by dragging them along Γ, so that the function $(\Lambda^{\#}\omega)\,(\xi, \eta)$ will be constant along the curve. Then ξ^a, Ξ^a satisfy the equations

$$\frac{\mathrm{d}\xi^a}{\mathrm{d}t} - \Xi^a = 0, \quad \frac{\mathrm{d}\Xi^a}{\mathrm{d}t} + \frac{\partial\Gamma^a_{cd}}{\partial x^b}\xi^b\dot{\gamma}^c\dot{\gamma}^d + 2\Gamma^a_{bc}\,\Xi^b\dot{\gamma}^c = 0,$$

with similar equations for η^a, H^a. Substitution for Ξ^a from the first equation in the second and rearrangement gives

$$\frac{D^2\xi^a}{\mathrm{d}t^2} + R^a_{bcd}\,\dot{\gamma}^b\dot{\gamma}^c\xi^d = 0,$$

where

$$\frac{D\xi^a}{\mathrm{d}t} = \frac{\mathrm{d}\xi^a}{\mathrm{d}t} + \Gamma^a_{bc}\,\xi^b\dot{\gamma}^c,$$

and R^a_{bcd} are the components of the curvature tensor. In other words, the projected vector fields $\pi_{\#}\xi, \pi_{\#}\eta$, defined along the geodesic γ, are Jacobi fields, and

$$(\Lambda^{\#}\omega)(\xi, \eta) = g_{ab}\left(\eta^a \frac{D\xi^b}{\mathrm{d}t} - \xi^a \frac{D\eta^b}{\mathrm{d}t}\right)$$

is constant along γ. This result is known as Lagrange's identity.

It remains to be shown how the twistor symplectic structure is related to the natural symplectic structure exhibited above.

The starting point for twistor theory[3] is the fact that zero-rest-mass field equations, suitably formulated, are invariant not only under Lorentz transformations, but under conformal transformations. In general relativity this invariance appears as the conformal invariance of null geodesics. In flat space-time an arbitrary twistor defines a shear-free congruence of null geodesics, and in particular a null twistor defines a congruence of null geodesics which all intersect a single null geodesic. Thus a null twistor defines a null geodesic. On the other hand a null twistor may be represented, relative to a given space-time origin, by a pair of two-spinors. This pair of spinors has a simple geometrical interpretation. Let $(\omega^A, \pi_{A'})$ represent the null twistor Z^α. Then if p_a is the null cotangent vector corresponding to $\bar{\pi}_A\pi_{A'}$, $g^{ab}p_b$ is the tangent vector to the null geodesic γ defined by Z^α, and points x on γ are determined by the equation $\omega^A - \mathrm{i}x^{AA'}\pi_{A'} = 0$, where $x^{AA'}$ is the spinor equivalent of x^a. The null geodesic is well defined if and only if $\omega^A\bar{\pi}_A$ is real, which is the condition that the corresponding twistor be null.

The twistor $\bar{Z}_\alpha$, the complex conjugate of Z^α, is represented by the spinor pair $(\bar{\pi}_A, \bar{\omega}^{A'})$; if moreover $(\lambda^A, \mu_{A'})$ represents a twistor W, the scalar product $\bar{Z}_\alpha W^\alpha$ is $\bar{\pi}_A\lambda^A + \bar{\omega}^{A'}\mu_{A'}$. In particular, $\bar{Z}_\alpha\,\mathrm{d}Z^\alpha = \bar{\pi}_A\,\mathrm{d}\omega^A + \bar{\omega}^{A'}\,\mathrm{d}\pi_{A'}$, so for a null twistor

$$\begin{aligned}\bar{Z}_\alpha\,\mathrm{d}Z^\alpha &= \mathrm{i}\bar{\pi}_A(\pi_{A'}\,\mathrm{d}x^{AA'} + x^{AA'}\,\mathrm{d}\pi_{A'}) - \mathrm{i}x^{AA'}\bar{\pi}_A\,\mathrm{d}\pi_{A'}\\ &= \mathrm{i}\bar{\pi}_A\pi_{A'}\,\mathrm{d}x^{AA'} = \mathrm{i}p_a\,\mathrm{d}x^a,\end{aligned}$$

and so $\mathrm{d}\bar{Z}_\alpha \wedge \mathrm{d}Z^\alpha = \mathrm{i}\mathrm{d}p_a \wedge \mathrm{d}x^a$: the twistor two-form $\mathrm{d}\bar{Z}_\alpha \wedge \mathrm{d}Z^\alpha$, which defines Penrose's symplectic structure, agrees, apart from the factor i, with the natural two-form ω.

This shows that the behaviour of null geodesics, elucidated by Penrose in the special case of an impulsive plane wave as the unfolding of a contact transformation of null twistors, is precisely the behaviour described by the unfolding of the contact transformation defined by the ordinary Hamiltonian $\frac{1}{2}g^{ab}p_a p_b$ on the cotangent bundle over space-time.

REFERENCES

1. R. Penrose, *International J. of Th. Physics*, **1**, 61 (1968).
2. Readers who wish to pursue these matters may consult R. Abraham, *Foundations of Mechanics*, Benjamin, New York, 1966; L. H. Loomis and S. Sternberg, *Advanced Calculus*, Addison-Wesley, Reading, Mass., U.S.A., especially Chapter 13.
3. For a detailed exposition see R. Penrose, *J. Math. Phys.*, **8**, 345 (1967) or [1].

PAPER 12

Static electromagnetic fields in general relativity

BIDYUT KUMAR DATTA
Surendra Nath College, Calcutta 9, India

ABSTRACT

The equations of Rainich's "already-unified field theory" are studied in the case of static electromagnetic fields and electromagnetic fields are obtained for two space-time metrics each of which admits a group G_4 of automorphisms. It turns out that in one case there exists an electromagnetic field for which there is a uniform charge along the axis of density, while in the other field there is a flow of current along the axial direction which produces a magnetic field in the angular direction.

1 INTRODUCTION

As is well known, for a non-null electromagnetic field in an otherwise empty Riemann space, one can present the entire content of the Einstein–Maxwell theory in a unified form[1,2,3]

$$R = 0 \tag{1.1}$$

$$R^{\mu}_{\nu}R^{\nu}_{\lambda} = \delta^{\mu}_{\lambda}\,(\tfrac{1}{4}R^{\beta}_{\alpha}R^{\alpha}_{\beta}) \tag{1.2}$$

$$R^{0}_{0} < 0 \tag{1.3}$$

and

$$\omega_{[\mu\nu,\lambda]} = 0, \tag{1.4}$$

where $\omega_{\mu\nu}$ is a self-dual antisymmetric tensor[4*] and a comma denotes an ordinary partial derivative. Here Greek letters take the values 0, 1, 2, 3 and the signature of the metric is $(+ - - -)$.

Geometrical relations (1.1)–(1.3) are applicable to any gravitational field whose source is a divergence-free Maxwell field. By virtue of geometrical relations (1.1)–(1.3) one can show that the Ricci tensor R_μ^ν has two real null eigenvectors, k_μ and l_μ, having the same real eigenvalue. Hence k_μ and l_μ can be found from the relations where

$$k^\mu l_\mu = \tfrac{1}{2}A \tag{1.5}$$

and the eigenvalue A is positive. One can then calculate the antisymmetric tensor $\omega_{\mu\nu}$ from the relation[3]

$$\omega^{\mu\nu} = 2\,(l^\mu k^\nu - k^\mu l^\nu - g^{-1/2}\varepsilon^{\mu\lambda\nu\sigma}k_\sigma l_\lambda)/A^{1/2}. \tag{1.6}$$

It can be shown that relation (1.6) does not uniquely determine $\omega_{\mu\nu}$. By setting

$$\omega'_{\mu\nu} = \omega_{\mu\nu}e^{i\alpha}, \tag{1.7}$$

where α is an arbitrary real function of space-time, one can show that $\omega'_{\mu\nu}$ satisfies relations (1.1)–(1.3).

Moreover, it satisfies Maxwell's field Eq. (1.4) if and only if

$$\alpha_{\beta,\gamma} - \alpha_{\gamma,\beta} = 0, \tag{1.8}$$

where the vector α_β is defined by

$$\alpha_\beta \equiv (-g)^{1/2}\,\varepsilon_{\beta\lambda\mu\nu}R^{\lambda\gamma;\mu}R_\gamma^\nu/R_{\sigma\delta}R^{\sigma\delta}, \tag{1.9}$$

g being the determinant of the metric tensor. The Levi–Civitá symbol $\varepsilon_{\beta\lambda\mu\nu}$ is skew-symmetric in all pairs of indices and ε_{0123} has the value unity. α is evidently a scalar invariant and is rightly called the "complexion" of the electromagnetic field.

When the field is non-null, the geometric relations (1.1), (1.2), (1.3), (1.8) and (1.9) are entirely equivalent to Einstein's original description of gravitation and electromagnetic radiation, in the absence of charges. Thus the geometrization of the gravitational field and electromagnetism is accomplished and the "already-unified theory" of Rainich emerges. As is seen an arbitrary field may be obtained from $\omega_{\mu\nu}$ by a duality rotation α and a scale factor. Thus, one cannot uniquely specify the Maxwell field but can determine it up to a constant phase factor.

In case the electromagnetic field is null in the sense

$$R_\nu^\mu R_\lambda^\nu = 0, \tag{1.10}$$

Eq. (1.8) falls off and one has yet to accomplish complete geometrization.

If there exists a non-vanishing charge-current vector density, j^μ within spatially limited regions in space-time, one can write Maxwell's equations in the integral form[3]

$$\iint \omega_{\mu\nu}\, \mathrm{d}\,(x^{(\mu)}, x^{(\nu)}) = \iiint g^{1/2}\varepsilon_{\mu\nu\sigma\tau}\, j^\mu\, \mathrm{d}\,(x^{(\nu)}, x^{(\sigma)}, x^{(\tau)}), \tag{1.11}$$

where the surface integral is taken over a closed two dimensional surface, and the volume integral is taken over the three dimensional volume enclosed by that surface. Maxwell's equations in the form (1.4) and its geometric counterpart (1.8), however, hold everywhere outside these regions within which there exists a non-vanishing charge-current vector.

In a series of papers Witten[3,5,6], Raychaudhuri[7] and the author[8,9] have discussed the solutions of the Rainich equations in the case of static fields. In the present paper the case of a static field is investigated when the space-time metric admits a group G_4 of automorphisms with the group structure:

$$\begin{gathered}[X_i, X_j] = 0 \quad (i, j = 1, 2, 3), \quad [X_1, X_4] = lX_2, \\ [X_2, X_4] = mX_1, \quad (l = m = 0, \pm 1), \quad [X_3, X_4] = 0.\end{gathered} \tag{1.12}$$

We note that the group G_4 includes the Abelian subgroups G_3. Two special types of metrics have been considered. It turns out that in one case there exists an electromagnetic field for which there is a uniform charge along the axis of density, while in the other field there is a flow of current along the axial direction which produces a magnetic field in the angular direction.

It is well known that any type of group transformation, if it is intransitive, results in the space-time points being reflected on a point lying on some certain surface of transitiveness. The whole space-time thus divides into such surfaces and physically they represent invariant images.

2 THE SOLUTIONS FOR THE FIRST TYPE OF METRIC

First, we consider the space-time metric which admits a group G_4 of automorphisms and has the group structure (1.12) to be of the form

$$\mathrm{d}s^2 = \nu\, \mathrm{d}t^2 - \mu\,(\mathrm{d}x^2 + \mathrm{d}y^2) - \mathrm{d}z^2, \tag{2.1}$$

with

$$\mu \equiv \mu(z), \quad \nu \equiv \nu(z).$$

Here t is the time coordinate, x, y, z are space coordinates, and μ, ν are both positive; t, x, y, z are numbered respectively as 0, 1, 2, 3. The group of motions admitted by this metric is intransitive.

As all the nondiagonal components of the Ricci tensor R^ν_μ vanish for line element (2.1), Eq. (1.2) reduces to

$$(R^1_1)^2 = (R^2_2)^2 = (R^3_3)^2 = (R^0_0)^2, \tag{2.2}$$

while Eq. (1.1) is expressible as

$$R^1_1 + R^2_2 + R^3_3 + R^0_0 = 0. \tag{2.3}$$

By virtue of Eqs. (2.2) and (2.3) one can easily see that the diagonal components of the Ricci tensor are all of the same magnitude with a pair of opposite signs and that there arise three possible cases:

case (i): $$R^1_1 = R^2_2 = -R^3_3 = -R^0_0; \tag{2.4}$$

case (ii): $$R^1_1 = -R^2_2 = R^3_3 = -R^0_0; \tag{2.5}$$

case (iii): $$R^1_1 = -R^2_2 = -R^3_3 = R^0_0. \tag{2.6}$$

For the line element (2.1) the equation

$$R^1_1 = R^2_2 \tag{2.7}$$

is satisfied identically and hence the last two possibilities lead to trivial cases. The Rainich equations thus yield case (i) as the only admissible case for metric (2.1) and we have at our disposal in case (i) the two distinct equations

$$R^3_3 = R^0_0 \tag{2.8}$$

and

$$R^1_1 = -R^0_0 \tag{2.9}$$

to determine μ and ν.

With line element (2.1), Eq. (2.8) reduces to

$$\alpha'' + \tfrac{1}{2}\alpha'^2 - \tfrac{1}{2}\alpha'\beta' = 0, \tag{2.10}$$

where we have set

$$\left.\begin{aligned} \alpha &\equiv \log \mu \\ \beta &\equiv \log \nu \end{aligned}\right\} \tag{2.11}$$

and where a dash indicates differentiation with respect to z.

Similarly Eq. (2.9) gives, in view of (2.11)

$$3\alpha'' + 2\alpha'^2 + \beta'' + \tfrac{1}{2}\beta'^2 + \tfrac{1}{2}\alpha'\beta' = 0. \tag{2.12}$$

As is well known, the "complexion" vector α_β vanishes for static fields and hence Eq. (1.8) is automatically satisfied. Thus our problem reduces to solving Eq. (2.10) and (2.12) for μ and ν subject to condition (1.3).

The solutions of Eqs. (2.10) and (2.12) divide naturally into two types according to whether α' vanishes or not. Let us first consider the case

$$\alpha' = 0. \tag{2.13}$$

Eq. (2.10) is then automatically satisfied, while Eq. (2.12) reduces to

$$\beta'' + \tfrac{1}{2}\beta'^2 = 0, \tag{2.14}$$

which gives either

$$\beta' = 0 \tag{2.15}$$

or,

$$\exp\left(\tfrac{1}{2}\beta\right) = az + b, \tag{2.16}$$

a, b being arbitrary constants of integration. We disregard both the cases as they lead to empty flat space.

Next, we consider the case when α' does not vanish. Eq. (2.10) then gives on integration with respect to z

$$\beta = \log\left(A\alpha'^2\right) + \alpha, \tag{2.17}$$

where A is an integration constant.

Eliminating β between (2.12) and (2.17), we get

$$2\alpha''' + 7\alpha'\alpha'' + 3\alpha'^3 = 0 \tag{2.18}$$

which reduces to

$$2\xi'' + 7\xi\xi' + 3\xi^3 = 0, \tag{2.19}$$

where

$$\xi = \alpha'. \tag{2.20}$$

Eq. (2.19) can be presented in the form

$$2\left(\xi'' + a\xi\xi'\right) + \xi\left(b\xi' + 3\xi^2\right) = 0, \tag{2.21}$$

where a and b are constants connected by two relations, one of which is

$$2a + b = 7. \tag{2.22}$$

The satisfaction of Eq. (2.21) is ensured if the equations

$$\xi'' + a\xi\xi' = 0 \tag{2.23}$$

and

$$b\xi' + 3\xi^2 = 0 \tag{2.24}$$

are satisfied simultaneously. This requirement yields

$$ab = 6, \tag{2.25}$$

which is the second relation between a and b. As this involves the vanishing of the integration constant obtained by integrating Eq. (2.23) once with respect to z, we shall in our later consideration use, without loss of generality, Eq. (2.24) only in solving for ξ. Eqs. (2.22) and (2.25) give either

$$a = 2, \quad b = 3 \tag{2.26}$$

or else

$$a = \tfrac{3}{2}, \quad b = 4. \tag{2.27}$$

Case I: If $a = 2$, $b = 3$, Eq. (2.24) reduces to

$$\xi' + \xi^2 = 0, \tag{2.28}$$

which gives on integration with respect to z

$$\xi = (z + C)^{-1}, \tag{2.29}$$

where C is an arbitrary constant.

On further integration, we get in view of (2.11) and (2.20)

$$\mu = B(z + C), \tag{2.30}$$

B being the constant of integration.

By an obvious transformation, we can write

$$\mu = Bz. \tag{2.31}$$

Next, from (2.17) we get by virtue of (2.11), (2.20) and (2.31)

$$\nu = AB/z. \tag{2.32}$$

Line element (2.1) thus takes the form

$$ds^2 = \frac{AB}{z}\, dt^2 - Bz\,(dx^2 + dy^2) - dz^2. \tag{2.33}$$

By the substitution

$$\left.\begin{aligned} \tau &= (AB)^{1/2}\, t, & \varrho &= z, \\ \phi &= B^{1/2}x, & \zeta &= B^{1/2}y \end{aligned}\right\}, \tag{2.34}$$

metric (2.33) can be transformed to

$$ds^2 = \varrho^{-1}\, d\tau^2 - \varrho\,(d\phi^2 + d\zeta^2) - d\varrho^2, \tag{2.35}$$

where ϱ, ϕ and ζ may be considered as radial, angular and axial coordinates respectively.

$\tau, \phi, \zeta, \varrho$ are numbered as 0, 1, 2, 3 according to our previous order.

Condition (1.3) is satisfied for the line-element (2.35) everywhere except for $\varrho = 0$ and $\varrho \to \infty$. The metric (2.35) is regular everywhere except for $\varrho = 0$ and $\varrho \to \infty$.

It appears that there exists a singularity at the origin and that at infinite distance from the axis $\varrho = 0$ one gets a completely empty flat space.

Next, we consider the case (II) where $a = \frac{3}{2}$, $b = 4$ and get as before

$$\left.\begin{aligned} \mu &= B_1 z^{4/3} \\ \nu &= (16AB_1)/(9z^{2/3}), \end{aligned}\right\} \tag{2.36}$$

where B_1 is an arbitrary constant of integration.

With metric (2.36),

$$R_0^0 = 0$$

and consequently this leads to empty flat space.

3 THE ELECTROMAGNETIC FIELD FOR THE FIRST TYPE OF METRIC

For the sake of complete analysis we venture to exhibit the electromagnetic field explicitly. One can find the null eigenvectors of the Ricci tensor R_μ^ν and their eigenvalues from relations (1.5) and then construct the self-dual electromagnetic field tensor $\omega_{\mu\nu}$ employing relation (1.6).

By virtue of Eq. (2.4) and recalling that $k_0 \neq 0$, one can easily see from relations (1.5) that

$$\left.\begin{aligned} k_1 &= k_2 = 0 \\ l_1 &= l_2 = 0 \end{aligned}\right\} \tag{3.1}$$

and

$$A = -R_0^0. \tag{3.2}$$

The two null eigenvectors are thus given by

$$k_\mu = \left[k_0, 0, 0, \left(-\frac{g^{00}}{g^{33}}\right)^{1/2} k_0\right], \tag{3.3}$$

$$l_\mu = \left[-\frac{1}{4}\frac{R_{00}}{k_0}, 0, 0, \frac{1}{4}\frac{R_{00}}{k_0}\left(-\frac{g^{00}}{g^{33}}\right)^{1/2}\right]. \tag{3.4}$$

One then easily sees that all components of the self-dual electromagnetic field tensor $\omega_{\mu\nu}$ except ω_{03} and ω_{12} vanish and one gets by virtue of (1.6) and (1.7)

$$\omega_{03} = -\tfrac{1}{2}\varrho^{-3/2}\mathrm{e}^{i\alpha} \tag{3.5}$$

and

$$\omega_{12} = \frac{i}{2}\,\mathrm{e}^{i\alpha}. \tag{3.6}$$

It is easy to verify that the electromagnetic field thus obtained satisfies Maxwell's equations in the differential form (1.4). Next, we turn to Maxwell's equations in the integral form (1.11) in order to exhibit the electromagnetic field explicitly and to find a physical meaning for the solution. For this purpose we consider an electromagnetic field with vanishing current

$$j^1 = j^2 = j^3 = 0. \tag{3.7}$$

We further consider an ordinary cylinder of height ζ and radius ϱ in the ordinary three-space (ϱ, ϕ, ζ) which enclose the axis $\varrho = 0$ and get from (1.11)

$$\int_0^{2\pi}\int_0^{\zeta}\omega_{12}\,\mathrm{d}\phi\,\mathrm{d}\zeta = \int_0^{\zeta}\int_0^{\varrho}\int_0^{2\pi} g^{1/2}\,j^0\,\mathrm{d}\zeta\,\mathrm{d}\varrho\,\mathrm{d}\phi. \tag{3.8}$$

Next, defining the charge density by

$$Q \equiv 2\pi\int_0^{\varrho}(-g)^{1/2}j^0\,\mathrm{d}\varrho, \tag{3.9}$$

we get from (3.6) and (3.8)

$$Q = +\pi, \tag{3.10}$$

where we have chosen $\alpha = \pi$ in order to make the charge density real. Thus we see that there exists a uniform charge along the axis of density given by (3.10). For this choice of α, the only non-vanishing component of the electromagnetic field tensor is given to be

$$f_{30} = -\tfrac{1}{2}\varrho^{-3/2}, \tag{3.11}$$

which, in the terminology of flat space, would correspond to a component of the electric field in the radial direction. It may be noted that the electromagnetic field decreases with ϱ and vanishes for large values and hence at infinite distance from the axis $\varrho = 0$ one gets a completely empty flat space.

4 THE SOLUTIONS FOR THE SECOND TYPE OF METRIC

Next, we consider the space-time metric

$$ds^2 = \mu(dt^2 - dx^2) - \nu\, dy^2 - dz^2, \tag{4.1}$$

where μ, ν are functions of z alone and both positive. The group of motions admitted by this metric is also intransitive.

Arguing as before, one can see that for line element (4.1) the Rainich equations give as the only admissible case three independent relations between the Ricci tensor components:

$$R_1^1 = R_0^0, \tag{4.2}$$

$$R_2^2 = R_3^3, \tag{4.3}$$

and

$$R_1^1 = -R_2^2. \tag{4.4}$$

With line element (4.1), Eq. (4.2) is satisfied identically and Eq. (4.3) gives

$$\alpha'' + \tfrac{1}{2}\alpha'^2 - \tfrac{1}{2}\alpha'\beta' = 0, \tag{4.5}$$

where α, β are given by (2.11).

Also Eq. (4.4) yields

$$3\alpha'' + 2\alpha'^2 + \beta'' + \tfrac{1}{2}\beta'^2 + \tfrac{1}{2}\alpha'\beta' = 0. \tag{4.6}$$

We note in passing that Eq. (1.8) is automatically satisfied in this case and our problem reduces to solving Eqs. (4.5) and (4.6) for μ and ν subject to condition (1.3).

Now proceeding in a similar manner to that in the previous case, and disregarding the case where α' vanishes as it leads to empty flat space, we have

$$\left.\begin{aligned} \mu &= Bz \\ \nu &= ABz^{-1} \end{aligned}\right\}, \tag{4.7}$$

and

$$\left.\begin{aligned} \mu &= B_1 z^{4/3} \\ \nu &= (16/9)\, AB_1 z^{-2/3} \end{aligned}\right\}, \tag{4.8}$$

where A, B and B_1 are constants of integration.

With metric (4.8),

$$R_0^0 = 0,$$

and hence this leads to empty flat space.

By the substitution

$$\left.\begin{aligned} \tau &= B^{1/2}t, \quad \varrho = z, \\ \phi &= B^{1/2}x, \quad \zeta = (AB)^{1/2}y \end{aligned}\right\}, \tag{4.9}$$

metric (4.7) can be transformed to

$$ds^2 = \varrho\,(d\tau^2 - d\phi^2) - \varrho^{-1}\,d\zeta^2 - d\varrho^2, \tag{4.10}$$

where ϱ, ϕ and ζ may be considered as radial, angular and axial coordinates respectively.

Condition (1.3) is satisfied for line element (4.10) everywhere except for $\varrho = 0$ and $\varrho \to \infty$. Metric (4.10) is regular everywhere except for $\varrho = 0$ and $\varrho \to \infty$. It appears that there exists a singularity at the origin and that at infinite distance from the axis $\varrho = 0$ one gets a completely empty flat space.

5 THE ELECTROMAGNETIC FIELD FOR THE SECOND TYPE OF METRIC

Now we exhibit the structure of the electromagnetic field corresponding to metric (4.10). By virtue of Eq. (4.2)–(4.4) and recalling that $k_0 \neq 0$, one can find from relations (1.5) the two null eigenvectors to be

$$k_\mu = [k_0, k_0, 0, 0], \tag{5.1}$$

$$l_\mu = \left[-\frac{1}{4}\frac{R_{00}}{k_0}, \frac{1}{4}\frac{R_{00}}{k_0}, 0, 0\right], \tag{5.2}$$

with

$$A = -R_0^0. \tag{5.3}$$

One can then easily see that all components of the self-dual electromagnetic field tensor $\omega_{\mu\nu}$ except ω_{01} and ω_{23} vanish and hence, by virtue of (1.6) and (1.7), we get

$$\omega_{01} = \tfrac{1}{2}\,e^{i\alpha} \tag{5.4}$$

and

$$\omega_{23} = -\frac{i}{2}\,\varrho^{-3/2}\,e^{i\alpha}. \tag{5.5}$$

It is easy to verify that the electromagnetic field thus obtained satisfies Maxwell's equations in the differential form (1.4). In order to find a meaningful physical interpretation for the solution arrived at we next turn to Maxwell's equations in the integral form (1.11). For this purpose we consider an electromagnetic field characterized by

$$j^0 = j^1 = j^3 = 0. \tag{5.6}$$

We further consider a cylinder of finite length, with its central axis along the time axis, in the three-space (ϱ, ϕ, τ) and get from (1.11)

$$\int_0^T \int_0^{2\pi} \omega_{01} \, d\tau \, d\phi = \int_0^T \int_0^{2\pi} \int_0^{\varrho} g^{1/2} j^2 \, d\tau \, d\phi \, d\varrho . \tag{5.7}$$

Next, defining the total current flowing along the axial direction by

$$I \equiv 2\pi \int_0^{\varrho} (-g)^{1/2} j^2 \, d\varrho , \tag{5.8}$$

we get from (5.4) and (5.7)

$$I = +\pi , \tag{5.9}$$

where we have chosen $\alpha = -\pi/2$ in order to make the current real. It is thus seen that there exists a current parallel to the axial direction of magnitude given by (5.9). For this choice of α, the only non-vanishing component of the electromagnetic field tensor is

$$f_{23} = -\tfrac{1}{2}\varrho^{-3/2}. \tag{5.10}$$

In the terminology of flat space, we have the case of a current along the axial direction which produces a magnetic field in the angular direction. It may be noted that the electromagnetic field decreases with ϱ and vanishes for large values and hence at infinite distance from the axis $\varrho = 0$ one gets a completely empty flat space.

REFERENCES

1. G. Y. Rainich, *Trans. Am. Math. Soc.*, **27**, 106 (1925).
2. C. W. Misner and J. A. Wheeler, *Ann. Phys.*, **2**, 525 (1957).
3. L. Witten, *Colloque sur la Théorie de la Relativité*, Centre Belge de Recherches Mathématiques, 1959, pp. 59–77.
4. We define

$$\omega_{\mu\nu} = f_{\mu\nu} + {}^*f_{\mu\nu},$$

where ${}^*f_{\mu\nu}$ is the dual of $f_{\mu\nu}$, defined as

$${}^*f^{\mu\nu} \equiv g^{-1/2}\varepsilon^{\mu\nu\lambda\sigma} f_{\lambda\sigma},$$

and

$$f_{\mu\nu} \equiv (2G/c^4)^{1/2} F_{\mu\nu},$$

$F_{\mu\nu}$ being the antisymmetric electromagnetic tensor.

5. L. Witten, *Phys. Rev.*, **115**, 1, 206 (1959).
6. L. Witten, *Phys. Rev.*, **120**, 635 (1960).
7. A. K. Raychaudhuri, *Ann. Phys.*, **11**, 501 (1960).
8. B. K. Datta, *Ann. Phys.*, **12**, 295 (1961).
9. B. K. Datta, *Ann. Phys.*, **15**, 403 (1961).

PAPER 13

Gravitational–scalar field coupling*

S. DESER** and J. HIGBIE†
Brandeis University, Waltham Massachusetts, U.S.A.

ABSTRACT

Spherically symmetric initial solutions of the system gravitation plus massless scalar field, generated by a physical "particle" source of both, are obtained. These solutions exhibit unusual behavior in terms of the source parameters (bare mass m_0, Einstein constant $\varkappa$, scalar coupling constant f, and source size ε): There are no solutions for some ranges of these parameters, while two (non-singular) branches exist elsewhere. In each of the limits $f \to 0$, $\varkappa \to 0$ or $\varepsilon \to \infty$ there is a normal and an anomalous branch, the latter involving essential singularities in f, $\varkappa$, ε^{-1} through factors $\sim f^2 \varepsilon^{-1} \exp [\varepsilon^2 f^{-2} \varkappa^{-1}]$ in the mass and the fields. There is a minimum finite particle extension for all solutions. The analogous system: particle source of the scalar-tensor gravitational field is also treated in the scalar-tensor theory of gravitation, where the results are qualitatively as in general relativity.

1 INTRODUCTION

An isolated electrically (e) and gravitationally (m_0) charged spherically symmetric distribution ("particle"), together with its Coulomb and "Newtonian" self fields is the simplest example of the three-part system: general relativity—Maxwell field—source of both. The initial value problem for this system was solved some time ago[1]. The result was a perfectly non-singular, unique solu-

* Work supported by the USAF OAR under Grant AFOSR 368–67; Reprinted from *Ann. Phys.*, **58**, 56 (1970) with the editor's kind permission.

** and NORDITA, Copenhagen, Denmark.
† *National Science Foundation* Predoctoral Fellow.

tion for every set of parameters (m_0, e, ε), as expected for the solution of the Cauchy problem of a physically reasonable matter source interacting with the gravitational field. In this paper[2], we deal with an apparently analogous and simpler problem, in which the "particle's" electric charge and Maxwell self-field are replaced by a coupling constant f to a long-range scalar field ψ. We shall see that the situation is actually considerably more complicated, due especially to the fact that there is no longer a conserved "charge", nor a unique "scalar Coulomb" self field. Unlike the vector field, a spherically symmetric scalar field is capable of radiating, so that the notion of a (classical) one-particle state becomes ambiguous. Thus the initial value of the field amplitude ψ has no necessary connection, from the field equations, to the source density.

We shall analyze this question, and choose as our initial situation one which is physically the best candidate for a one-particle state. It will then be possible to obtain general solutions to the problem in an appropriate, non-singular, coordinate frame. These will have a very anomalous character in comparison to the electric case. In general, there will be either no solution for a given set (m_0, f, ε) of input parameters, or two solutions. Thus, some perfectly normal (from the flat-space viewpoint) matter configurations do not have general relativistic counterparts. Others involve the opposite problem that they give rise to two different total masses (and solutions). Indeed, as either f or the Einstein constant $\varkappa$ vanish, or the source size becomes large for fixed parameters, one branch yields the physically desired limit, while the other exhibits an essential singularity, which manifests itself in the total mass m, and in the various field strengths, through factors $\sim f^2\varepsilon^{-1}\exp[f^{-2}\varkappa^{-1}\varepsilon^2]$. This branch is perfectly non-singular in its spatial behavior, but clearly the limiting values (and in particular the order of $r \to \infty$ and f or $\varkappa \to 0$) are not well-defined. Thus the relation between a structureless bare dust source distribution and the exterior metric it supports is quite unusual. It would appear that the chief source of the anomaly lies in the mentioned lack of charge conservation together with a peculiar degeneracy between the scalar field and Newtonian part of the metric field; the negative nature of the energy in the (attractive) scalar field is also involved.

The general spherical solution obtained here will also be compared to a recent[3] purely external static solution. While the latter also exhibited an anomaly in the limits of one of the integration constants, it turns out, from our point of view, to correspond to an infinitely extended and "dilute" source, for which the notion of "exterior" itself is not clear. Its static nature is likewise related to this limiting character; in general, we shall show that our initial configurations are not static. Contraction might be expected from the attractive nature of both the scalar and gravitational forces, but the negative character of part of the source energy complicates the question. From the form of the time development equations, one finds initial contraction for dilute systems at least. The question of collapse is still more difficult. We can show that at the initial instant, there is a non-zero lower bound on the coordinate (and invariant) extension of the system, so that for finite (m_0, f) we cannot have $\varepsilon = 0$. This is also in contrast to the Maxwell case.

The scalar-tensor Brans–Dicke theory[4] is well-known to be formally similar to the above Einstein-plus-scalar-field system, and it is natural to investigate the corresponding problem there. The scalar field is now of quite different origin, being an integral part of the description of gravitation; the appropriate problem is then one in which the source has only its gravitational coupling m_0, but no separate scalar coupling. The structure of the theory itself determines the relative couplings to tensor and scalar fields of m_0; indeed the famous deviations from Einstein's predictions for bending of light and precession are due to this division of coupling strength, for reasonably dilute sources like the sun. We will solve the initial problem with a structureless dust source in this theory, and express the observed mass in terms of m_0, ε (and of course ω, the scalar-tensor coupling parameter). Here the situation is qualitatively like that of a neutral particle in general relativity. There is a unique, non-singular solution for every set (m_0, ε) which reduces correctly to the Einstein case as $\omega \to \infty$. As $\varepsilon \to 0$, the total mass vanishes, as in general relativity (but somewhat faster), while the opposite (dilute), $\varepsilon \to \infty$, limit yields the Brans–Dicke equivalent of the Newtonian limit. The theory is somewhat more "Machian" than general relativity in that the (coordinate) point limit really corresponds to a vanishing invariant particle extension or volume, i.e. to an effective decoupling of the metric (but not of the scalar field) and source. The purely exterior static solution given in [4] stands in the same relation to the general case here as did that of [3] to the earlier system: It is a particular dilute limit.

2 METRIC-SCALAR FIELD-PARTICLE SYSTEM

We formulate the action for this system in the canonical framework[2] of [1], with identical notation (except for the explicit appearance of the Einstein constant $\varkappa$). We have

$$\begin{aligned}\varkappa\mathscr{L} &\equiv {}^4\mathscr{R} + \varkappa\mathscr{L}_\psi(\pi, \psi) + \varkappa\mathscr{L}_P(\mathbf{p}, \mathbf{x}) + \varkappa\mathscr{L}_I(\mathbf{x}, \psi) \\ &= -\dot{\pi}^{ij}g_{ij} + \varkappa\pi\dot{\psi} + \varkappa p_i\dot{x}^i\varrho(r) - N_\mu\bar{R}^\mu + \mathscr{D}\end{aligned} \tag{2.1}$$

$$\begin{aligned}\bar{R}^0 \equiv &-g^{1/2}({}^3R) + g^{-1/2}(\pi^{ij}\pi_{ij} - \tfrac{1}{2}\pi^i_i\pi^j_j) + \tfrac{1}{2}\varkappa g^{-1/2}\pi^2 \\ &+ \tfrac{1}{2}\varkappa g^{1/2}\psi^{,i}\psi_{,i} + \varkappa(p_ip^i + m_0^2)^{1/2}\varrho(r) - \varkappa f\psi(p_ip^im_0^{-2} \\ &+ 1)^{-1/2}\varrho(r)\end{aligned} \tag{2.2a}$$

$$\bar{R}^i \equiv -2\pi^{ij}{}_{|j} + \varkappa\pi\psi^{,i} - \varkappa p^i\varrho(r) \tag{2.2b}$$

$$\mathscr{D} \equiv -2(\pi^{ij}N_j - \tfrac{1}{2}\pi^j_jN^i + g^{1/2}N^{|i})_{,i}. \tag{2.3}$$

Here π^{ij} (essentially $\dot{g}_{ij}$) is the momentum conjugate to g_{ij}, π is the scalar field momentum conjugate to ψ, where (p_i, x^i) describe the "particle", which is a spherically symmetric three-scalar density distribution $\varrho(r)$. We shall take it to be a shell, $\varrho \sim \delta(r - \varepsilon)/4\pi r^2$, for simplicity. In addition to the kinetic energy terms, the initial value equations $\bar{R}^\mu = 0$ (constraints) of the theory determine the dependent metric variables as functions of the true degrees of freedom. [The constraints are basically the $G^0_\mu = \varkappa T^0_\mu$ field equations, obtained upon variation of $N_0 \equiv N = (-g^{00})^{-1/2}$ and $N_i \equiv g_{0i}$.] All operations are with respect to the three-metric g_{ij} (thus g is its determinant, and "$|$" denotes covariant differentiation with respect to g_{ij}). The total derivative $\mathscr{D}$ does not affect the field equations at all, but is kept here because it will contribute in Brans–Dicke theory where ${}^4\mathscr{R}$ is multiplied by the Brans–Dicke scalar field ϕ. The initial data are specified by giving the values of the unconstrained degrees of freedom of the system at $t = 0$. These include p, r, π and ψ as well as the two pairs of gravitational degrees of freedom (for this spin-2 massless field) from among the six pairs (π^{ij}, g_{ij}). The remaining gravitational variables in (π^{ij}, g_{ij}) are four initial coordinate conditions and four constraint variables, to be solved for in principle through Eq. (2.2). The N_μ are just Lagrange multipliers which may be assigned arbitrarily initially, or more fundamentally, they are determined by the choice of coordinates off the $t = 0$ surface. Their initial values, in particular, are determined by the coordinate choice at $t = \mathrm{d}t$. In addition to the initial data, there are of course the input parameters (m_0, f, ε) which describe the bare mass (i.e. the particle's mass in the absence of gravitation), the scalar coupling strength, and the source's (coordinate) extension respectively.

We wish to study the pure self-field situation in which no external gravitons or scalar excitations are present. In the electromagnetic case, this is particularly simple, since spherical symmetry insures that no transverse tensor or vector components are present, leaving only the respective longitudinal Newtonian and Coulomb fields. The latter are determined entirely by $\bar{R}^{\mu} = 0$ and the Gauss equation $\nabla \cdot \mathscr{E} = e\varrho$. Thus, spherical symmetry automatically provides the desired conditions. This is still the case for the present system as far as the gravitational variables are concerned. Likewise, we obviously set $p_i = 0$ (and $x^i = 0$ for convenience). However, the scalar field does not provide us with a Gauss equation or a conserved "charge". While it is clear that we should set the field momentum π to zero to avoid any irrelevant "velocity" contributions, the field strength $\psi(r)$ is in principle entirely arbitrary. Thus the field equations read

$$\pi = g^{1/2}(N^{-1}\dot{\psi} - N^i N^{-1}\psi_{,i}),$$
$$\dot{\pi} = (Ng^{1/2}g^{ij}\psi_{,j} + N^i\pi)_{,i} + fN(p_i p^i m_0^{-2} + 1)^{-1/2}\varrho(r) \tag{2.4}$$

and the $\dot{\pi}$ equation then gives the value of $\dot{\pi}$ once ψ is specified. Thus we could initially set $\psi = f/4\pi r$, as in flat space, or even $\psi = 0$. The latter choice clearly indicates the unusual nature of this system. For in that case, the total energy, determined by the constraints, would be *independent* of f (since neither f nor π appear in them), and would in fact be that of a neutral particle m_0 in pure Einstein theory (Eq. 3.3a below). Since energy is conserved, this value would remain the same at later times when ψ no longer vanishes[6]. Also, this choice would mean that $\dot{\pi}(0)\,\mathrm{d}t = \pi\,(t = \mathrm{d}t)$ would have a discontinuity at the shell. Clearly, the most static initial configuration is one where ψ satisfies the covariant Poisson equation suggested by Eq. (2.4), so that[7] $\dot{\pi}(0) = 0$, and the rest of our discussion follows this choice. We are now faced with having to specify the $g_{0\mu}$ initially since they enter in this Poisson equation. Again, we could, whatever our initial coordinate choice, pick the $g_{0\mu}$ arbitrarily, e.g. set $g_{0\mu} = -\delta_{0\mu}$. But then we would find, as will emerge below, that the coordinates dt later would have a singularity at the shell. This difficulty with an *a priori* choice of $g_{0\mu}$ has been noted earlier[1] in a similar context. Instead, we shall keep the same coordinate choice at the next instant, which will determine the initial $g_{0\mu}$, all in a non-singular way. The coordinate choice is, as in [1], the isotropic-minimal surface one, determined by the requirement

$$g_{ij,jkk} - \tfrac{1}{4}g_{kj,kji} - \tfrac{1}{4}g_{jj,kki} = 0, \quad \pi^i_i \equiv g_{ij}\pi^{ij} = 0. \tag{2.5}$$

With this choice, and spherical symmetry, g_{ij} takes the particularly simple isotropic form, and the curvature the form of a (flat space) Laplacian:

$$g_{ij} = \chi^4 \delta_{ij}, \quad {}^3\mathscr{R} \equiv g^{1/2} g^{ij} R_{ij} = -8\chi \nabla^2 \chi. \tag{2.6}$$

It then also follows from symmetry and the other initial data that the remaining components of π^{ij}, determined by $\bar{R}^{\mu} = 0$, vanish initially. Thus $t = 0$ is an instant of time symmetry, at which all momenta vanish. It is then easy to see, from the g_{ij} equations obtained by varying (2.1) with respect to π^{ij}, that g_{0i} vanish initially, while N is determined when the $\pi^i_i = 0$ requirement is inserted into the $\dot{\pi}^{ij}$ equation.

At this point, we may write down a simple effective Lagrangian for the initial value situation, by simply omitting both the kinetic terms and all momenta in the action (2.1), (2.2).

We then find at $t = 0$,

$$\mathscr{L}_{\text{eff}} = -N\,[8\chi\nabla^2\chi + \tfrac{1}{2}\varkappa\chi^2\,(\nabla\psi)^2 + \varkappa\,(m_0 - f\psi)\,\varrho(r)] \tag{2.7}$$

where N, χ and ψ are to be varied independently. Outside the shell ($r = \varepsilon$), the source term disappears. It is clear from (2.7) that there is an exterior quantity which is divergence-free, namely $N\chi^2\nabla\psi$, and we will see that the same is true of $\chi^2\nabla N$. Because of this fact and because of spherical symmetry, it will be convenient to use the new independent variable $u \equiv r^{-1}$, and all differentiation will be with respect to it. Hence $N\chi^2\psi'$ and $\chi^2 N'$ are exterior "constants of the motion". [A greater advantage of using the variable u is that matching conditions at the shell are simply read from the coefficients of $\nabla^2 u \sim \varrho(r)$ since we choose ϱ to be a shell distribution. Consequently the independent variable will not appear in the matching conditions or in the field equations.] With the two above "constants of the motion", the problem becomes one of determining the function $\psi(u)$, in terms of which the two others are expressible. The initial equations read as follows:

$$\delta N: \qquad -8\chi\chi'' = \tfrac{1}{2}\varkappa\chi^2\,(\psi')^2, \qquad u < \varepsilon^{-1} \tag{2.8a}$$

$$\delta\psi: \qquad \psi'' + 2\chi^{-1}\chi'\psi' + N^{-1}N'\psi' = 0, \qquad u < \varepsilon^{-1} \tag{2.8b}$$

$$\delta\chi: \quad 8N'' + 16N\chi^{-1}\chi'' + 16N'\chi^{-1}\chi' + \varkappa N\,(\psi')^2 = 0, \quad u < \varepsilon^{-1}. \tag{2.8c}$$

Thus (2.8b) has the immediate integral:

$$N\chi^2\psi' = \text{const.} \tag{2.9a}$$

while use of (2.8a) in (2.8c) yields:

$$N'\chi^2 = \text{const.} \tag{2.9b}$$

It will be convenient to introduce dimensionless constants as follows:

$$N\chi^2\psi' = (\varkappa^{1/2}m_0/8\pi)\, K_1\,, \tag{2.10a}$$

$$N'\chi^2 = (-\varkappa m_0/16\pi)\, K_2\,. \tag{2.10b}$$

Using the boundary conditions at $u = 0: \psi(0) = 0,\ N(0) = 1,\ \chi(0) = 1$, Eq. (2.10) yield:

$$N = \exp\left(-\tfrac{1}{2}\Lambda\varkappa^{1/2}\psi\right), \qquad \Lambda \equiv K_2K_1^{-1}\,, \tag{2.11a}$$

$$\chi^2 = (\varkappa m_0/16\pi)\, K_1\left(\tfrac{1}{2}N\varkappa^{1/2}\psi'\right)^{-1}. \tag{2.11b}$$

The equation for ψ follows then from (2.8a) and (2.11)

$$\varkappa\,(1 + \Lambda^2)\,(\psi')^2 + 12\,(\psi''/\psi')^2 - 8(\psi'''/\psi') = 0\,, \tag{2.12}$$

whose first integral is

$$2\,(\psi''/\psi') = \pm\left[\varkappa\,(1 + \Lambda^2)\,(\psi')^2 - (\varkappa^{3/2}m_0/8\pi)\, K_3\psi'\right]^{1/2}. \tag{2.13}$$

Using $\chi(0) = 1$ and $\psi(0) = 0$, (2.13) is integrated to the explicit form

$$\frac{1}{2}\varkappa^{1/2}\psi = \frac{f}{|f|}\,(1 + \Lambda^2)^{-1/2} \times$$

$$\times\ \ln\left|\frac{1 + K_1(f/|f|)\,(1 + \Lambda^2)^{1/2}\,(1 + C)\,(\varkappa m_0 u/32\pi)}{1 - K_1(f/|f|)\,(1 + \Lambda^2)^{1/2}\,(1 - C)\,(\varkappa m_0 u/32\pi)}\right|. \tag{2.14}$$

Here we have written $C \equiv \mp\left[1 - K_3K_1^{-1}\,(1 + \Lambda^2)^{-1}\right]^{1/2}$ with the $\mp$ sign correlated to the $\pm$ sign in (2.13). We shall see that the positive (lower) sign must hold to prevent singularities in the exterior region. We now find the matching conditions at the shell, equating coefficients of $\delta\,(r - \varepsilon)/4\pi r^2$ in the equations corresponding to (2.8):

$$\delta N: \qquad -\chi\chi' + (\varkappa/32\pi)\,(m_0 - f\psi) = 0\,, \qquad u = \varepsilon^{-1} \tag{2.15a}$$

$$\delta\psi: \qquad (f/4\pi)\, N - N\chi^2\psi' = 0\,, \qquad u = \varepsilon^{-1} \tag{2.15b}$$

$$\delta\chi: \qquad 2N\chi' + N'\chi = 0\,, \qquad u = \varepsilon^{-1}. \tag{2.15c}$$

Eqs. (2.15b), (2.10a) and (2.11a) yield:

$$K_1 = (2f/\varkappa^{1/2}m_0)\exp\left[-\tfrac{1}{2}\Lambda\varkappa^{1/2}\psi(\varepsilon^{-1})\right]; \tag{2.16}$$

(2.15c) and (2.11) yield:

$$\psi''(\varepsilon^{-1})/\psi'(\varepsilon^{-1}) = 0; \tag{2.17}$$

with (2.13) this gives

$$\psi'(\varepsilon^{-1}) = (\varkappa^{1/2}m_0/8\pi)\,K_1\,(1 - C^2); \tag{2.18}$$

with (2.14):

$$(\varkappa m_0/32\pi\varepsilon) = C\,(1 + \Lambda^2)^{-1/2}\,(|K_1|\,[1 - C^2])^{-1}; \tag{2.19}$$

so that

$$\frac{1}{2}\varkappa^{1/2}\psi(\varepsilon^{-1}) = \frac{f}{|f|}(1 + \Lambda^2)^{-1/2}\ln\left|\frac{1 + C}{1 - C}\right|. \tag{2.20}$$

To avoid singularities for $u < \varepsilon^{-1}$, this requires $0 \leqslant C < 1$. We now have, defining

$$\Lambda\,(1 + \Lambda^2)^{-1/2} \equiv \lambda,$$

$$K_1 = (2f/\varkappa^{1/2}m_0)\left(\frac{1 + C}{1 - C}\right)^{-\lambda}. \tag{2.21}$$

Finally (2.15a) yields

$$\frac{\Lambda}{\lambda}\left(\frac{\varkappa^{1/2}m_0}{2|f|} - \Lambda\right) = \ln\left(\frac{1 + C}{1 - C}\right). \tag{2.22a}$$

Inserting (2.21) in (2.19) gives

$$\frac{\Lambda}{\lambda}\left(\frac{\varkappa^{1/2}|f|}{16\pi\varepsilon}\right)(1 - C^2) = C\left(\frac{1 + C}{1 - C}\right)^{\lambda}. \tag{2.22b}$$

This fixes the constants of integration (K_1, C, Λ) in terms of the input parameters (m_0, f, ε). We could also rewrite (2.22b) in terms of the invariant particle radius

$$\bar{\varepsilon} \equiv \int_0^{\varepsilon}(g_{rr})^{1/2}\,dr = \varepsilon\chi^2(\varepsilon^{-1}) = \varepsilon\,(1 - C^2)^{-1}\left(\frac{1 + C}{1 - C}\right)^{\lambda}.$$

Then (2.22b) gives:

$$\Lambda\lambda^{-1}\,(\varkappa^{1/2}|f|/16\pi\bar{\varepsilon}) = C. \tag{2.22b'}$$

The total mass, defined as the coefficient of $\varkappa u/32\pi$ in the asymptotic expansion of χ, is obtained now from (2.11), (2.14) and (2.21):

$$\varkappa^{1/2}m/2|f| = \Lambda\,(1 + C\lambda^{-1})\,(1 + C)^{-\lambda}\,(1 - C)^{\lambda}. \tag{2.23}$$

When solved for (K_1, Λ, C) in terms of (m_0, f, ε), Eq. (2.21), (2.22), (2.23) together with the fields (2.11) and (2.14) constitute in principle the full solution to our initial value problem. For future reference we write the fields in terms of the final parameters (Λ, C):

$$\frac{1}{2}\varkappa^{1/2}\psi = \frac{f}{|f|}\Lambda^{-1}\lambda \ln\left[\frac{1 + a(1 + C)u}{1 - a(1 - C)u}\right], \tag{2.24a}$$

$$N = \left[\frac{1 + a(1 + C)u}{1 - a(1 - C)u}\right]^{-\lambda}, \tag{2.24b}$$

$$\chi^2 = [1 + a(1 + C)u]^{1+\lambda}\,[1 - a(1 - C)u]^{1-\lambda}, \tag{2.24c}$$

where
$$a \equiv (\varkappa^{1/2}\,|f|/16\pi)\Lambda\lambda^{-1}\left(\frac{1 + C}{1 - C}\right)^{-\lambda}.$$

For f negative, ψ changes sign but N, χ and m are unchanged. Henceforth we will assume f, ψ positive, with the understanding that there is a physically identical solution with these quantities changing sign. The solutions are guaranteed to be everywhere non-singular for $0 \leqslant C < 1$. In the next section we discuss their form and exhibit the peculiarities mentioned in Section I.

3 PROPERTIES OF THE SOLUTIONS

The general form of the allowed, non-singular, solutions with the initial data examined in the previous section is determined by $\psi(r)$, the metric components N, χ being conveniently expressed in terms of ψ. The latter function depends on two parameters (Λ, C) which are determined by the dimensionless quantities $\alpha \equiv \varkappa^{1/2}m_0/2f$ and $\beta \equiv \varkappa^{1/2}f(16\pi\varepsilon)^{-1}$. The total mass is likewise determined in the combination $\varkappa^{1/2}m/2f$ by α and β.

For orientation, we begin with the normal limits of small $\varkappa$, small $|f|$, or large ε, which should check with known results for the total mass. The anomalous solution discussed in Section I will not appear in these limits as its behavior is essentially singular, and will be treated later.

As the gravitational constant becomes negligible we expect to recover the usual flat space energy of the system. In the limit, with $\nabla^2\phi \sim f\varrho(r)$ so that $\phi \sim (f/4\pi r)$, the flat space Hamiltonian is

$$\begin{aligned}\mathscr{H} &= m_0 + \tfrac{1}{2}\textstyle\int(\nabla\psi)^2 - f\int\psi\varrho \rightarrow m_0 + \tfrac{1}{2}(f^2/4\pi\varepsilon) - (f^2/4\pi\varepsilon)\\ &= m_0 - (f^2/8\pi\varepsilon)\end{aligned} \tag{3.1}$$

the total sign depending on which term predominates. We recover this result from (2.22), (2.23) by taking the limit $\alpha \to 0$, $\beta \to 0$, and assuming C does not $\to 1$. Then we get

$$\alpha - \Lambda \approx \ln\left(\frac{1+C}{1-C}\right) \approx 2C \tag{3.2a}$$

$$\beta \approx C \tag{3.2b}$$

$$\varkappa^{1/2} m/2f \approx C + \Lambda \approx \alpha - \beta$$

so that

$$m \approx m_0 - (f^2/8\pi\varepsilon) \tag{3.2c}$$

in agreement with (3.1).

Consider next the $f \to 0$ limit, which means $\alpha \to \infty$, $\beta \to 0$, with a fixed product. Assuming C does not tend to 1, (2.22) and (2.23) give

$$\Lambda \approx \alpha \tag{3.3a}$$

$$C \approx (1/2\alpha\beta) + 1 - [(1/2\alpha\beta)^2 + (1/\alpha\beta)]^{1/2} \tag{3.3b}$$

$$\varkappa^{1/2} m/2f \approx -\frac{1}{2\beta} + [(1/2\beta)^2 + (\alpha/\beta)]^{1/2}$$

$$m \approx -(16\pi\varepsilon/\varkappa) + [(16\pi\varepsilon/\varkappa)^2 + m_0\,(32\pi\varepsilon/\varkappa)]^{1/2}. \tag{3.3c}$$

Eq. (3.3c) gives the mass of a neutral source in general relativity[1]. As ε becomes large relative to $\varkappa m_0$ (dilute source), this reduces further to the Newtonian limit:

$$m \approx m_0 - (\varkappa m_0^2/32\pi\varepsilon). \tag{3.3d}$$

More generally, if we let $\varepsilon \to \infty$, keeping f finite, so that $\beta \to 0$ (and, again requiring that C not $\to 1$), we get:

$$C \approx \beta\,(1 + \alpha^2)^{1/2} \tag{3.4a}$$

$$\Lambda \approx \alpha - 2\beta \tag{3.4b}$$

$$\varkappa^{1/2} m/2f \approx \alpha - \beta\,(\alpha^2 + 1)$$

$$m \approx m_0 - (\varkappa m_0^2/32\pi\varepsilon) - (f^2/8\pi\varepsilon), \tag{3.4c}$$

the expected Newtonian plus scalar energy result. Thus the solutions contain all the physically required limits as the interactions become weak because of small coupling constants or because of diluteness of the source.

It may be shown by the same limiting considerations, that the $f \to \infty$ or $\varepsilon \to 0$ limits do not exist. The former is interesting as a symptom of the fact that for certain ranges of the input parameters (m_0, f), for given ε, there will be no solutions, i.e., if f is large enough, though finite. This is a very surpris-

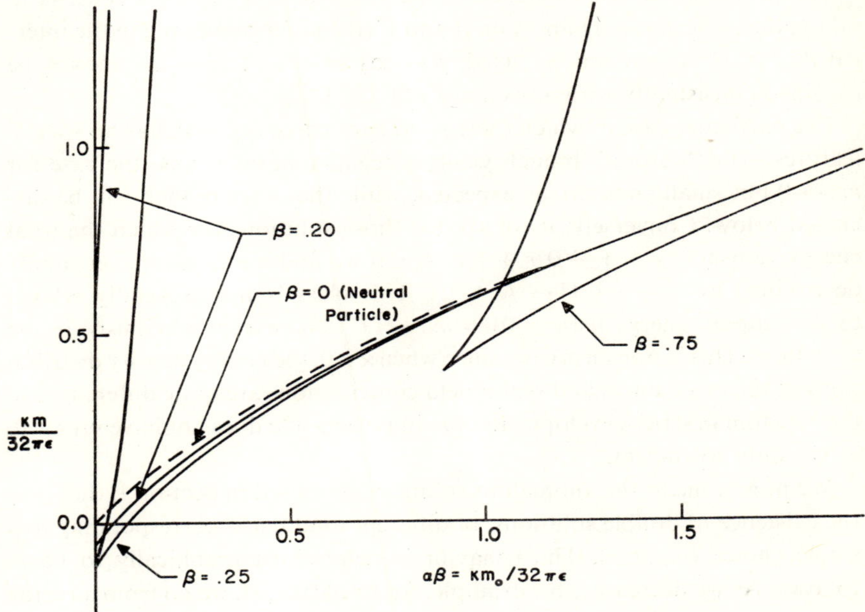

Figure 1. The masses of solutions are plotted against bare mass, each "pair" of curves corresponding to fixed scalar coupling constant f. This exhibits the ranges for which there are no or two solutions; note the sharp rise of the anomalous (upper) branch. The dashed curve is the solution corresponding to a neutral $(f = 0)$ particle

ing result; for example if m_0 vanishes altogether, β has an upper bound which is computed numerically to be $\beta \approx 0.25$. Figure 1 plots $m\varkappa\,(32\pi\varepsilon)^{-1}$ vs. $m_0\varkappa\,(32\pi\varepsilon)^{-1} = \alpha\beta$ for various values of $\varkappa^{1/2}f\,(16\pi\varepsilon)^{-1} = \beta$, and illustrates the regions of no solution and of double solutions. The impossibility of a "point" shell (for $f \neq 0$) is also amusing, since it stands in contrast with the electromagnetic or neutral $(e = f = 0)$ results. There, for any values of (m_0, e) a unique $\varepsilon \to 0$ limit existed. Here, there is a lower bound, $\varepsilon > 0$, for finite (m_0, f), which turns out to be the point at which the two "branches" coincide; below this bound is the region of no solutions. Of course, even in the electromagnetic or purely neutral case, the *invariant* radius remains finite[1] as $\varepsilon \to 0$ [$32\pi\varkappa^{-1}\bar{\varepsilon}\ (e = 0) = m_0^2/(m_0 - m)$, or alternately $m = m_0 - (\varkappa m_0^2/32\pi\bar{\varepsilon})$], as it does here. Here, (2.22b′) shows that $\bar{\varepsilon} \geqslant \varkappa^{1/2} f/16\pi$ for all solutions. We will see that in Brans–Dicke theory, in contrast, $\bar{\varepsilon}_{\min} = \varepsilon_{\min} = 0$ for any m_0.

The proof of the non-existence of the $\varepsilon \to 0$ or $f \to \infty$ limits consists in taking the corresponding limits on Λ and C (the latter remaining in the interval (0, 1)). It is seen that (α fixed, $\beta \to \infty$) or ($\alpha \to 0, \beta \to \infty$) cannot be obtained consistently with any choice of (Λ, C).

The particular case in which there is no bare mass, $m_0 = 0$, has no special features: The "natural" branch yields a negative m (as is also the case for $m_0 \to 0$ but small enough), as expected, while the other branch will be discussed below. Conversely, if we ask for those solutions for which the total energy vanishes ($\chi = 1 + 0/r + b/r^2 + \cdots$) we find a one parameter family determined by $\Lambda = -C(1 - C^2)^{-1/2}$. This class is not especially related to the "input" energy $m\,(\varkappa = 0) = m_0 - (f^2/8\pi\varepsilon)$ except, obviously, in the $\varkappa = 0$ limit. This is not surprising since when $\varkappa \neq 0$, the energy density distributions of the mechanical and scalar field contributions are quite different, and the Newtonian series development of m in powers of $\varkappa$ does not have $m\,(\varkappa = 0)$ as the only coefficient.

We now come to the anomalous solutions promised in Section I, that is to the existence of double solutions, of different total mass, corresponding to a single choice (m_0, f, ε). The behavior is determined graphically to be as follows: As we decrease f, for example, for fixed (m_0, ε), we go from a region of no solutions, through a point where there is just one, to a branching with two solutions. In the algebra, the latter fact is most dramatically evidenced by looking at the limiting cases in (α, β) as reflected by (Λ, C) and seeing that they can be fulfilled in two different ways. From among the various fashions in which α and β may each tend to zero or infinity, we pick out some interesting cases. Consider first $\varepsilon \to \infty$, which means $\beta \to 0$, α finite. This is accomplished by the natural choice $C \to 0, \Lambda \to \alpha$ treated earlier, or by

$$\Lambda \approx -(4\beta)^{-1} \to -\infty, \tag{3.5a}$$

$$C \approx 1 - 2\exp(-1/16\beta^2) \to 1. \tag{3.5b}$$

Then (2.23) gives

$$m \approx (f^2/4\pi\varepsilon)\exp[(4\pi\varepsilon)^2/\varkappa f^2] \tag{3.5c}$$

which has an essential singularity as $\varepsilon \to \infty$. The characteristic parameters here are the flat-space scalar field energy ($f^2/4\pi\varepsilon$) and the exponent, which is the inverse of the ratio: Schwarzschild radius $\varkappa(f^2/4\pi\varepsilon)$ of this energy to extension ε.

In the preceding, we had $\Lambda \to -\infty$, so that $\alpha - \Lambda \to -\Lambda$ for α finite. Since α appears nowhere else in (2.22) or (2.23), Eqs. (3.5) hold for $\alpha \to 0$, $\beta \to 0$, i.e. for $\varkappa \to 0$, as well as for $\varepsilon \to \infty$.

Finally the $f \to 0$ ($\alpha \to \infty$, $\beta \to 0$) limit is slightly different. Again the anomalous solution is found by looking at the $C \to 1$ limit and again $\Lambda \approx -(1/4\beta)$, but $\alpha\beta$ is finite here and Λ/α stays finite. The only change is therefore to retain $(\alpha - \Lambda)$ in the exponential:

$$C \approx 1 - 2\exp\left[-\alpha(\alpha - \Lambda)|\Lambda|\right]$$

$$\approx 1 - 2\exp\left[-(1/16\beta^2) - (\alpha/4\beta)\right] \tag{3.6a}$$

and

$$m \approx (f^2/4\pi\varepsilon)\exp\left[(4\pi\varepsilon)^2(\varkappa f^2)^{-1}(1 + \varkappa m_0/8\pi\varepsilon)\right]. \tag{3.6b}$$

Again this is an essential singularity as $f \to 0$. In (3.5, 6) the mass is also expressible in terms of $\bar{\varepsilon}$ by replacing ε with $4\bar{\varepsilon}$.

It should be emphasized that the metric and ψ components are in no way singular or anomalous for this branch (except of course *at* $\beta = 0$); they will contain the exponential of (3.5c) or (3.6b) through the factor $(1 - C)^{-1}$. This means of course that the $r \to \infty$ and the $f \to 0$, $\varkappa \to 0$ or $\varepsilon \to \infty$ limits are not interchangeable as e.g. $\chi \sim 1 + (\varkappa m/32\pi r)$, and indeed the latter limits simply do not exist for this branch.

We have been discussing so far the initial properties of our system, especially its energy, which is of course conserved, and so valid at later times as well. We now investigate whether the system is ever static, possibly with some particular choice of input parameters, and if not whether it begins to contract or expand from the moment of time symmetry. Normally one would expect contraction for a dust cloud, which has no other interactions to compensate the gravitational attraction. At first sight, the negative energy aspects of the scalar field might seem to alter this conclusion; on the other hand we know that the scalar interaction itself is attractive, as is the gravitational force between two negative energy regions (but not between positive and negative).

In the electromagnetic case, the motion's direction depends on whether the Coulomb repulsion dominates the Newtonian attraction, and there is even a particular choice of (m_0, e) for which they balance out. Here things are more complicated, and we must turn to the equations of motion.[8] The time-development field equations at $t = 0$ simplify when we recall that the choice of coordinates $g_{ij} = \chi^4\delta_{ij}$, $\pi^i_i = 0$ is retained at the later instant, so that the time derivative of Eq. (2.5) also vanishes. The original equations read

$$\dot{g}_{ij} = 2Ng^{-1/2}(\pi_{ij} - \tfrac{1}{2}g_{ij}\pi^k_k) + N_{i|j} + N_{j|i} \tag{3.7a}$$

$$\dot{\pi}^{ij} = -Ng^{1/2}(R^{ij} - \tfrac{1}{2}g^{ij}R) + g^{1/2}(N^{|ij} - g^{ij}N^{|k}{}_{|k}) + \tfrac{1}{2}\varkappa\mathcal{T}^{ij}_{\mathrm{M}} \tag{3.7b}$$

where we have dropped all N_i and π^{ij} terms in (3.7b), since they vanish initially and will not be further differentiated; $\mathcal{T}^{ij}_{\mathrm{M}}$ is the matter spatial stress $(\delta\mathcal{L}_M/\delta g_{ij})$. From (3.7a), we learn that $\dot{g}_{ij}\,(t = 0) = 0$ while clearly

$$\ddot{g}_{ij}\,(t = 0) = 2Ng^{-1/2}g_{ik}g_{jl}\dot{\pi}^{kl} + \dot{N}_{i|j} + \dot{N}_{j|i}. \tag{3.8}$$

The $\dot{\pi}^{ij}$ equation is subject to the requirement that

$$\dot{\pi}^i_i = 0 = \tfrac{1}{2}Ng^{1/2}R - 2g^{1/2}N^{|i}{}_{|i} + \tfrac{1}{2}\varkappa\,\mathcal{T}^i_{\mathrm{M}i}. \tag{3.9}$$

Thus, we have

$$\dot{\pi}^{ij} = -Ng^{1/2}R^{ij} + g^{1/2}(N^{|ij} + g^{ij}N^{|k}{}_{|k}) + \tfrac{1}{2}\varkappa(\mathcal{T}^{ij}_{\mathrm{M}} - g^{ij}g_{kl}\mathcal{T}^{kl}_{\mathrm{M}}). \tag{3.10}$$

We now insert the following forms into (3.10)

$$\begin{aligned}
\tfrac{1}{2}R_{ij} &= \chi^{-2}[3\chi_{,i}\chi_{,j} - \chi_{,k}\chi_{,k}\delta_{ij}] - \chi^{-1}[\chi_{,ij} + \chi_{,kk}\delta_{ij}] \\
g^{1/2}N^{|ij} &= \chi^{-2}[N_{,ij} - 2\chi^{-1}(N_{,i}\chi_{,j} + N_{,j}\chi_{,i} - N_{,k}\chi_{,k}\delta_{ij})] \\
\varkappa\chi^2\mathcal{T}^{ij}_{\mathrm{M}} &= \tfrac{1}{2}\varkappa N[\psi_{,i}\psi_{,j} - \tfrac{1}{2}\psi_{,k}\psi_{,k}\delta_{ij}].
\end{aligned}$$

These may be verified from the definitions and the simple isotropic form of g_{ij}. Note that the source does not contribute to $\mathcal{T}^{ij}_{\mathrm{M}}$. We next use the initial value equations (2.8) and the matching equations (2.22), having converted (3.10) by spherical symmetry and use of $u \equiv r^{-1}$ to find simply

$$u^{-4}N^{-1}\chi^2\dot{\pi}^{ij} = -\tfrac{1}{2}\varkappa^{1/2}\psi'\varepsilon C\,\Lambda\lambda^{-1}(\delta_{ij} - 3\hat{r}_i\hat{r}_j) \tag{3.11}$$

which exhibits the tracelessness of $\dot{\pi}^{ij}$. We must next discuss, in (3.8) the $\dot{N}_i$ terms, which read

$$\dot{N}_{i|j} + \dot{N}_{j|i} \equiv \dot{N}_{i,j} + \dot{N}_{j,i} - 2\dot{N}_k\Gamma^k_{ij}. \tag{3.12a}$$

By spherical symmetry, we may set

$$\dot{N}_i \equiv \tfrac{1}{2}f(u)\,x^i \qquad \text{3.12b)}$$

and express the result as the sum of a traceless part and a trace:

$$\dot{N}_{i|j} + \dot{N}_{j|i} = (\tfrac{1}{3}\delta_{ij} - \hat{r}_i\hat{r}_j)\,u(f' - 4\chi'f/\chi) + \tfrac{1}{3}\delta_{ij}(3f - uf' - 2uf\chi'/\chi). \tag{3.13}$$

Since $\ddot{g}_{ij}$ is a pure trace due to our continuation of the initial isotropic coordinates off $t = 0$, the form of $\dot{N}_i$ is determined by the vanishing of the coefficient of $(\frac{1}{3}\delta_{ij} - \hat{r}_i\hat{r}_j)$ in the $\ddot{g}_{ij}$ equation, while the trace of this equation then determines $\ddot{\chi}$ $(t = 0)$:

$$0 = f' - (4f\chi'/\chi) - 3\varkappa^{1/2}\psi'\varepsilon C\Lambda\lambda^{-1}N^2 \tag{3.14a}$$

$$4\chi^3\ddot{\chi} = -\tfrac{1}{3}u^4\chi^{-2}(\chi^2u^{-3}f)'. \tag{3.14b}$$

The first equation states that

$$(f\chi^{-4})' = 3u^3N^2\chi^{-4}\varkappa^{1/2}\psi'\varepsilon C\Lambda\lambda^{-1} \tag{3.15}$$

while we need the sign of $(u^{-3}\chi^2 f)'$ for (3.14b). It is easily seen that

$$-\tfrac{1}{3}u^4\chi^{-2}(fu^{-3}\chi^2)' = -u^4\chi^{-2}\varkappa^{1/2}\varepsilon C\Lambda\lambda^{-1}\left[u^{-3}\chi^6\int_0^u u^3\chi^{-4}N^2\psi'\,du\right]'. \tag{3.16}$$

The function being differentiated on the right is clearly positive, since N, χ and ψ' (which properly is $|\psi'|$ here) are, and it vanishes at infinity (at $u \sim 0$, $\psi' \to$ finite). Thus it is increasing with positive (u) derivative near $u = 0$. Since C is positive, $\ddot{\chi}$ is negative near $u = 0$ ("infinity"). We have not calculated whether this persists down to ε or whether there is a turning point in the differentiation due to the $u^{-3}\chi^6$ factor. [In that case, one might define such static configurations as "one-particle states", although they are still extended.] For fairly dilute sources or sufficiently weak f, we can verify that there will be an initial contraction as expected. We have not investigated the point $C = 1$, which is always delicate. However, we note that for $C \to 0$, corresponding to an infinitely dilute distribution, the invariant radius $\varepsilon\chi^2$ $(u = \varepsilon^{-1})$ does not decrease, and this limit is static. We shall return to it in the next section, but only to remark that in this case there is no "exterior" either, since the shell is at infinity. The question of whether gravitational collapse eventually occurs requires a finite-time integration of the field equations, which we have not attempted.

4 PURELY EXTERIOR SOLUTIONS

In the previous section, we have given the general nonsingular solutions of the posed initial value problem. The exterior solutions were "supported" by a physically sensible source which was free of any interactions but those being investigated—"dust" with parameters m_0, f and ε. We also noted the generally non-static character of the system. In this section, we compare our results with the earlier, purely exterior solution of JNW[3]. The latter is static, with metric, in the JNW coordinate system

$$ds^2_{\text{JNW}} = (R_+/R_-)^{1/\mu}(\mathrm{d}R^2 + R_+R_-\,\mathrm{d}\Omega) - (R_+/R_-)^{-1/\mu}\,\mathrm{d}t^2$$
$$R_\pm \equiv R \pm \tfrac{1}{2}r_0(\mu \pm 1), \quad \mu \equiv (1 + 4\varkappa A^2 r_0^{-2})^{1/2} \tag{4.1a}$$

and field

$$\psi_{\text{JNW}} = 2^{1/2}A\,(\mu r_0)^{-1}\ln(R_-/R_+). \tag{4.1b}$$

Here (r_0, A) are independent integration parameters. The total mass will be seen to be $m = 8\pi\varkappa^{-1}r_0$. The coefficient of $1/r$ in ψ is $-A\sqrt{2}$ for JNW whereas this coefficient was

$$(f/4\pi)([1 + C]/[1 - C])^{-\lambda}$$

in our solution, Eq. (2.24). Note that there is effectively only *one* parameter of integration to be related to a possible source, since $8\pi\varkappa^{-1}r_0$ is just the total resulting mass, while A alone must account for all input parameters (m_0, f, ε).

To relate (4.1) to our general solution, we transform from Schwarzschild to isotropic coordinates[9] r, and obtain in terms of $z \equiv \frac{1}{4}r_0\mu$

$$N = ([1 + zu]/[1 - zu])^{-1/\mu} \tag{4.2a}$$

$$\psi = -2^{1/2}Az^{-1}\ln([1 + zu]/[1 - zu]) \tag{4.2b}$$

$$\chi^2 = [(1 + zu]/[1 - zu])^{1/\mu}(1 - z^2u^2). \tag{4.2c}$$

The coefficient of $(\varkappa u/16\pi)$ in χ^2 is $8\pi\varkappa^{-1}r_0$. Comparing (4.2) to our solution (2.24), we see that the combinations $a(1 \pm C)$ must both be equal to z, which means that either $C = 0$, or $C = 1$ and $z = 0$. For $C = 0$, the source is infinitely dilute and $m = m_0$. For $C \to 1$, $z = 0$ requires $N = \chi = 1$, $\psi = 0$ throughout the exterior, and $m = 0$. In [3], it is remarked that the limit $\mu \to 1$ (corresponding to $|A| \to \infty$) has different behavior according to whether $r = z$ or $r > z$. We see from the realistic source point of view that this problem evaporates since the distribution is always infinitely extended ($\varepsilon = \infty$) and the exterior solution in question does not extend down to $r =$ any finite z. Alternately, the JNW solution is not generated by any bounded finite source. The static nature of this metric is likewise due to its infinite extension, as remarked in our discussion of time development in the previous section.

5 BRANS–DICKE THEORY

In this section, we investigate a physically different but mathematically analogous system, namely the initial value problem in the scalar-tensor theory proposed by Brans and Dicke. This theory is formally akin to the Einstein plus scalar field system, as we shall see; the scalar is, however, part of the gravitational field system and as such is coupled to matter only through (the trace of) the latter's stress tensor; the coupling constant f is replaced by the constant ω determining the relative strength of tensor to scalar in the action. Thus the source here is just a shell with m_0, the general action having the form

$$\begin{aligned} I &= \int [{}^4\mathscr{R}\phi - \omega\phi^{-1}\phi_{,\mu}\phi^{,\mu} (-{}^4g)^{1/2} + \mathscr{L}_M(g)]\, d^4x \\ &= \int [\phi\, ({}^4\mathscr{R} - \omega\varkappa\psi_{,\mu}\psi^{,\mu} (-{}^4g)^{1/2} + \phi^{-1} \mathscr{L}_M(g))]\, d^4x \end{aligned} \tag{5.1}$$

where ϕ is the scalar field, $\varkappa^{1/2}\psi \equiv \ln(\varkappa\phi/16\pi)$, and $\mathscr{L}_M$ the usual "minimally coupled" $(\eta_{\mu\nu} \to g_{\mu\nu})$ matter action. Although one could use conformal transformations to eliminate the ϕ coefficient of ${}^4\mathscr{R}$, it is most straightforward for our problem to proceed as in Section II. The only differences are that $\mathscr{L}_M$ is simpler, being just the $Nm_0\varrho$ mass term, that in the curvature ${}^4\mathscr{R}$ the divergence $\mathscr{D}$ must be retained since it is now multiplied by the field ϕ, and that the new term $\varkappa\omega \int\phi\psi_{,\mu}\psi^{,\mu}(-{}^4g)^{1/2}$ replaces the old scalar action. It is easy to see that the effective initial Lagrangian now becomes

$$\begin{aligned} \mathscr{L}_{\text{eff}} = 2\bar{N}\, [&-2\nabla^2 L + (\nabla L)^2 L^{-1} - \tfrac{1}{2}\omega\varkappa L\, (\nabla\psi)^2 + \tfrac{1}{2}m_0\varrho \exp(-\varkappa^{1/2}\psi) \\ &- \varkappa^{1/2}\nabla L \cdot \nabla\psi - \varkappa^{1/2} L\nabla^2\psi - \varkappa L\, (\nabla\psi)^2] \end{aligned} \tag{5.2}$$

$$\bar{N} \equiv (N\varkappa\phi/16\pi), \quad L \equiv \chi^2.$$

Once again we introduce $u \equiv r^{-1}$, vary $\bar{N}$, ψ and χ independently and obtain three exterior equations along with their matching conditions. Note that we have also taken the field and particle momenta to vanish, the same coordinate conditions for the physical metric (which is *not* rescaled in any way), while the initial ψ is again specified such that its $\dot{\pi} = 0$ initially. The last three terms in (5.2) arise from the $\mathscr{D}$ term in (2.3), while the first two are just ${}^3\mathscr{R}N\varkappa\phi/16\pi \equiv -8\bar{N}\chi\nabla^2\chi$ written in terms of L. The exterior equations $(u < \varepsilon^{-1})$ read

$$-2L''/L + (L'/L)^2 - (1 + \tfrac{1}{2}\omega)\,\varkappa(\psi')^2 - \varkappa^{1/2}\psi'' - \varkappa^{1/2}\psi'\,(L'/L) = 0 \tag{5.3a}$$

$$\begin{aligned} -2\,(\bar{N}''/\bar{N}) - 2\,(L''/L) - 2\,(\bar{N}'L'/\bar{N}L) + \varkappa^{1/2}\psi'\,(\bar{N}'/\bar{N}) \\ + (L'/L)^2 - \varkappa(\psi')^2\,(1 + \tfrac{1}{2}\omega) = 0. \end{aligned} \tag{5.3b}$$

Using (5.3a) to simplify it, the $\delta\psi$ equation reads:

$$\varkappa^{1/2} (2 + \omega) [\psi'' + \psi' ([\bar{N}'/\bar{N}] + [L'/L])] - (\bar{N}'L'/\bar{N}L) - (\bar{N}''/\bar{N}) = 0. \tag{5.3c}$$

[Eq. (5.3a) would have the form $-8\bar{\chi}\nabla^2\bar{\chi} = (\omega + \frac{3}{2})\varkappa (\nabla\chi)^2$ in terms of a rescaled $\bar{\chi} \equiv \chi (\varkappa\phi/16\pi)^{1/4}$.] It is easy to show that the same two first integrals found in Section II exist, but in terms of $\bar{N}$:

$$L\bar{N}\psi' = (\varkappa^{1/2}m_0/16\pi) K_1, \quad L\bar{N}' = -(\varkappa m_0/16\pi) K_2 \tag{5.4}$$

where again we have made K_1, K_2 dimensionless for convenience. Also, the ψ equation may be put into the form (2.14) with $(1 + \Lambda^2)$ replaced by

$$4\left[\left(\frac{K_2}{K_1}\right)^2 + \frac{K_2}{K_1} + \left(1 + \frac{\omega}{2}\right)\right] \equiv 4B^2$$

so that

$$\varkappa^{1/2}\psi = B^{-1} \ln ([1 + a (1 + C) u]/[1 - a (1 - C) u]) \tag{5.5a}$$

$$a \equiv (\varkappa m_0/32\pi) K_1 B.$$

Then (5.4) gives as before

$$\bar{N} = ([1 + a (1 + C) u]/[1 - a (1 - C) u])^{-K_2/K_1B} \tag{5.5b}$$

$$L = [1 + a (1 + C) u]^{(K_2/K_1B)+1} [1 - a (1 - C) u]^{-(K_2/K_1B)+1}. \tag{5.5c}$$

The three conditions at $r = \varepsilon$,

$$(\varkappa m_0/8\pi) \exp (-\varkappa^{1/2}\psi) - \varkappa^{1/2}L\psi' - 2L' = 0 \tag{5.6a}$$

$$\bar{N}'L + \bar{N}L' = 0 \tag{5.6b}$$

$$(2 + \omega) \varkappa^{1/2}\bar{N}L\psi' - (\varkappa m_0/8\pi) \bar{N} \exp (-\varkappa^{1/2}\psi) - L\bar{N}' = 0 \tag{5.6c}$$

give

$$(\varkappa m_0/32\pi\varepsilon) = C (1 - C^2)^{-1} (BK_1)^{-1}$$

$$(K_2/K_1) = \omega + 1, \quad \therefore B^2 = (\omega + 2) (\omega + \tfrac{3}{2}) \tag{5.7a}$$

$$K_1 = (\omega + \tfrac{3}{2})^{-1} (1 + C)^{-\tau} (1 - C)^{\tau} \tag{5.7b}$$

$$\tau \equiv (\omega + 2)^{1/2} (\omega + \tfrac{3}{2})^{-1/2}.$$

Thus there is only one effective parameter C and one condition for $C(m_0)$ at ε, namely

$$d_0 \equiv (\varkappa m_0/32\pi\varepsilon) = C\tau^{-1} (1 + C)^{\tau-1} (1 - C)^{-\tau-1}. \tag{5.8}$$

Since $1 \leqslant \tau^2 \leqslant 4/3$ and $0 \leqslant C \leqslant 1$, we see that (5.8) gives m_0 as a monotonic function of C. Hence, unlike the previous case, to a given set of input parameters (m_0, ε) there corresponds a unique set of integration constants and a unique solution. This is analogous to the neutral Einstein situation.

We now need to find an expression for the total energy of the system. Here, in contrast to the Einstein case, m is no longer given by the u coefficient of χ^4, but rather by that of $\chi^4\varphi$. The reason for this is that the overall φ factor in (5.1) multiplies, in particular, the time translation generator (Hamiltonian). Alternately, if the φ coefficient is removed by a conformal transformation of the metric, $g_{\mu\nu} \to g_{\mu\nu}\phi$, the new action looks like the Einstein one with respect to this rescaled metric, which is just $\chi^4\phi$. Details of the Hamiltonian reduction of the theory, along the lines of reference [10], may be found in the work of Toton[11]. From (5.5a, c) and the boundary conditions (5.7), we find directly that

$$d \equiv \varkappa m/32\pi\varepsilon = (\tau^{-1} + C)(1 - C^2)^{-1} C \tag{5.9}$$

which determines $m(m_0, \varepsilon)$ when the solution of (5.8) for C is inserted. (Had we used χ, rather than $\chi\phi^{1/4}$ to define m, τ^{-1} would have been replaced by $\tau^{-1} - [(2\omega + 4)(2\omega + 3)]^{-1/2}$ in (5.9).)

We first check that the limit to general relativity, $\omega \to \infty$, is correct. Here, $\tau = 1$ and

$$d_0 = C(1 - C)^{-2}, \quad d = C(1 - C)^{-1} \tag{5.10}$$

which is equivalent, upon eliminating C, to the neutral point particle solution (3.3c) in Einstein theory, as are the metric components. Next, consider the dilute limit, $d_0 \ll 1$. Here C is small and we find

$$m \sim m_0 - (\tau^2 \varkappa_{\text{BD}})(m_0^2/32\pi\varepsilon) + \cdots \tag{5.11}$$

which differs from the Einstein (or Newtonian) expression by the τ^2 factor. (Note, incidentally, that the "wrong" m would lead to the clearly wrong result $m \sim m_0\tau^2$ here.) At the same time, the $1/r$ coefficient of N, which gives the Newtonian force and red shift, is in this limit, $\sim\tau^2\varkappa_{\text{BD}}m_0$ so that the τ^2 must be absorbed into $\varkappa_{\text{BD}}$ to define the Newtonian constant $\gamma = \tau^2\varkappa_{\text{BD}}/16\pi$. On the other hand, the bending of light, determined by $dr/dt \sim N\chi^{-2}$ has a $1/r$ coefficient which is $\sim\varkappa_{\text{BD}}m_0$, independent of τ, as in Einstein theory (since light is unaffected by the scalar field). Thus the light bending prediction is, as given in [4], different by a factor τ^2, which is $\approx 6\%$ for $\omega \approx 6$. For general "density" (m_0/ε), the relation between the two theories is of course not so simple.

More interesting is the comparison of point particle limits (with m_0 fixed) in the two theories. Here $1 - C \equiv \mu$ is small by (5.8), which yields $d_0 \sim \mu^{-(\tau+1)}$. But (5.9) reads $d \sim \mu^{-1}$. Consequently, omitting numerical coefficients, we have

$$m_{\text{BD}} \underset{\varepsilon \to 0}{\sim} [m\varepsilon^{\tau}]^{(1+\tau)^{-1}}. \qquad (5.12a)$$

Since τ exceeds unity, the mass vanishes *faster*, i.e., as a higher power of ε than the Einstein result

$$m_E \underset{\varepsilon \to 0}{\sim} (m_0 \varepsilon)^{1/2}. \qquad (5.12b)$$

In this sense, the theory is somewhat more "Machian"[12] than general relativity. Likewise, the invariant radius $\bar{\varepsilon} \equiv \chi^2(\varepsilon)\,\varepsilon$ vanishes in this limit, whereas $\bar{\varepsilon} \to \varkappa m_0 \neq 0$ in Einstein theory. This is established by use of (5.5c), with the subsequent values of the parameters; χ^2 is seen to go as the following power of ε^{-1}: $[1 + (\omega + 1)\,B^{-1}]\,(1 + \tau)^{-1}$. This power is always <1 for $\omega < \infty$, while at $\omega = \infty$, $\chi^2 \sim \varepsilon^{-1}$ as stated above.

6 SUMMARY

We have investigated the coupling of gravitation to a long range scalar field with a realistic source of both. The complete solutions thus obtained displayed a number of very surprising features, primarily the fact that for a given set of coupling parameters (m_0, f) either no solutions or two solutions exist in general. In the weak coupling ($\varkappa$ or $f \to 0$) or equivalently, in the dilute ($\varepsilon \to \infty$) limits, one branch reduced to the expected value *at* $\varkappa = 0$ or $f = 0$ while the other gave a mass roughly proportional to

$$(f^2/4\pi\varepsilon) \exp\,[(4\pi\varepsilon)^2/\varkappa f^2],$$

which has no perturbation theoretic counterpart. These results would seem to be due to the peculiar relation between the spin zero field and the Newtonian (spin zero) part of the Einstein field[13], the absence of a unique scalar "Coulomb" self-field and the negative character of the scalar field energy.

The solution and self-energy of a massive source in the Brans–Dicke theory was also obtained in general, and seen to behave qualitatively as in general relativity, with unique solutions for all values of the source parameters and vanishing mass in the "point" limit.

Added in proof:

Two generalizations of the gravitation-scalar field system have been investigated subsequently:

1) Non-minimal coupling $\sim(\varkappa\psi^2/12)\, {}^4\mathscr{R}$, to render the scalar field conformally invariant, and 2) making the source electrically charged.

The conformally invariant system retains the two-branch structure of the solutions, including essential singularity limits, but the anomalous branch is restricted to the region $\varkappa^{1/2}m_0 < 2\sqrt{3}\,|f|$. For greater $m_0/|f|$ ratios, there is exactly one solution and it is on the well-behaved branch, so that in this case the system behaves completely "sensibly": Every set $m_0\ (>2\sqrt{3}\,\varkappa^{-1/2}|f|), f, \varepsilon$ gives a unique initially static solution. None are static.

An electrically charged source leads to a discontinuity in the $m\ (m_0, f, e, \varepsilon)$ relation. The mass curves are similar to those in figure 1, but with a missing segment which may be on either branch, or straddle the "vertex" on both branches. Curiously, taking the $f \to 0$ limit of these truncated curves only yields the expected $f \equiv 0$ relation[1] for $m\ (m_0, e, \varepsilon)$ for the region $\varkappa^{1/2}m_0 \geqslant 2|e|$; no $f \to 0$ solutions exist for $\varkappa^{1/2}m_0 < 2|e|$. There are again no static solutions for $|f| > 0$. Similar results obtain in Brans-Dicke theory when the source is charged: $\varkappa^{1/2}m_0$ is constrained to be $\geqslant 2|e|$ and no static solutions are possible. Finally, a charged source in the conformally invariant model was studied. It was found that in the unique solution region $\varkappa^{1/2}m_0 > 2\sqrt{3}\,|f|$ the discontinuity remains, so the $f \to 0$ limit again requires $\varkappa^{1/2}m_0 \geqslant 2|e|$. No non-trivial static solutions are possible. These results will be discussed in detail elsewhere.

REFERENCES

1. R. Arnowitt, S. Deser and C. W. Misner, *Phys. Rev.*, **120**, 313 (1960) and *Ann. Phys.* (N.Y.), **33**, 88 (1965).
2. We use the approach, notation and units of [1]. For a general description of the canonical formulation, see [10].
3. A. I. Janis, E. T. Newman and J. Winicour, *Phys. Rev. Letters*, **20**, 878 (1968), as corrected in *Phys. Rev.*, **176**, 1507 (1968); hereafter referred to as JNW.
4. C. Brans and R. H. Dicke, *Phys. Rev.*, **124**, 925 (1961).
5. The scalar field-particle interaction we have chosen is $\mathscr{L} \sim -fm_0^{-1}\psi\mathscr{T}^{\mu}_{P\mu}$. This is kinematically the simplest covariant coupling to a particle; however, we do not imply that ψ need couple universally to all matter $\mathscr{T}^{\mu\prime}_{\mu}\,s$, as in a scalar-tensor gravitational theory. We are merely ensuring that the scalar (f) and gravitational (m_0) spatial distribution of the particle are the same, $\varrho\ (r)$, just as was assumed for the charge, e, in [1]. The scalar field is taken to be minimally coupled to gravitation (see note added in proof).

6. Of course, the size of the particle, which enters in the mass formula, varies in time and f dependence will be introduced in this indirect way (except for zero initial extension).
7. Since ψ cannot be uniquely divided into self and radiation field at one instant, it is still not strictly meaningful to say that this choice avoids free scalar field excitations, except perhaps for static solutions, if such exist.
8. The initial acceleration of the shell may also be treated in terms of the equations of motion for the shell's radial momentum change, $\dot{\mathbf{p}}(0)$, which is specified by the "particle" equation of motion, $\dot{\mathbf{p}}(0) \sim f\nabla(N\psi) + m_0\nabla N$, evaluated at ε.
9. The unsuitability of Schwarzschild coordinates for this sort of problem has been remarked on in R. PENNEY, *Phys. Rev.*, **174**, 1578 (1968), and discussed also in ref. 1. The former reference also discusses the JNW metric as a particular limit of an axially symmetric exterior solution.
10. R. ARNOWITT, S. DESER and C. W. MISNER, in *Gravitation*, edited by L. WITTEN, Wiley, New York, 1962, Chapter 7.
11. This exterior solution agrees with that given for a charged particle in Brans-Dicke theory by E. Toton, Phys. Rev. Letters **21**, 1401 (1968) and Ph. D. Thesis, University of Maryland, 1969.
12. Properties of m_{BD} as compared to m_{E} are treated in a rather different fashion in ref. 11.
13. A formally similar essential singularity was found in a different context in M. FIERZ and W. PAULI, *Proc. Roy. Soc.*, **173**, 211 (1939), where a scalar ψ is coupled to the spin zero part of a vector field, with $L_{\mathrm{I}} \sim \psi(\partial_\mu f^\mu)^2$. We thank A. S. WIGHTMAN for this reference.

PAPER 14

Kinetic theory of gases in general relativity

JÜRGEN EHLERS*
University of Texas at Austin, U.S.A.

1

For several reasons, the kinetic theory of gases in general relativity is of interest:

i) Kinetic theory provides a simple model of matter which takes into account the particle structure of matter, in contrast to the more usual hydrodynamical description.

ii) It provides a way to complete the Einstein field equation

$$G^{ab} = T^{ab}$$

by an additional equation, the Liouville or Boltzmann equation, such that a deterministic model for a gravitating material system is obtained without the need for "equations of state".

iii) It gives a basis for relativistic thermodynamics and, in particular, permits the treatment of transport processes.

iv) It has proved to be a useful tool for a variety of astrophysical and cosmological investigations.

v) It is far from complete and suggests further work which might be of astrophysical and cosmological relevance.

* This work was supported by Aerospace Research Labs. OAR, AF–33 (615) 1029.

2

A brief historical survey, which is not meant to be complete, indicates the rapid growth of interest in the subject in recent years, and gives some idea of the problems that have been considered.

The (special-) relativistic analogues of the Boltzmann, Fermi and Bose equilibrium distributions of classical and quantum gases, respectively, have been determined by Jüttner[1,2].

In 1934 Synge began to consider collections of particles in flat and curved space-time, introduced an invariant distribution function, and used it to express particle and 4-momentum currents, N^a and T^{ab}, in terms of it[3]. In 1936, Walker established the Liouville equation for a collision-free gas in curved space-time[4]. The first model of a self-consistent system of particles moving in their average gravitational field was constructed by Einstein[5]. In 1940, Lichnerowicz and Marrot formulated the (special-)relativistic Boltzmann equation for a classical simple gas[6].

In 1957 Synge[7] gave an elegant survey of the treatment of a classical gas in Minkowski space-time. He did not use the Boltzmann equation, but quickly made the transition to the hydrodynamics of perfect fluids, assuming that the most probable distribution is everywhere attained without delay. An attempt to derive Eckart's relativistic transport equations for a simple gas from the Boltzmann equation was made by Sasaki in 1958[8].

The general-relativistic Boltzmann equation, for both Fermions and Bosons, was formulated and used to derive an H-theorem and several other results by Tauber and Weinberg[9]. Independently, I established the H-theorem in 1961[10] and Tauber, Weinberg and I pointed out that, in contrast to a classical theorem of Boltzmann's in the nonrelativistic case, a relativistic gas of particles with positive proper masses can be in thermal equilibrium only if space-time is stationary and if the mean motion of the gas is a rigid one.

In 1963, Israel adapted the Chapman–Enskog approximation method to the relativistic theory of a simple gas and showed that such a gas has a bulk viscosity[11]; this result "explains" the preceding result.

In deriving the momentum-dependence of equilibrium distributions from the requirement of vanishing entropy production, one needs to use Grad's theorem[12] that the general additive collision invariant in elastic binary collisions is a linear combination of a constant, the energy, and the components of linear momentum. In the relativistic case, this theorem has been proven by Bichteler[13] and by Boyer[14]. The theorem has also been established by

Chernikov[35] and by Marle[36]. Bichteler also succeeded in establishing the existence and uniqueness of solutions to the Cauchy problem for the relativistic Boltzmann equation in a given space-time[15].

The basic equations for radiative transport were derived in the general-relativistic case by Lindquist (1966)[16]. Zel'dovich and Podurets[17], and Fackerell[18] gave several results on general-relativistic, static, spherically symmetric, collisionless star clusters. A rederivation and new characterization of the Robertson–Walker cosmological models from the point of view of kinetic theory was given by Geren, Sachs and myself[19]. Additional results extending this work are due to Rienstra[20]. Lyle developed a systematic perturbation theory for the coupled Einstein–Liouville equations for a photon gas, the background being an Einstein–de Sitter model[21].

Ipser and Thorne[22] investigated the stability of relativistic star clusters.

Misner[23] was able to show that neutrino viscosity and anisotropic pressure of collisionless radiation in the early stages of an expanding universe are capable of damping out spatial anisotropies; this work has been extended by Matzner[24].

Applications of the Liouville equation for photons in curved space-time are contained, e.g., in the paper by Sachs and Wolfe on the anisotropy of the 3 °K—radiation due to density perturbations in an Einstein–de Sitter universe[25], and in the analysis of the optical appearance of a collapsing star by W. L. Ames and K. S. Thorne[26].

Anderson and Stewart have developed a relativistic version of Grad's method of moments* and applied it to transport processes; they confirmed Israel's result that a relativistic gas has a bulk viscosity[27].

Systematic expositions of relativistic kinetic theory have been given by Chernikov[35,37], Ehlers and Sachs[28], Marle[36], and the author[29]. For more references and critical remarks on relativistic statistical mechanics see also Havas[30].

3

I shall now outline the concepts and assumptions of the theory, and state some theorems. (For details and proofs, see [28], [29] or [36].)

A gas is represented in space-time as a broken complex of world lines, corners representing collisions. The particles are assumed to move like test

* See also Marle[36].

particles in an external or a self-consistent gravitational field, except during point collisions caused by short-range forces.

There is, then, an invariant distribution function $f(x^a, p^a)$ such that, for a system of particles of proper mass m,

$$\mathrm{d}N = fp^a\sigma_a \wedge \pi \tag{1}$$

is the (average) number of particles intersecting the hypersurface element σ_a with a 4-momentum contained in the cell

$$\pi = \frac{\mathrm{d}^3 p}{E} \tag{2}$$

of the mass shell. For any local observer, f equals the ordinary density in phase.

If no collisions occur, f satisfies the Liouville equation

$$L(f) = p^a \frac{\partial f}{\partial x^a} - \Gamma^{\lambda}_{bc} p^b p^c \frac{\partial f}{\partial p^{\lambda}} = 0 \tag{3}$$

$(a, b, \ldots = 1, \ldots, 4; \lambda, \mu, \ldots = 1, 2, 3)$ which expresses that f is constant on any particle orbit in (x^a, p^a)-phase space. More generally, $L_m(f)$ measures the phase space density of collisions.

If the "*Stosszahlansatz*" is accepted, one has, in the case of a classical simple gas with binary collisions, the Boltzmann equation

$$L(f_1) = \tfrac{1}{2} \int (f_3 f_4 - f_1 f_2)\, \delta(\Delta p)\, R_{12;34} \pi_2 \wedge \pi_3 \wedge \pi_4, \tag{4}$$

where 1, 2, 3, 4 stand for momentum values, the factor $\delta(\Delta p) = \delta(p_1 + p_2 - p_3 - p_4)$ takes care of 4-momentum conservation, and $R\ldots$ is, except for some kinematical factors, the differential scattering cross section.

Similar collision integrals can be written down for emission and absorption processes, and for Bose and Fermi gases.

By means of the formulae

$$N^a = \int p^a f \pi, \quad T^{ab} = \int p^a p^b f \pi, \quad S^a = -\int p^a f \log f \pi \tag{5}$$

the particle 4-current N^a, the stress energy tensor T^{ab}, and the entropy 4-current S^a are defined.

If there are no collisions or if the cross sections in (4) satisfy conservation laws, these (and similar) currents will obey continuity equations; e.g., the δ-factor implies that

$$T^{ab}_{;b} = 0. \tag{6}$$

More generally, one can derive balance equations from (4).

If the interaction responsible for the collisions is invariant under the space-time reflection PT, then $R\ldots$ has the symmetry property

$$R_{12;34} = R_{34;12}, \tag{7}$$

and (4) implies the H-theorem,

$$S^{a}_{;a} \geqslant 0. \tag{8}$$

Since (6) is a consequence of the Boltzmann equation (4), the self-consistent equations

$$\left.\begin{aligned} G^{ab} = T^{ab} &= \int p^a p^b f \,\pi_m \\ L_m(f) &= \text{collision integral} \end{aligned}\right\}, \tag{9}$$

for a gravitating system have the involution property: If the constraints $G^4_a = T^4_a$ hold on a space-like initial hypersurface $\Sigma: x^4 = \text{const.}$, and if the evolution equations $R_{\lambda\mu} = T_{\lambda\mu} - \frac{1}{2}g_{\lambda\mu}T$, $L_m(f) = \ldots$ are satisfied everywhere, then the constraints will also hold true off Σ. Presumably, then, Eq. (9) define a deterministic dynamical system, though no existence or uniqueness theorems seem to be known for the system (9).

The H-theorem (8) indicates that an isolated gaseous body tends toward a state in which the total entropy $S = {}_{\Sigma}\int S^a \sigma_a$ (Σ: spatial cross section of the system) is maximal; hence in that state one will have

$$S^{a}_{;a} = 0. \tag{10}$$

Eq. (10) is satisfied for precisely those solutions f of the Boltzmann Eq. (4) for which

(a) $\log f$ (or, for Bosons and Fermions, $\log(rh^{-3}f^{-1} \pm 1)$ is an additive collision invariant, or, alternatively, for which

(b) $L(f) = 0$, which means detailed balancing between all collisions and their inverse collisions.

These statements hold for simple gases and, *mutatis mutandis*, for mixtures; they hold for Boltzmann particles, Bosons, and Fermions, and for all types of collisions; one can also include electromagnetic fields.

If elastic binary collisions occur (among others), then (a) and Chernikov's analogue of Grad's theorem lead to the familiar equilibrium distribution

$$f(x, p) = \frac{r}{h^3}\left(e^{-\alpha(x) - \beta(x)\cdot p} \mp \varepsilon\right)^{-1} \tag{11}$$

where

$$\varepsilon = \begin{cases} 0 & \text{for classical particles,} \\ 1 & \text{for Bosons,} \\ -1 & \text{for Fermions,} \end{cases}$$

and r is the spin degeneracy of a particle. Here α is a scalar field, and β^a a future-directed, time-like 4-vector which is split according to

$$\beta^a = \frac{u^a}{T}, \quad u^2 = -1, \quad T > 0. \tag{12}$$

Eq. (11) implies

$$\left. \begin{aligned} N^a &= nu^a, \quad S^a = su^a \\ T^{ab} &= (\mu + p)\, u^a u^b + pg^{ab} \end{aligned} \right\}, \tag{13}$$

whence u^a is identified as the mean 4-velocity of the gas, and it implies the familiar thermostatic relations

$$\mathrm{d}\mu = T\,\mathrm{d}s + T\alpha\,\mathrm{d}n \tag{14}$$

and

$$\mu + p = Ts + \alpha Tn. \tag{15}$$

Hence T is the (thermodynamic) temperature and αT is the chemical potential (per particle).

Condition (b) determines the space-time dependence of α, T, and u^a:

u^a/T must be a Killing vector for a gas of particles with $m > 0$, and a conformal Killing vector if $m = 0$. (16)

If there is no mean electric field, $F_{ab}u^b = 0$, α must be constant (otherwise, see [9] or [29]). (17)

Condition (16) shows that (for $m > 0$) equilibrium is possible only in a stationary space-time; conversely, stationarity (i.e., invariance of g_{ab}, f under a one-parameter group with time-like orbits) implies equilibrium, $S^a_{;a} = 0$. Moreover, (16) and (17) give the dependence of temperature and chemical potential on the scalar gravitational potential, defined for stationary space-times by $U = \frac{1}{2}\log(-\xi^2)$ where ξ is the group generator.

For non-equilibrium situations not too far removed from equilibrium, one will have to modify (13) by transport terms. Let us consider a simple Boltzmann gas with $m > 0$. Following Anderson and Stewart[27] one can associate uniquely with the actual non-equilibrium distribution f an equilibrium distribution f_0 of the form (11) (with $\varepsilon = 0$) by requiring that N^a and T^a_a give the same values, whether computed by means of f or by means of f_0. One can then decompose N^a, T^{ab} according to

$$N^a = nu^a,$$
$$T^{ab} = T_0^{ab} + \varepsilon u^a u^b + \tfrac{1}{3}\varepsilon (g^{ab} + u^a u^b) + \pi^{ab} + 2u^{(a}q^{b)} \tag{18}$$

with $u_a q^a = u_a \pi^{ab} = \pi_a^a = 0$, and with T_0^{ab} computed from f_0. Then

$$\left.\begin{aligned} \mu &= T_{ab}u^a u^b = \mu_0 + \varepsilon = \text{energy density} \\ p &= p_0 + \tfrac{1}{3}\varepsilon = \text{kinetic pressure} \\ \varrho &= mn = \text{rest mass density} \end{aligned}\right\}, \tag{19}$$

and we will call

$$\left.\begin{aligned} &q^a \quad \text{the heat flux} \\ &\pi^{ab} \quad \text{the shear stress} \end{aligned}\right\}. \tag{20}$$

The equilibrium pressure which would belong to μ and ϱ from Eq. (19) is called the thermal pressure p_{th} of the gas, and we put $\bar{p} = p - p_{\text{th}}$. Anderson and Stewart proceed to expand f/f_0 into a series of generalized Hermite polynomials (depending on f_0) of the momenta p^a, derive differential equations for the moments of f from (4), and thereby deduce transport laws for normal solutions of the form

$$\left.\begin{aligned} \pi^{ab} &= -2\eta\sigma^{ab} \quad (\sigma^{ab} = \text{shear velocity}), \\ q^a &= -\lambda\,(\delta_b^a + u^a u_b)\,(T\dot{u}^b + T^{,b}) \\ (\dot{u}^a &= u^a{}_{;b}u^b = \text{acceleration}) \\ \bar{p} &= -\zeta\Theta \quad (\Theta = \text{dilation rate}), \end{aligned}\right\} \tag{21}$$

with positive coefficients η, λ, ζ. It turns out that $\zeta \neq 0$, which means that the gas has a bulk viscosity. These results establish the relativistic Navier–Stokes equation and the irreversible thermodynamics of a simple gas.

4

Now I wish to describe briefly a few applications of the theory.

Considering radiation as a photon gas, one immediately obtains from Eq. (1) the relation

$$I_\nu = h^4 \nu^3 f \tag{22}$$

between the distribution function f and the specific intensity I_ν, as judged by an observer who measures the frequency $\nu = (2\pi)^{-1}\,|u \cdot p|$. Since f is observer-independent, so is (I_ν/ν^3), which contains several kinematic effects important, e.g., in cosmology.

If the photons do not interact with matter between source S and observer O, Liouville's Eqs. (3) and (22) give the relation

$$I_{\nu_0} = \frac{I_{\nu_s}}{(1 + z)^3} \tag{23}$$

between I_ν near the source and at the observer, where z is the redshift. This equation is basic for the derivation of the m, z relation not only in Robertson–Walker universes, but in general models. Using (an approximation of) Eq. (4) rather than Eq. (3) one can derive scattering and absorption corrections to (23).

If the radiation is thermal, with a temperature T and a mean velocity u^a, (11) and (22) predict that an (arbitrary) observer will measure a Planck energy distribution in each direction with an effective temperature T_e depending on the velocity v which the observer has relative to the radiation and on the angle ϑ between the direction of observation and the direction with which he moves through the radiation field:

$$T_e = T \frac{\sqrt{1 - v^2}}{1 - v \cos \vartheta}. \tag{24}$$

This relation is now being used to find the velocity of the earth relative to the 3 °K-radiation[31].

If one assumes that the 3 °K-radiation was emitted thermally from the "recombination hypersurface $T \sim 3000$ °K", one obtains from (23) the observed intensity distribution in each direction in an arbitrary (anisotropic) model universe, provided one can compute z from the null geodesics. This idea was used by Sachs and Wolfe to estimate the influence of material "lumps" on the radiation[25]. The same method yields the optical appearance of a collapsing star for a distant observer (see Ames and Thorne[26]).

Due to results obtained in [9] and [19], a collision-free gas with a distribution function which is isotropic with respect to a geodesic mean 4-velocity field, can exist only in a Robertson–Walker space-time. Since the microwave background radiation is observed to be highly isotropic, this theorem gives a much better empirical motivation for using these models than arguments based on (rather uncertain) galaxy counts[32].

Very few exact solutions of the self-consistent Eq. (9) are known so far. Besides Fackerell's static, spherically symmetric solutions[18] and the kinetic theory version of the Robertson–Walker models[19] only some special $m = 0$ solutions with Gödel-type metrics or plane wave metrics have been constructed by Sachs[33]. The Robertson–Walker models have been characterized as the only solutions of (9) with a locally-isotropic distribution function, if the particles have positive proper masses[19]. The problem whether locally-isotropic solutions with zero mass (i.e., ultrarelativistic) particles and with rotation exist is still unsolved, although Rienstra[20] was able to show that the answer is negative in a number of kinematially specialized cases.

Stability analyses of various spherical, static, relativistic star clusters (with both isotropic and anisotropic velocity distributions, and with truncated Maxwell distributions or polytropic pressure density relations) by Ipser and Thorne[22] have led to the result that no stable models with central redshifts markedly larger than 0.5 seem to exist. This result is of interest in connection with the problem of the quasar redshifts[34]. If, in a quasi-static contraction, the central redshift exceeds 0.5, the cluster becomes unstable against gravitational collapse.

5

Finally, I want to point out that there are many unsolved problems connected with the topics discussed above. Firstly, the foundation on which the Boltzmann equation (or gravitational Vlasov equation) rests is not understood at all from a statistical mechanical point of view[30]. Neither the separation between long-range and short-range forces assumed in Eq. (4) nor the randomness assumptions which enter the "Stosszahlansatz" are well understood. There is no treatment of finite-range interactions (except via the mean field). The influence of curvature on the collisions is not analyzed. In treating stellar clusters, no collective interactions have been taken into account; the effects of bodily collisions and of encounters have not been treated. So far, no post-Newtonian approximation for the system (9) has been worked out. Irreversible processes in heterogeneous gases, such as nuclear reactions in a hot big bang, have not been treated systematically in analogy to the theory of the simple gas sketched at the end of Section 3. No internal structure of the particles (spin) has been taken into account. Also there are very few models so far of selfgravitating systems, and few results on stability.

REFERENCES

1. F. Jüttner, *Ann. Phys.*, **34**, 856; **35**, 145 (1911).
2. F. Jüttner, *Z. Physik*, **47**, 542 (1928).
3. J. L. Synge, *Trans. Soc. Canada*, III, **28**, 127 (1934).

4. A. G. WALKER, *Proc. Edinburgh Math. Soc.*, **4**, 238 (1936).
5. A. EINSTEIN, *Ann. of Math.*, **40**, 922 (1939).
6. A. LICHNEROWICZ and R. MARROT, *C.R. Acad. Sc. Paris*, **210**, 759 (1940).
7. J. L. SYNGE, *The Relativistic Gas*, North Holland Publ. Co. Amsterdam, 1957.
8. M. SASAKI, in: *Max Planck Festschrift 1958* (VEB Deutscher Verlag der Wissenschaften, Berlin 1958), p. 129.
9. G. E. TAUBER and J. W. WEINBERG, *Phys. Rev.*, **122**, 1342 (1961).
10. J. EHLERS, *Akad. Wiss. Mainz Abh., math.-naturwiss.*, Kl. 1961, Nr. 11.
11. W. ISRAEL, *J. Math. Phys.*, **4**, 1163 (1963).
12. H. GRAD, *Comm. Pure Appl. Math.*, **2**, 331 (1949).
13. K. BICHTELER, *Z. Physik*, **182**, 521 (1965).
14. R. H. BOYER, *Amer. Journ. of Physics*, **33**, 910 (1965).
15. K. BICHTELER, *Commun. math. Phys.*, **4**, 352 (1967).
16. R. W. LINDQUIST, *Annals of Physics*, **37**, 487 (1966).
17. YA. B. ZEL'DOVICH and M. A. PODURETS, *Astr. Zh.*, **42**, 963 (1965); Engl. transl. in Soviet Astronomy-AJ, **9**, 742 (1966).
18. E. D. FACKERELL, unpubl. Ph.D. thesis, University of Sidney (1966); see also *Ap. J.*, **153**, 643 (1968) and *Proc. Astron. Soc. Australia*, **1**, 86 (1968).
19. J. EHLERS, P. GEREN and R. K. SACHS, *J. Math. Phys.*, **9**, 1344 (1968).
20. W. RIENSTRA, unpublished Ph.D. thesis at the University of Texas at Austin (1969).
21. R. LYLE, unpublished Ph.D. thesis at the University of Texas at Austin (1968).
22. J. R. IPSER and K. S. THORNE, *Ap. J.*, **154**, 251 (1968); see also forthcoming papers by J. R. IPSER in the same journal.
23. C. W. MISNER, *Ap. J.*, **151**, 431 (1968).
24. R. A. MATZNER, *Ap. J.*, **157**, 1085 (1969). See also a forthcoming paper by C. W. MISNER and R. MATZNER, to appear in *Ap. J.*
25. R. K. SACHS and A. M. WOLFE, *Ap. J.*, **147**, 73 (1967).
26. W. L. AMES and K. S. THORNE, *Ap. J.*, **151**, 659 (1968).
27. J. L. ANDERSON and J. M. STEWART (1969), to be published. I am indebted to Dr. Stewart for informing me about this work prior to publication.
28. J. EHLERS and R. K. SACHS, *Kinetic Theory and Cosmology*, to appear in the *Proceedings of the Brandeis Summer Institute for Theoretical Physics* (1968).
29. J. EHLERS, *General Relativity and Kinetic Theory*, to appear in *Rendiconti della Scuola Internazionale di Fisica "Enrico Fermi"*, Course XLVII (1969).
30. P. HAVAS, in J. MEIXNER (Ed.), *Statistical Mechanics of Equilibrium and Non-Equilibrium*, North-Holland Publishing Company, Amsterdam, 1965.
31. E. K. CONKLIN, *Nature*, **222**, 971 (1969).
32. W. KUNDT, in G. HÖHLER (Ed.), *Springer Tracts in Modern Physics*, **47**, 111 (1968).
33. R. K. SACHS, communication at Cincinnati-Meeting on General Relativity (1969).
34. F. HOYLE and W. A. FOWLER, *Nature*, **213**, 373 (1967).
35. N. A. CHERNIKOV, *Acta Phys. Polon.*, **26**, 1069 (1964).
36. C. MARLE, *Ann. Inst. Henri Poincaré* A, **10**, 67 and 127 (1969).
37. N. A. CHERNIKOV, *Acta Phys. Polon.* **23**, 629 (1963), *ibid.* **27**, 465 (1964).

PAPER 15

A new technique for the analysis of singularities

DAVID FEINBLUM

The Weizmann Institute of Science, Rehovot, Israel and the State University of New York at Albany, U.S.A.

It has been shown[1] that the study of a suspected singularity in a geometry may be reduced to that of adding singular "boundary" points to a well behaved but incomplete geometry. If the metric tensor for the geometry is positive definite, the addition of boundary points is quite simple; in fact it is often trivial. The techniques evolved over several years for adding such points are seldom simple to use, and may give incorrect results in complex cases. The reason for the difference caused by metric indefiniteness is that the notion of "Cauchy-ness" of a sequence is available if the metric is definite, but apparently not otherwise. I say "apparently" because mathematicians have developed the concept of a *uniform* space[2], which is more general than that of a metric space, and they have succeeded in extending the notion of a Cauchy sequence to all of these spaces. All metrizable topological spaces may be given a "uniform structure" which makes them uniform spaces. In particular every indefinite metric space may be made into a uniform space; a unique uniform structure is determined in a rather natural way by the metric. With this structure it is possible to show that a broad class of indefinite metric spaces are each related to a definite metric space in such a way that they have the "same" completion. More precisely, the following theorem has been proven:

Let M be a space-time containing points $(x^0, \mathbf{x})$ which is the union of a finite number of pieces $P_0 \cdots P_n$ such that: (1) P_0 has a time-orthogonal differentiable metric tensor with bounded components and determinant bounded away from zero. (2) Each P_i $(i = 1, \ldots n)$ has a time-orthogonal coordinate system such that the metric tensor is continuous and either one curve

$\mathbf{x}$ = constant, or one space-like slice x^0 = constant, is compact. Then each piece may be completed by (1) changing the sign of g_{00} to get a related piece with definite metric, (2) complete the related pieces, and (3) form the union of the related pieces. The resulting space is related to the desired completion in the following way: (1) The points contained in the spaces are the same. (2) The two spaces have equivalent topologies. (3) The two spaces have equivalent uniform structures. (4) The two spaces do not have equivalent metrics.

The technique of completion based on this result passes some very severe tests some of which are not passed by any other technique which I know. I shall restrict myself, however, to a particular example, chosen because the discrepancy between the results of the new technique and the most refined previous techniques is about as extreme as possible; also, the given geometry may be isometrically embedded in a flat space, and this embedding yields a completion. I think that most would agree that the answer given by the embedding should be regarded as the "correct" one; it is therefore of interest to compare the result of this last technique with the other results. The example in question has the metric

$$\mathrm{d}s^2 = \cos x\,(\mathrm{d}y^2 - \mathrm{d}x^2) + 2\sin x\,\mathrm{d}x\,\mathrm{d}y$$

with the point (x, y) identified with $(x + 2m\pi, y + 2m\pi)$ for integral m and n. This looks like an analytic metric for a torus, but there is strange behavior at $x = \pi/2$ and $3\pi/2$. Geroch[1] describes the analysis done by himself in consultation with several others. The result is that the piece of the "torus" $0 < x \leqslant \pi/2$ should be regarded as a *non-Hausdorff*[3] manifold with boundary; the boundary being the disjoint union of *two circles and a point*. The present techniques require a time-orthogonal coordinate system which is achieved by the transformations

$$\bar{y} - \bar{y}_0 = \ln|\cos x| + y$$

$$\bar{x} = \int_{\pi/2}^{x} (\cos x)^{-1/2}\,\mathrm{d}x \equiv F^{-1}(x).$$

The metric is then

$$\mathrm{d}s^2 = -\mathrm{d}\bar{x}^2 + \cos F\,\mathrm{d}\bar{y}^2.$$

One deduces that all points of the form $(\pi/2, \bar{y})$ are equivalent. The piece in question becomes a Hausdorff manifold with (singular) boundary; the boundary being a *single point*.

This geometry may be isometrically embedded in a flat space with cylindrical coordinates (T, ϱ, θ) and signature ($-$ $+$ $+$). The embedding is given by

$$\theta = \bar{y}$$

$$\varrho = (\cos F)^{1/2}$$

$$T = \int_0^{\bar{x}} \mathrm{d}\xi \left[1 + \left(\frac{\mathrm{d}\cos F}{\mathrm{d}\xi}\right)^2\right]^{1/2}$$

Near the singularity (here at $T = 0$), the surface is given by

$$\varrho = \frac{|T|}{\sqrt{5}}\left(1 - \frac{2T^4}{625}\right).$$

The second term is negligible for $-1 < T < 0$. In this region the surface is essentially a cone; clearly one wants to add only the origin at $T = 0$. Consideration of the region $x > \pi/2$ leads to another cone. This result is in complete agreement with the new technique, and in flagrant disagreement with the older one.

REFERENCES

1. R. Geroch, *J. Math. Phys.*, **9**, 450 (1968).
2. See e.g. J. Kelley, *General Topology*, Van Nostrand, Princeton, N.J. An outline will be given in a forthcoming paper by the author expanding the subject of this talk.
3. That is, there is a pair of points p and q, such that for each open neighborhood U of p, and each open neighborhood V of q, U intersects V. This behavior cannot be observed in any regular space-time.

PAPER 16

Quantum space-time and general relativity*

DAVID FINKELSTEIN

Young Men's Philanthropic League and Belfer Graduate School of Science, Yeshiva University, U.S.A.

This little report is written as an affectionate birthday greeting to Professor Nathan Rosen. Appropriately, it deals with the modifications made necessary by general relativity in a recent attempt to overcome the divergency problem of quantum field theory by introducing a quantum theory of space-time.

We begin with a sketch of the proposed system of the world.

In the history of the physics of extended systems, two avenues have lead from the old classical theory to the new quantum theory. We could call them the electrodynamic treatment and the water treatment after their most conspicuous examples. They may also be labeled the formal and the physical methods of quantization.

In the example of electrodynamics, we start from the classical system of equations and apply a purely formal prescription to obtain a quantum theory, a prescription based either on the classical canonical theory, the classical Lagrangion theory, or some other standard system of equations.

In the case of water, however, such a procedure does not lead us to the true microscopic nature of water, although it may provide a useful approximate or phenomenological description of such things as quantum vortices. Instead, to understand the true microscopic nature of water, we must add new physical information such as the existence of water molecules, and then apply the methods of quantum theory to these new entities. The atomic nature of water can never be discovered from the Navier–Stokes equations of hydrodynamics.

---

* Supported in part by the National Science Foundation.

In the case of electrodynamics, it is reasonable to suppose that the quantum equations are quite similar in form to the classical ones, because the classical ones are nearly linear. In the case of hydrodynamics, the classical equations are highly non-linear and there is no reason *a priori* to suppose that they resemble the underlying quantum theory at all.

It is well known that general relativity, the theory of space-time, has strong resemblences both to electrodynamics and to hydrodynamics. There have been many applications of the electrodynamic treatment to the theory of space-time.

We are pursuing a hydrodynamic treatment of space-time.

Our hunt for the quantum of space-time, the chronon, is guided by the following scheme.

Physical theories are generally couched in differential equations while more and more of late differential equations serve as an intermediate step in the formation of a digital computer program from which consequences of the theory are then drawn. Indeed, if a theory is convergent it is practically indistinguishable from some such discrete system and if not convergent it is not really a theory at all. In our search for a convergent theory, why should we not omit the treacherous middle step and directly formulate physical theories in digital-computer language? Thus we come naturally to the extreme form of this thought which may be put dogmatically as follows:

The world is a digital computer. Physical events are computational events, and the causal relation which defines most of the structure of space-time is the relation of computational dependence. (Here we say event as in statistics for the most general kind of happening, not as in special relativity for a happening at one space-time point, a troublesome concept. Events have elsewhere been called propositions, questions, and classes, and form an ortholattice.)

The world as a computer is evidently highly *asynchronous* or parallel and *quantum-mechanical.* These granted, there is no difficulty in constructing models which are exactly Lorentz-invariant, although Poincaré invariance can only be approximate due to the assumed digital nature.[1]

We now take up general-relativistic invariance.

The conceptual changes suggested by general relativity in this world scheme are quite obvious and to us quite compelling. We may consider the question from the side of computer theory as well as physics. The computers we have considered so far are in one important aspect much simpler then those of present-day technology. Their flow chart as specified by the relation of computational dependence is given absolutely. Only the specific information content is subject to determination by initial conditions. This is the state of affairs in adding machines. It is the state of the computer art prior to such functions as conditional instructions, address modification, and the ability to compute instructions to be carried out. Let us then distinguish between *special* and *general* computers, between those with absolutely specified flow charts and those whose flow charts are potentially self-modified during the course of computation.

But this is just the expression in computer terms of one familiar and basic distinction between special relativity and general relativity: the former has an absolute causal structure, the latter a conditional, self-influencing causal structure.

We infer that if the world is a digital computer then the world computer must be a *general* asynchronous quantum-mechanical digital computer.

REFERENCE

1. DAVID FINKELSTEIN, The Space-Time Code, *Phys. Rev.*, **184**, 1261 (1969); *Coral Gables Conference on Fundamental Interactions at High Energy*, Gordon and Breach, New York, N.Y., 1969, p. 338.

PAPER 17

New experimental tests of relativity

R. FOX and J. SHAMIR
Technion—Israel Institute of Technology, Haifa, Israel

ABSTRACT

It is shown that the Michelson–Morley experiment performed in a solid transparent medium is capable of distinguishing between special relativity and rival theories based on a preferred frame of reference and physical Lorentz contractions. This experiment was performed and its negative result enhances the experimental basis of special relativity. An experiment suggested by Yilmaz some time ago to test the equivalence principle for photons is also described. Theoretically, a violation of the equivalence principle could lead to an anisotropy in the velocity of light with respect to the sun of the order of $\Delta c/c \approx 9 \times 10^{-9}$. An experimental upper limit of 3×10^{-11} was found.

1 INTRODUCTION

Some time ago one of us (R.F.) suggested an experimental program for making tests of the relativity theory using a highly sensitive Michelson-type interferometer[1]. It was emphasized then that the sensitivity of such an instrument for making relativity tests was very great. The program is being carried out and some recent work is reported here.

Two of the basic ideas of Einstein's theory of relativity are the lack of a preferred reference frame and the equivalence principle. Though most scientists believe that the special theory of relativity (STR) is fully verified experimentally, *this is not so*. A number of theories based on the existence of a preferred reference frame are also in agreement with the existing experiments confirming the Lorentz contraction. The best known of these rival theories is that of Lorentz himself[2]. H.E. Ives[3-5], L. Janossy[6-8] and C.N. Gordon[9] also obtain the Lorentz contraction based on a preferred reference frame (PRF). The discovery of the cosmic microwave background radiation[10] makes the existence of a PRF even more of a possibility. This radiation can serve as a PRF since it should be possible to detect motion through it.[11] If there is a PRF one would expect a coupling between it and matter.

The velocity of light, for example, measured in a solid medium of an inertial frame (e.g. the earth) should show a small anisotropy as a function of the direction of the velocity between the inertial reference frame and the preferred reference frame. Such an experiment is described in Section II (for greater detail, see ref. 12). The equivalence principle as tested by the Eötvös–Dicke experiment for material bodies showed that to a high accuracy different masses of different constituencies have no relative motion when placed in the gravitational field of the earth or sun[13,14]. The question arises whether anomalous motion arises between *photons and matter* in a gravitational field. An experiment of this type is described in Section III (for greater detail, see ref. 15).

2 INTERFEROMETER TEST FOR THE EXISTENCE OF A PREFERRED REFERENCE FRAME

The Michelson–Morley experiment (MME) did not yield a strictly zero result[16]. The non-zero result might have been real and due to the fact that the experiment was performed in air and not in vacuum. The effect of the lengthened optical path due to the presence of air, in contrast to vacuum, would not be cancelled by a physical Lorentz contraction. The MME would then yield a zero result only if performed in vacuum. We performed the MME in a solid transparent medium which would enhance the possible effect of the refractive index.

We make three assumptions here which are included in most of the rival theories to STR.

a) There exists a preferred frame of reference with respect to which we can (in principle, at least) measure velocities and accelerations.

b) The Lorentz contraction of length is a real physical process.

c) The Fresnel drag coefficient is given exactly by

$$b = 1 - \frac{1}{n^2}, \tag{1}$$

where n is the refractive index of a transparent material.

Assumptions (a) and (b) are always needed in order to ensure a negative result of the MME. Assumption (c) is usually dealt with only for the special case when light travels parallel to the direction of motion of the transparent medium. Assumption (c) is known very accurately by experiment[17]. The result of assumption (c) is that if a light ray travels through a transparent moving medium, in any direction, it is dragged along in the direction of motion and the change in the velocity of light, $\Delta \mathbf{c}$ is given by

$$\Delta \mathbf{c} = \mathbf{v}b, \tag{2}$$

where $\mathbf{v}$ is the velocity of the transparent medium through the PRF.

We analyze the MME performed in a transparent medium on the basis of these three assumptions, describing the experiment as seen by an observer stationary with respect to the PRF. An observer moving with the system (laboratory frame) observes the same final measured effect which is a permanent record of the fringe shift.

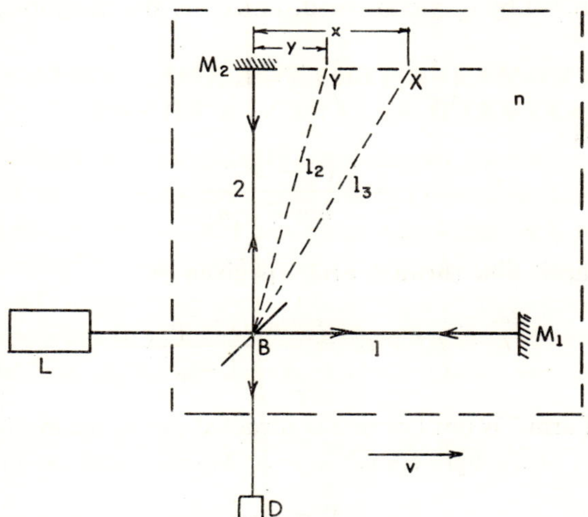

Figure 1. Diagram of the experiment. L—laser, B—beam splitter, M_1 and M_2—mirrors, D—detecting system

Figure 1 describes the experiment: An observer stationary with respect to the PRF sees a Michelson interferometer immersed in a transparent medium of index n moving to the right with a velocity v. Assume the interferometer arms are each of length l, as measured when stationary with respect to the PRF. We wish to calculate the transit times of a light ray in each arm in the moving system. The length of arm 1 in Figure 1 is contracted by the Lorentz contraction:

$$l_1 = l(1 - \beta^2)^{1/2}, \quad \beta = v/c, \tag{3}$$

where c is the velocity of light in vacuo. The velocity of light in this arm is

$$u_{\pm} = \frac{c}{n} \pm vb. \tag{4}$$

Here we used Eq. (2); the (+) sign refers to the velocity parallel to v while the (−) sign refers to the velocity anti-parallel to v.

During the light transit, mirror M_1 and beam splitter B move to the right, so that the light travels distances:

$$l_{\pm} = l_1 \pm vt_{\pm}, \tag{5}$$

where $t_{\pm}$ refers to the respective transit times in the two directions. Multiplication of Eq. (4) by $t_{\pm}$ yields the distance $l_{\pm}$. We thus obtain

$$\left(\frac{c}{n} \pm vb\right) t_{\pm} = l_1 \pm vt_{\pm} \tag{6}$$

or, using Eq. (1),

$$t_{\pm} = l \frac{(1 - \beta^2)^{1/2}}{c/n \mp v/n^2}. \tag{7}$$

The overall transit time through arm 1 is given by

$$t_1 = t_+ + t_- = \frac{2ln}{c} \frac{(1 - \beta^2)^{1/2}}{1 - \beta^2/n^2}. \tag{8}$$

The length l of arm 2 is not Lorentz-contracted due to the motion. However, during the time of the light transit, mirror M_2 moves to the point X, where

$$X = vt \tag{9}$$

(t is the transit time of the light from B to M_2).

The light ray striking M_2 at X, has thus travelled a distance l_3. From Figure 2, we see that the velocity along l_3 has two components. One is the drag velocity directed to the right with magnitude vb; the second has a magnitude c/n.

Then we have

$$\mathbf{c}' = \mathbf{c}/n + \mathbf{v}b. \tag{10}$$

Multiplying this equation by t we obtain the relation between the distances l_2, l_3 and $(x - y)$ that appear in Figure 1:

$$\mathbf{l}_3 = \mathbf{l}_2 + (\mathbf{x} - \mathbf{y}). \tag{11}$$

Using Eq. (1), we can express y by

$$y = vt - vbt = \frac{v}{n^2}\, t. \tag{12}$$

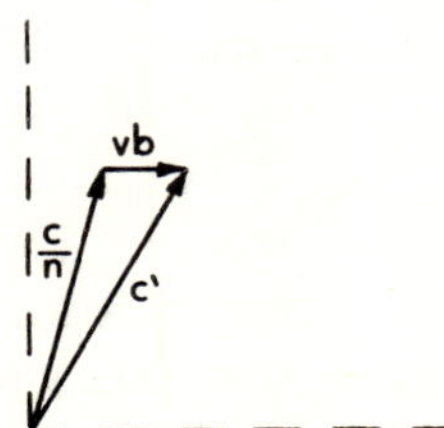

Figure 2. Demonstration of the dragging effect

Again, from Figure 1:

$$l_2^2 = l^2 + y^2$$

or

$$\left(\frac{c}{n}\, t\right)^2 = l^2 + \left(\frac{v}{n^2}\, t\right)^2 \tag{13}$$

We obtain then that the overall transit time through arm 2 is

$$t_2 = 2t = \frac{2ln}{c}\left[\frac{1}{(1 - \beta^2/n^2)^{1/2}}\right]. \tag{14}$$

The difference in transit times between the two arms is given by

$$\Delta t = t_2 - t_1 = \frac{2ln}{c}\left[\frac{1}{(1 - \beta^2/n^2)^{1/2}} - \frac{(1 - \beta^2)^{1/2}}{1 - \beta^2/n^2}\right]. \tag{15}$$

Expanding and retaining only terms of second order in β we have

$$\Delta t \approx \frac{ln\beta^2}{c}\left[1 - \frac{1}{n^2}\right] = \frac{lnb}{c}\,\beta^2. \tag{16}$$

If the frequency of light is ν and its wavelength in vacuum λ, the phase difference of the two beams will be

$$\delta = \nu\,\Delta t = \frac{l}{\lambda}\, nb\beta^2. \tag{17}$$

A rotation of 90° will interchange arms 1 and 2, but will not effect the frequency ν of the light source (a laser), due to our assumption that the MME gives negative result in vacuum. We thus have for the total observed fringe shift for a 90° rotation of the system,

$$\Delta = 2\delta = 2 \frac{l}{\lambda} nb\beta^2. \tag{18}$$

The Michelson-interferometer arms consisted of perspex rods, and the light source was a He-Ne laser. For sensitive detection of fringe shifts, the fringes were projected on to a pair of photoresistors that consisted of two arms of a Wheatstone bridge. Such a set-up is capable of measuring fringe shifts with a sensitivity up to 10^{-5} fringe[18]. The whole system rested on a heavy turntable (about 3.5 tons), which floated on mercury. The output of the fringe sensing system was connected to the y-input of an x-y recorder, the x-input being a voltage proportional to the sine of the angle of rotation of the table.

The final result obtained was that the velocity v, of the earth with respect to the preferred reference frame was

$$v \leqslant 7 \text{ km sec}^{-1}. \tag{19}$$

This is much smaller than the orbital velocity of the earth around the sun (~ 30 km sec^{-1}).

3 INTERFEREOMETER TEST OF THE EQUIVALENCE PRINCIPLE FOR PHOTONS

The Schwartzschild metric as interpreted by Yilmaz has a physical meaning and leads to an anisotropy in the velocity of light in a gravitational field[19]. This anisotropy is given for a mass M at a distance R from its center by the relation

$$\frac{\Delta c}{c} = \frac{GM}{Rc^2} \equiv \gamma \tag{20}$$

where c is the velocity of light in vacuum, and G the gravitational constant. A difference Δc should occur when the velocity of light parallel to the lines of gravitational force is compared to the velocity of light perpendicular to these lines.

The equivalence principle, as tested by the Eötvös–Dicke experiment for material bodies ensures that no mechanical strains are induced in an interferometer support. Yilmaz suggested the comparison of the velocity of light parallel to the gravitational field of the sun with the velocity of light perpendicular to it. Such an experiment can be performed by rotating a Michelson interferometer in a horizontal plane, when the sun is close to the horizon

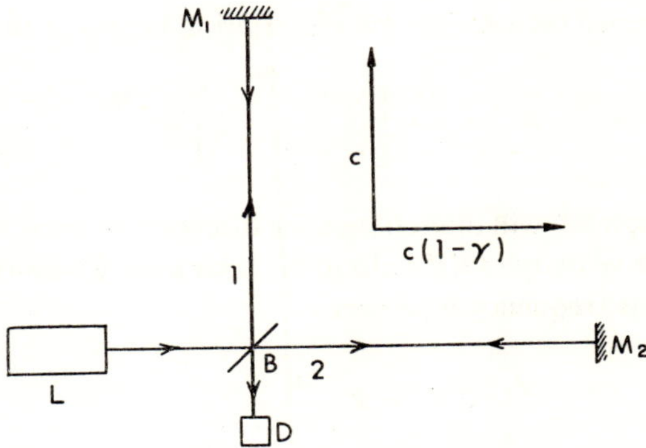

Figure 3. Diagram of the experiment. L—laser, M_1 and M_2—mirrors, B—beam splitter, D—detecting system

(morning and evening hours). If the earth was stationary with respect to the sun, we would expect to observe a $(\Delta c)/c$ given by (20). However, since the earth is in free fall with respect to the sun this experiment becomes a test of the equivalence principle for photons as also suggested by Yilmaz. According to the equivalence principle, the free fall of the earth along its orbit creates a change in the velocity of light $(\Delta c)/c$ with respect to the earth, which is exactly equal and opposite to (20). No net $(\Delta c)/c$ should then be observed. If however, the equivalence principle held for matter, but not for photons, we would expect an effect of the order of (20)

$$[\Delta c/c]_{\mathrm{sun}} \cong 9 \times 10^{-9}. \tag{21}$$

A schematic diagram of the experiment is presented in Figure 3. A laser light source (L) with a frequency ν illuminates a Michelson interferometer with arms each of length l. The velocity of light in arm 1 is c, while in arm 2 it is $c(1 - \gamma)$, (arm 2 is in the direction of the sun). The wavelength of the light beam in each arm is respectively:

$$\lambda_1 = \frac{c}{\nu}; \quad \lambda_2 = \frac{c(1-\gamma)}{\nu}. \tag{22}$$

The number of wavelengths in each arm is

$$N_1 = \frac{2l}{\lambda_1} = \frac{2l\nu}{c}, \quad N_2 = \frac{2l}{\lambda_2} = \frac{2l\nu}{c\,(1-\gamma)}. \tag{23}$$

The phase difference between the two interfering beams at the detector D is

$$\phi \equiv N_2 - N_1 = \frac{2l\nu}{c}\left[\frac{1}{1-\gamma} - 1\right] = \frac{2l\nu}{c}\,\frac{\gamma}{1-\gamma}. \tag{24}$$

A rotation through 90° will interchange the direction of arms 1 and 2. Due to the change in the velocity of light along the laser axis, which rotates with the interferometer, its frequency ν becomes

$$\nu' = \nu\,\frac{1}{1-\gamma}. \tag{25}$$

As a result of the rotation through 90°, the numbers of wavelengths, N_1, N_2 of Eq. (23) become

$$N_1' = \frac{2l\nu'}{c\,(1-\gamma)} = \frac{2l\nu}{c}\,\frac{1}{(1-\gamma)^2} \quad N_2' = \frac{2l\nu'}{c} = \frac{2l\nu}{c}\,\frac{1}{1-\gamma}. \tag{26}$$

The new phase difference is then

$$\phi' \equiv N_2' - N_1' = \frac{2l\nu}{c}\left[\frac{1}{1-\gamma} - \frac{1}{(1-\gamma)^2}\right] = -\frac{2l\nu}{c}\,\frac{\gamma}{(1-\gamma)^2}. \tag{27}$$

Thus the rotation introduced a phase shift (or fringe shift) of magnitude

$$\delta \equiv \phi - \phi' = \frac{2l\nu\gamma}{c\,(1-\gamma)}\left[1 + \frac{1}{1-\gamma}\right]. \tag{28}$$

To first order in γ, we have

$$\delta \approx 4\,\frac{l\nu}{c}\,\gamma. \tag{29}$$

We obtained experimentally that

$$\delta \leqslant 10^{-4}. \tag{30}$$

In the interferometer used l was 47 cm, and the frequency of the laser light, ν, was 5×10^{14} Hz. Putting these numbers into (29) we obtain from (20)

$$\frac{\Delta c}{c} \leqslant 3 \times 10^{-11}. \tag{31}$$

This is smaller by two orders of magnitude than (21), the estimate of $(\Delta c)/c$ obtained from (20).

Acknowledgments

We wish to thank Professor Rosen for his helpful comments throughout this work and take great pleasure in dedicating this paper to him on the occasion of his 60th birthday.

REFERENCES

1. R. Fox, Talk given at the Tenth Einstein Symposium in Theoretical Physics, Technion, 1965, unpublished.
2. H. A. Lorentz, *Proc. of the Acad. of Sci. of Amsterdam*, 6 (1904).
3. H. E. Ives, *J. Opt. Soc. Amer.*, **27**, 263 (1937).
4. H. E. Ives, *J. Opt. Soc. Amer.*, **27**, 310 (1967).
5. H. E. Ives, *Phil. Mag.*, **36**, 392 (1945).
6. L. Janossy, *Filozofion Szemle*, **6**, 153 (1962).
7. L. Janossy, *Acta Phys. Hung.*, **17**, 421 (1963).
8. L. Janossy, *Acta Phys. Hung.*, **21**, 1 (1966).
9. C. N. Gordon, *Proc. Phys. Soc.*, **80**, 569 (1962).
10. R. B. Partridge and D. T. Wilkinson, *Phys. Rev. Letters*, **18**, 557 (1967) and references there.
11. C. V. Heer and R. H. Kohl, *Phys. Rev.*, **174**, 1611 (1968).
12. J. Shamir and R. Fox, *Nuovo Cimento* **62** B, 258 (1969).
13. R. V. Eötvös, D. Pekar and E. Fekete, *Ann. Physik*, **68**, 11 (1922).
14. P. G. Roll, R. Krotkov and R. H. Dicke, *Ann. Phys.*, **26**, 442 (1964).
15. J. Shamir and R. Fox, *Phys. Rev.*, **184**, 1303 (1969).
16. A. A. Michelson, *Studies in Optics*, University of Chicago Press, Chicago, Ill., 1927.
17. W. M. Macek, J. R. Schneider and R. M. Salamon, *J. Appl. Phys.*, **35**, 2556 (1964).
18. J. Shamir, R. Fox and S. G. Lipson, *Appl. Opt.*, **8**, 103 (1969).
19. Huseyin Yilmaz, *Phys. Rev. Letters*, **3**, 320 (1959).

PAPER 18

On the possibility of cosmological foundations for classical and relativistic thermodynamics

BENJAMIN GAL-OR*

Technion–Israel Institute of Technology, Haifa, Israel

The constant search towards better thermodynamic theories that are based on more unified and universal foundations than present-day thermodynamics is not merely an academic, aesthetic and logical necessity, but is mainly due to the failure of statistical, classical, and special-relativistic thermodynamics to solve the existing problems, inconsistencies, and paradoxes. In previous papers[1,2] we reviewed some of these problems whereby we illustrated not only the need for better foundations but also the reason why the solution may eventually come from general-relativistic cosmology[6].

The fundamental difference between classical thermodynamics and general relativity was a constant source of concern to Einstein who wrote in 1949[3]:

> "The problem disturbed me already at the time of building of the general theory of relativity, without my having succeeded in clarifying it. What is essential in this is the fact that the sending of a signal is, in the sense of thermodynamics, an irreversible process, a process which is connected with the growth of entropy (whereas, according to our present knowledge, all elementary processes are reversible)."

This problem is also known as the paradox posed by the "arrow of time"[6], or Loschmidt's paradox[2].

* Also at the University of Pittsburgh, Pittsburgh, Pa., U.S.A.

In the present note the argument is extended to show that this fundamental problem of "irreversibility" may be solved, in principle, if one takes into account the cosmological fact that *space itself* is expanding[6]. Therefore, in considering the process of sending of a signal one must consider first the effect of space expansion on the signal intensity. In this way one can show that as far as the universe is expanding the contribution of all kinds of radiation in space is weakened "irreversibly" due to the expansion phenomenon itself. Such "loss" or "degradation" of energy in the depth of the intergalactic expanding space may then be considered as a *universal* "*sink*" for all the radiation flowing out of material bodies. In such a context one may then take the cosmological fact of the expansion of the universe as the point of departure in laying new foundations for thermodynamics[2].

This approach is in a sense similar to the Mach principle since it claims that one cannot isolate a thermodynamic system completely from the rest of the expanding universe. From this point of view cosmology must be the basic framework for thermodynamics[6]. (It would be of particular interest to compare the results and conclusions drawn in this approach with those of statistical thermodynamics. This, however, is outside the scope of this note. Thus we employ here the continuum concepts of macroscopic physics and develop explanations for the origin of such "laws" as the "second law" *without* employing or referring to the set of postulates and axioms underlying statistical or classical thermodynamics. There exists also an intimate connection between the principle of wave retardation and entropy[2] increase which we will not try to treat here. To judge then which approach is more logical one may compare the total number of independent postulates and axioms underlying each theory and conclude to what extent the existing problems and paradoxes are solved and eliminated.)

How is it possible that such remote expansions affect all irreversible processes on, say, the Earth? To answer this question let us first refer to Olbers' well known paradox according to which the night sky should be intensly brilliant in a static universe. This is, of course, avoided in an expanding universe because the radiation flowing into space from distant sources is greatly *weakened* by the *red-shift effect*. This phenomenon of "weakening", "degradation", or "sink" of the radiation energy in the depth of the expanding space is thus becoming a *result* of the expansion process. Various energy sources can exist in the galactic material but according to this cosmological mechanism, one finds only one "*universal sink*" which is the depth of the intergalactic expanding space[4-6].

The mechanism of this transfer and "sink" of energy in a non-uniform expanding universe must be then divided into two submechanisms. The first transfer is carried out by radiation from the surface of the material bodies into the universal sink. The mechanism of the second transfer is carried out within the material bodies and is involving *non-expanding* bodies. But according to this mechanism the energy released in any of these intermediate stages should, *eventually*, find its way out into the universal sink.[2]

Once the principles of such an approach are established one would like to formulate them in a more quantitative manner. For this aim one must first choose a quantitative theory and secondly in order to solve the equations, one needs a model. The first part is most likely to be supplied by general relativity and general-relativistic cosmology. The second choice is much more complicated due to various mathematical difficulties involved in treating non-uniform cosmological models.

The essential question here is: If the expansion of space is accepted as the cause for the existence of the "universal sink", and the time anisotropy[6], then perhaps the second law is not an *independent* postulate or "law?" Then perhaps it can be deduced and derived (in terms of energy units) from general relativistic models of the universal expansion. Adopting this approach, one may be able to reduce the total number of postulates, namely, the existence of entropy, the extensive properties of entropy, the second "law" and the third law. One must then build thermodynamics without the usual concept of entropy and redefine temperature. Instead of entropy one can use energy in agreement with an expanding cosmological model. The quantitative analysis of such an approach is available elsewhere[2] and will not be repeated here. The results show that due to the time symmetry of the equations the paradox posed by the "arrow of time" can be solved and eliminated. Furthermore, it has been shown[2] how a thermodynamic theory can be developed in terms of "dissipated radiation energy" or "geotropy" and intensive potentials (including temperature and chemical potentials). Such a formulation elucidates also other problems occurring in present-day thermodynamic theories such as the problems of thermal equilibrium in a stationary gravitational field (the "pocket" temperature), the correct Lorentz transformation of temperature, absolute negative temperature, and the use of Gibbs equation in irreversible thermodynamics. The entire superstructure of thermodynamics can thus be formulated by following the existing methods of thermodynamics. Most recently Narlikar[7] and Layzer[8] inferred also that origin of irreversibility in nature is cosmological rather than local.

REFERENCES

1. B. GAL-OR, A Gravitational Thermodynamic Theory, *Proceedings of Engineering Science in Space*, 5th Meeting of the Society of Engineering Science, Marshall Space Center, Huntsville, Alabama 1967, p. 67. (*Recent Advances in Engineering Science*, Vol. 4, pp. 111–131, Gordon & Breach, London (in press).)
2. E. B. STUART, B. GAL-OR, and A. J. BRAINARD (Editors) "*Critical Review of Thermodynamics*", Proceedings of International Conference on Thermodynamics, Pittsburgh, April 7–8, 1969. Mono Book Corp. Baltimore, Md. 1970.
3. A. EINSTEIN, Reply to criticisms, in *Albert Einstein, Philosopher-Scientist*, Harper, New York, 1959, p. 687.
4. R. C. TOLMAN, *Relativity Thermodynamics and Cosmology*, Oxford, The University Press, 1933.
5. G. C. MCVITTIE, *General Relativity and Cosmology*, Chapman Hall, London (1956).
6. T. GOLD, *Recent Developments in General Relativity*, Pergamon Press, New York, N.Y., 1962, pp. 225–234. See also L. D. LANDAU and E. M. LIFSHITZ "*Statistical Physics*", *2nd ed.* p. 29–30, Pergamon, London (1969).
7. J. V. NARLIKAR in *Proc. of International Conference on Thermodynamics*" (Edited by P. T. LANDSBERG) University College, Cardiff, April 1–4, 1970.
8. D. LAYZER, *ibid.* See also D. LAYZER in "*Relativity Theory and Astrophysics*" p. 237 (Edited by J. EHLERS) Am. Math. Soc., Providence, R.I. (1967).

PAPER 19

Gauge invariance and observer dependence

IRWIN GOLDBERG

Drexel Institute of Technology, Philadelphia, U.S.A.

1 INTRODUCTION

The problem of constraints in theories containing invariance groups is well known. The first problem to be faced in the quantization of such systems concerns the consistency between the operator algebra and the constraints. If the usual canonical quantization procedure is employed, the constraints do not commute with the dynamical variables. To solve the problem, many artifices have been tried with varying success, but the most straightforward procedure is the modification of the quantization procedure originated by Bergmann[1]. This technique introduces a new classical transformation group, and the quantization procedure is a homomorphism between the infinitesimal classical transformation group and the infinitesimal unitary transformations in Hilbert space. One of the major advantages of the Bergmann method is that it automatically provides a technique for finding the observables of the theory. I would like to discuss with you the insight I have gained in applying this method to electrodynamics and in particular the relationship between gauge invariance, manifest covariance, and observer dependence. My interest in the problem began when Gian-Carlo Wick asked whether it was possible to quantize electrodynamics without sacrificing part of the gauge invariance of the theory. Although the initial quantization was accomplished quickly, I spent the next ten years putting the theory in a form which would enable us to calculate by means of the Feynman method. The difficulties which I encountered all have their counterparts in the general theory of relativity.

2 BERGMANN TRANSFORMATIONS

Briefly, the Bergmann technique is to consider only those canonical transformations that map the hypersurface on which all the constraints are satisfied onto itself. The generator of the group-theoretical commutator of the infinitesimal transformations is called the Dirac bracket and determines the quantum-mechanical commutator of the observables. Formally the transformations, for a field theory, are given by

$$\int \varepsilon_{mn}(x, x')\, \delta y^n(x')\, \mathrm{d}^3x' = \frac{\delta \Gamma}{\delta y_m(x)}. \tag{1}$$

In Bergmann's notation $\xi^\mu(x)$ are the canonical field variables and momenta with $\mu = 1$ to N denoting field variables $\mu = N + 1$ to $2N$ denoting canonical momenta. The $y_m(x)$ are function of the field variables and momenta which determine points on the hypersurface $C^a = 0$ where C^a, $a = 1$ to p, are the constraints of the theory, and m runs from 1 to $2N - p$. The C^a and y^m taken together form a non-canonical coordinate system for the theory. The matrix ε_{mn} is given by

$$\varepsilon_{mn}(x, x') = \int \mathrm{d}^3x'' \,\frac{\delta \xi(x'')}{\delta y_m(x)}\, \varepsilon_{\mu\nu}\, \frac{\delta \xi^\nu(x'')}{\delta y_n(x')}, \tag{2}$$

where $\varepsilon_{\mu\nu}$ is the $2N$ by $2N$ matrix

$$\begin{pmatrix} 0 & -1 \\ 1 & 0 \end{pmatrix}. \tag{3}$$

When first-class constraints are present[2], the matrix ε_{mn} is singular and thus has null vectors (eigenvectors with eigenvalues zero). The null vectors provide a condition on the allowed generators. If we take the inner product of Eq. (1) with one of the null vectors $U_{(s)}$, then

$$\int \frac{\delta \Gamma}{\delta y_m(x)}\, U^m_{(s)}(x)\, \mathrm{d}^3x = 0. \tag{4}$$

Only those functions satisfying Eq. (4) can generate transformations of the type we are considering, and only the Γ's can be observables in the quantized theory. The Γ's are also the observables of the classical theory.

3 ELECTRODYNAMICS

When I first attempted to carry out the above procedure[3] for electrodynamics, the resulting theory was not manifestly covariant, which made it difficult to carry out calculations of the usual type. The lack of manifest covariance arose from the fact that in order to calculate the matrix ε_{mn}, the constraint equations must be solved. For electrodynamics the first constraint $\pi_0 = 0$ is simple, but the second constraint[4],

$$\pi^{s}_{,s} - e\psi^{+}\psi = 0, \tag{5}$$

is difficult. The straightforward method is to Fourier-analyze the theory and eliminate one of the components of the electric-field vector by Eq. (5). Although I was able to carry out this procedure and obtain a consistent gauge-invariant quantum theory the result was not entirely satisfactory. The first reason was obviously the lack of manifest covariance; the second was the fact that in order to carry out a perturbation calculation, one must separate the interaction term $e\psi\gamma^{\mu}\psi A_{\mu}$ from the free-field part. Then since neither of the two parts are gauge-invariant separately, this procedure seems to contradict the spirit of the approach.

The second problem can be solved by borrowing a technique invented by Dirac[5]. Dirac proposed a gauge-invariant electron operator

$$\left.\begin{aligned} \psi' &= e^{iC}\psi, \\ C &= e\int c_s(x, x')\, A_s(x)\, d^3x', \\ C_{s,s}(x, x') &= -\delta^3(x - x'). \end{aligned}\right\} \tag{6}$$

We note that, under a gauge transformation $\psi \rightarrow e^{i\Lambda}\psi$,

and

$$\left.\begin{aligned} A_s &\rightarrow A_s + \Lambda_{,s}, \\ \psi' &\rightarrow \psi'. \end{aligned}\right\} \tag{7}$$

Dirac used this technique to find electron creation operators which would create the electron together with its Coulomb field.

I applied this technique[6] to the Lagrangian for electrodynamics, and found that this transformation converted the Lagrangian to the form

$$L = \int d^3x\, [-\tfrac{1}{4}(F_{\mu\nu}F^{\mu\nu}) + \tfrac{1}{2}i\,[\overline{\psi}'\gamma^{\mu}\psi'_{,\mu} - \overline{\psi}'_{,\mu}\gamma^{\mu}\psi'] - m\overline{\psi}'\psi' + \\ + e\int d^3x'\, \overline{\psi}'(x)\,\gamma^{\mu}\,\psi'(x)\, C^{s}(x, x')\, F_{\mu s}(x')]. \tag{8}$$

Eq. (8) has the obvious advantage that the interaction term is manifestly gauge-invariant, but the formulation is not manifestly covariant. The lack of manifest covariance can be solved by generalizing Eq. (7), writing

$$C = \int d^4x' \, C_\mu (x, x') \, A_\mu(x'), \quad C_{\mu,\mu} (x, x') = \delta^4 (x - x'). \tag{9}$$

The Lagrangian then takes the form

$$L = \int d^3x \, [-\tfrac{1}{4} (F_{\mu\nu} F^{\mu\nu}) + \tfrac{1}{2} i \, (\bar{\psi}' \gamma^\mu \psi'_\mu - \bar{\psi}'_\mu \gamma^\mu \psi) - m \bar{\psi}' \psi' + e \int d^4x' \, \psi'(x) \, \gamma^\mu \psi \, (x) \, F_{\mu\nu}(x') \, C^\nu \, (x, x')]. \tag{10}$$

The Lagrangian of Eq. (10) is manifestly covariant and manifestly gauge-invariant, but at the sacrifice of locality. The interaction term is non-local and it is impossible to determine the Hamiltonian directly from the Lagrangian. The Hamiltonian was obtained in ref. 6 by the artifice of writing the Hamiltonian for the usual theory and then applying the transformation defined by Eq. (9). This must be the correct Hamiltonian because the two approaches differ only by a unitary transformation. The Lagrangians of Eq. (8) and (10) have the advantage that the secondary constraint is simplified and takes the form

$$\pi^s_{,s} = 0. \tag{11}$$

Physically this means that the momentum canonically conjugate to the field variable A_s is not the electric field component F_{s0}, but is the free part of the field. Then although the momenta π_s commute with the electron field variables ψ, ψ^+, the electric field F_{s0} does not commute with the electron field. In the quantized theory this means that the electrons are created together with their Coulomb field in contradistinction to the usual formulation.

We are faced with the problem of choosing between a local formulation of the theory and a manifestly covariant non-local theory. Marx[7] and I discovered a way of obtaining a formulation that is both local and manifestly covariant. This can only be obtained by explicitly introducing the coordinates of the observer into the formulation. The resulting theory is manifestly covariant, but some terms are explicitly observer dependent. This corresponds to the fact that varying combinations of electric and magnetic fields are measured depending on the state of motion of the observer. We set

$$x_\mu = \tilde{x}_\mu + \tau n_\mu, \tag{12}$$

where $\tau = x \cdot n$ is the observer time, while the space-like vector $\tilde{x}_\mu$ lies in a plane perpendicular to n. The surface element on this plane, $d\sigma$ is indicated by $d^3\tilde{x}$, and is invariant under Lorentz transformations. We now modify the transformation of Eq. (9) by writing

$$C = \int d^3\tilde{x} A_\mu (\tilde{x}', \tau)\, \tilde{C}^\mu (\tilde{x}, \tilde{x}') \tag{13}$$

and

$$\tilde{\partial}_\mu \tilde{C}^\mu(\tilde{x}, \tilde{x}') = \delta^3 (\tilde{x} - \tilde{x}'),$$

where $\delta^3(\tilde{x})$ is the Lorentz invariant extension[8] of $\delta^3(x)$ to the surface perpendicular to n, so that

$$\left.\begin{aligned} \int d^3\tilde{x}' f(\tilde{x}')\, \delta^2 (\tilde{x} - \tilde{x}') &= f(\tilde{x}) \\ \delta^3(\tilde{x})\, \delta(\tau) &= \delta^4(x). \end{aligned}\right\} \tag{14}$$

A possible choice for $\tilde{C}_\mu$ is

$$\left.\begin{aligned} \tilde{C}_\mu &= \tilde{\partial}_\mu G(\tilde{x}), \\ G(\tilde{x}) &= (1/4\pi)\, (-\tilde{x}^2)^{-1/2}. \end{aligned}\right\} \tag{15}$$

The action then becomes

$$\begin{aligned} I = \int d^4x\, [\tfrac{1}{2}i\, (\bar{\psi}'\gamma^\mu\psi'_{,\mu} - \bar{\psi}'_{,\mu}\gamma^\mu\psi') - m\bar{\psi}'\psi' - \tfrac{1}{4}F_{\mu\nu}F^{\mu\nu}] \\ + e \int d\tau\, d^3\tilde{x}\, d^3\tilde{x}'\, j_\mu (\tilde{x}, \tau)\, C_\lambda (\tilde{x}, \tilde{x}')\, F^{\mu\lambda} (\tilde{x}', \tau), \end{aligned} \tag{16}$$

where

$$j_\mu = \bar{\psi}'\gamma_\mu\psi'.$$

The equations of motion are:

$$\left.\begin{aligned} \partial^\nu F_{\mu\nu} &= j_\mu, \\ i\,\gamma^\mu\psi_{,\mu} - m\psi + e\gamma^\mu\psi \int d^3\tilde{x}'\, C^\lambda (\tilde{x}, \tilde{x}')\, F_{\mu\lambda} (\tilde{x}', \tau) &= 0, \\ -i\,\bar{\psi}_{,\mu}\gamma^\mu - m\bar{\psi} + e\bar{\psi}\gamma^\mu \int d^3\tilde{x}'\, C^\lambda (\tilde{x}, \tilde{x}')\, F_{\mu\lambda} (\tilde{x}', \tau) &= 0, \end{aligned}\right\} \tag{17}$$

where we have dropped the primes on the ψ, $\bar{\psi}$ which henceforth will refer to the gauge-invariant spinor-field variables.

The covariant canonical momenty for the electromagnetic field are

$$\left.\begin{aligned} \Pi_\mu &= n_\nu g_{\mu\alpha} \frac{\partial \mathcal{L}}{\partial A_{\alpha,\nu}}, \\ \Pi_\mu &= n^\nu F_{\mu\nu} - en^\nu \int d^3\tilde{x}'\, j_\nu (\tilde{x}', \tau)\, C_\mu (\tilde{x}' - \tilde{x}). \end{aligned}\right. \tag{18}$$

The constraint equations, the analogues to $\Pi_0 = 0$, and $\Pi^s_{,s} = j^0$ are

$$\left.\begin{aligned} n^\lambda \Pi_\lambda &= 0, \\ \partial_\lambda \Pi^\lambda &= \tilde{\partial}_\lambda \tilde{\Pi}^\lambda = 0. \end{aligned}\right\} \tag{19}$$

The second equation is the derivative with respect to τ of the first equation. We see that the electric field seen by the observer n is separated into a free-field part Π_μ and the part which arises from the sources.

The symmetrized canonical energy-momentum tensor is

$$\theta_{\mu\nu} = \tfrac{1}{4}\mathrm{i}\,(\bar{\psi}\gamma^\mu\psi_{,\nu} - \bar{\psi}_{,\nu}\gamma^\mu\psi + \bar{\psi}\gamma^\nu\psi_{,\mu} - \bar{\psi}_{,\mu}\gamma^\nu\psi) - F_{\lambda\mu}F^\lambda_\nu + \tfrac{1}{4}F_{\alpha\beta}F^{\alpha\beta}g_{\mu\nu}$$
$$+ \tfrac{1}{2}e\, j_\mu(x) \int \mathrm{d}^3\tilde{x}'\, C^\lambda(\tilde{x}, \tilde{x}')\, F_{\nu\lambda}(\tilde{x}', \tau). \tag{20}$$

Then the Hamiltonian density is defined by

$$\mathscr{H} = n^\alpha n^\beta \theta_{\alpha\beta}. \tag{21}$$

The Hamiltonian $H = \int \mathrm{d}^3\tilde{x}\, \mathscr{H}$ may be put into the form

$$H = \int \mathrm{d}^3\tilde{x}\, [\bar{\psi}\,(-\mathrm{i}\tilde{\partial}_\mu\gamma^\mu + m)\,\psi + \tfrac{1}{4}\tilde{F}_{\alpha\beta}\tilde{F}^{\alpha\beta} - \tfrac{1}{2}\tilde{\Pi}_\alpha\tilde{\Pi}^\alpha$$
$$+ \tfrac{1}{2}e^2 \int \mathrm{d}^3\tilde{x}\, \mathrm{d}^3\tilde{x}'\, n^\alpha j_\alpha(\tilde{x}, \tau)\, n^\lambda j_\lambda(\tilde{x}', \tau)\, G(\tilde{x} - \tilde{x}')$$
$$+ e \int \mathrm{d}^3\tilde{x}\, \mathrm{d}^3\tilde{x}'\, j_\mu(\tilde{x}, \tau)\, \tilde{F}^{\mu\lambda}(\tilde{x}', \tau)\, C_\lambda(\tilde{x} - \tilde{x}'), \tag{22}$$

where

$$\tilde{F}_{\alpha\lambda} = F_{\alpha\lambda} - n_\alpha n_\beta F^\beta_\lambda - n_\lambda n_\beta F^\beta_\alpha$$
$$= \tilde{\partial}_\lambda \tilde{A}_\alpha - \tilde{\partial}_\alpha A_\lambda. \tag{23}$$

The Hamiltonian (22) is tied to the observer and can be broken into a free-field part, the Coulomb field seen by the observer, and the interaction between the radiation field and charged spinor field.

We now can apply the Bergmann formalism. First we choose as parameters y_m:

$$a_\alpha(\tilde{k}, \tau),\quad \Pi^{(A)}(\tilde{k}, \tau),\quad b_\lambda(\tilde{k}, \tau),\quad b^*_\lambda,\quad d_\lambda(\tilde{k}, \tau),\quad d^*_\lambda,$$

which are related to the canonical field variables by

$$\left.\begin{aligned} A_\alpha(x) &= (2\pi)^{-3/2} \int \mathrm{d}^3\tilde{k}\, a_\alpha(\tilde{k}, \tau)\, \mathrm{e}^{-\mathrm{i}\tilde{k}\cdot\tilde{x}}, \\ \Pi^\alpha(x) &= (2\pi)^{-3/2} \int \mathrm{d}^3\tilde{k}\, \mathrm{e}^{\mathrm{i}\tilde{k}\cdot\tilde{x}}\, [\Pi^{(1)}(\tilde{k}, \tau)\, \varepsilon^\alpha_{(1)}(\tilde{k}) + \Pi^{(2)}(\tilde{k}, \tau)\, \varepsilon^\alpha_{(2)}], \\ \psi(x) &= (2\pi)^{-3/2} \sum_{\lambda=1}^{2} \int \mathrm{d}^3\tilde{k}\, (m/\varkappa)^{1/2}\, [u_\lambda(\tilde{k})\, b_\lambda(\tilde{k}, \tau)\, \mathrm{e}^{-\mathrm{i}\tilde{k}\cdot\tilde{x}} + v_\lambda d^*_\lambda\, \mathrm{e}^{\mathrm{i}\tilde{k}\cdot\tilde{x}}, \end{aligned}\right\} \tag{24}$$

where

$$\left.\begin{aligned} \varkappa &= +(m^2 - \tilde{k}^2)^{1/2}, \\ \varepsilon_\mu^{(A)} \cdot \tilde{k}^\mu &= \varepsilon_\mu^{(A)} \cdot n^\mu = 0, \\ \varepsilon_\mu^{(A)} \cdot \varepsilon_{(B)}^\mu &= \delta_B^A, \\ \varepsilon_\mu^{(2)} &= \varepsilon_{\mu\nu\lambda\alpha} n^\nu \hat{k}^\lambda \varepsilon_{(1)}^\alpha, \\ \hat{k}_\lambda &= \tilde{k}_\lambda (-\tilde{k}^2)^{1/2}. \end{aligned}\right\} \tag{25}$$

The shift to momentum space is particularly convenient because of the simplification of the constraint equation:

$$\partial_\mu \tilde{\Pi}^\mu \to \tilde{k}_\mu \tilde{\Pi}^\mu (\tilde{k}, \tau) = 0. \tag{26}$$

The parameters we have chosen are obviously independent of the constraints and we can readily calculate $\varepsilon_{mn} (\tilde{k}, \tilde{k}', \tau)$, the Fourier transform of Eq. (2). Then

$$\varepsilon_{mn} (\tilde{k}, \tilde{k}') = \delta^3 (\tilde{k} - \tilde{k}') \times$$

$$\left|\begin{array}{cccccccccccccc}
0 & 0 & 0 & 0 & 0 & 0 & 0 & 0 & 0 & 0 & 0 & 0 & -\varepsilon^{(1)0} & -\varepsilon^{(2)0} \\
0 & 0 & 0 & 0 & 0 & 0 & 0 & 0 & 0 & 0 & 0 & 0 & -\varepsilon^{(1)1} & -\varepsilon^{(2)1} \\
0 & 0 & 0 & 0 & 0 & 0 & 0 & 0 & 0 & 0 & 0 & 0 & -\varepsilon^{(1)2} & -\varepsilon^{(2)2} \\
0 & 0 & 0 & 0 & 0 & 0 & 0 & 0 & 0 & 0 & 0 & 0 & -\varepsilon^{(1)3} & -\varepsilon^{(2)3} \\
0 & 0 & 0 & 0 & 0 & 0 & -i & 0 & 0 & 0 & 0 & 0 & 0 & 0 \\
0 & 0 & 0 & 0 & 0 & i & 0 & -i & 0 & 0 & 0 & 0 & 0 & 0 \\
0 & 0 & 0 & 0 & i & 0 & 0 & 0 & 0 & 0 & 0 & 0 & 0 & 0 \\
0 & 0 & 0 & 0 & 0 & 0 & 0 & 0 & 0 & 0 & 0 & 0 & 0 & 0 \\
0 & 0 & 0 & 0 & 0 & 0 & 0 & 0 & 0 & 0 & i & 0 & 0 & 0 \\
0 & 0 & 0 & 0 & 0 & 0 & 0 & 0 & 0 & 0 & 0 & i & 0 & 0 \\
0 & 0 & 0 & 0 & 0 & 0 & 0 & 0 & -i & 0 & 0 & 0 & 0 & 0 \\
0 & 0 & 0 & 0 & 0 & 0 & 0 & 0 & 0 & -i & 0 & 0 & 0 & 0 \\
\varepsilon^{(1)0} & \varepsilon^{(1)1} & \varepsilon^{(1)2} & \varepsilon^{(1)3} & 0 & 0 & 0 & 0 & 0 & 0 & 0 & 0 & 0 & 0 \\
\varepsilon^{(2)0} & \varepsilon^{(2)1} & \varepsilon^{(2)2} & \varepsilon^{(2)3} & 0 & 0 & 0 & 0 & 0 & 0 & 0 & 0 & 0 & 0
\end{array}\right|, \tag{27}$$

The null vectors of ε_{mn} are

$$U^{(1)} = \begin{vmatrix} n_0 \\ n_1 \\ n_2 \\ n_3 \\ 0 \\ 0 \\ 0 \\ 0 \\ 0 \\ 0 \\ 0 \\ 0 \\ 0 \\ 0 \end{vmatrix}, \quad U^{(2)} = \begin{vmatrix} \tilde{k}_0 \\ \tilde{k}_1 \\ \tilde{k}_2 \\ \tilde{k}_3 \\ 0 \\ 0 \\ 0 \\ 0 \\ 0 \\ 0 \\ 0 \\ 0 \\ 0 \\ 0 \end{vmatrix}. \tag{28}$$

The simplest set of observables are then $a^{\alpha}\varepsilon_a^{(A)}, \Pi^{(A)}, b_\lambda, b_\lambda^*, d_\lambda, d_\lambda^*$ which are all gauge invariant. The Hamiltonian may now be expressed completely in terms of observables. We define

$$\begin{aligned} Q_\lambda &= \tfrac{1}{2}\varepsilon_{\alpha\beta\lambda\nu} n_\nu \tilde{F}_{\alpha\beta} = n_\nu \tilde{F}^D_{\lambda\nu} \\ &= (2\pi)^{-3/2}\, \mathrm{i} \int \mathrm{d}^3\tilde{k}\, \omega\, (\varepsilon_\lambda^{(1)} q^{(2)} - \varepsilon_\lambda^{(2)} q^{(1)}) \exp(-\mathrm{i}\tilde{k}\cdot\tilde{x}), \end{aligned} \tag{29}$$

where

$$\omega = \sqrt{-\tilde{k}^2}, \quad q^{(A)} = -a^{\alpha}\varepsilon_\alpha^{(A)}.$$

We note that

$$Q_\alpha n^\alpha = 0, \quad Q_\alpha Q^\alpha = -\tfrac{1}{2}\tilde{F}_{\alpha\beta}\tilde{F}^{\alpha\beta}. \tag{30}$$

Then

$$\begin{aligned} H = \sum_{\substack{\lambda \\ A=1}}^{2} \int \mathrm{d}^3\tilde{k}\, [\varkappa\, (b_\lambda b_\lambda^* - d_\lambda d_\lambda^*) + \tfrac{1}{2}(\Pi^{(A)}\Pi^{*(A)} + \omega^2 q^{(A)} q^{*(A)})] \\ + \tfrac{1}{2}e^2 \int \mathrm{d}^3\tilde{x}\, \mathrm{d}^3\tilde{x}\, n_\alpha j^\alpha(\tilde{x},\tau)\, n_\beta j^\beta(\tilde{x}',\tau)\, G(\tilde{x}-\tilde{x}') \\ + e \int \mathrm{d}^3\tilde{x}\, j^\mu(\tilde{x},\tau)\, A_\mu^T(\tilde{x},\tau), \end{aligned} \tag{31}$$

where

$$A_\mu^T = (2\pi)^{-3/2} \int \mathrm{d}^3\tilde{k}\, \varepsilon_\mu^{(A)} q^{(A)}(\tilde{k},\tau) \exp(-\mathrm{i}\tilde{k}\cdot\tilde{x}).$$

From its definition it is apparent that A^T is an observable and consequently gauge-invariant. The brackets of the observables may now be obtained by choosing an observable as the generator of a transformation and calculating the change in the other observables generated by the transformation. Then, since

$$\bar{\delta}_1 \theta_2 = \{\theta_2, \theta_1\}, \tag{32}$$

we obtain the brackets. The quantum commutators may be obtained unambiguously from the brackets, because the basic set of observables are all linear in the parameters y_n. Then, by straightforward calculation,

$$\left.\begin{aligned} \{q^{(A)}(\tilde{k},\tau), q^{(B)}(\tilde{k}',\tau)\} &= \{\Pi^{(A)}(\tilde{k},\tau), \Pi^{(B)}(\tilde{k}',\tau)\} = 0, \\ \{q^{(A)}(\tilde{k},\tau), \Pi^{(B)}(\tilde{k}',\tau)\} &= -\delta^{AB}\delta^3(\tilde{k}-\tilde{k}'), \\ \{b_\lambda(\tilde{k},\tau), b^*_{\lambda'}(\tilde{k}',\tau)\} &= \{d_\lambda(\tilde{k},\tau), d^*_{\lambda'}(\tilde{k}',\tau)\} = \delta_{\lambda\lambda'}\delta^3(\tilde{k}-\tilde{k}'), \end{aligned}\right\} \tag{33}$$

and all other brackets between observables in the basic set vanish.

In the quantized theory the observables become operators and the commutators are given by multiplying the classical bracket by i. For the fermion operators we must change the commutators to anticommutators to get the correct relation between spin and statistics. It may be pointed out, however, that in the quantum theory only bilinear combinations of the fermion operators are observable and that the commutators of the observables are independent of whether commutation or anticommutation relations are used for the fermions. Then

$$\begin{aligned} [q^{(A)}(\tilde{k},\tau), \Pi^{(B)}(\tilde{k}',\tau)]_- &= -\mathrm{i}\,\delta^{AB}\delta^3(\tilde{k}-\tilde{k}'), \\ [b_\lambda(\tilde{k},\tau), b^*_{\lambda'}(\tilde{k}',\tau)]_+ &= [d_\lambda(\tilde{k},\tau), d^*_{\lambda'}(\tilde{k},\tau)] = \delta_{\lambda\lambda'}\delta^3(\tilde{k}-\tilde{k}') \end{aligned} \tag{34}$$

are the basic non-vanishing commutators and anticommutators.

4 PROPAGATORS

In order to carry out Feynman-type calculations, we must obtain the non-equal-time commutators of the free-field operators. These then determine the propagators. After a short calculation[9] we find:

$$\begin{aligned} A^T_\alpha(x) &= (2\pi)^{-3/2} \int \mathrm{d}^3\tilde{k}\,(2\omega)^{-1/2} \sum_{A=1}^{2} \varepsilon^{(A)}_\alpha [q^{(A)} \mathrm{e}^{-\mathrm{i}k\cdot x} + q^{+(A)} \mathrm{e}^{\mathrm{i}k\cdot x}, \\ k\cdot x &= \omega(\tilde{k})\,\tau. \end{aligned} \tag{35}$$

Then

$$[A_\alpha^T(x), A_\beta^T(x')]$$

$$= (2\pi)^{-3/2} \int \left\{ (2\omega)^{-1} \left(g_{\alpha\beta} - n_\alpha n_\beta - \frac{\tilde{k}_\alpha \tilde{k}_\beta}{\tilde{k}^2} \right) 2\mathrm{i} \sin k \cdot (x - x') \right\} \mathrm{d}^3\tilde{k}. \quad (36)$$

We note that no indefinite metric is required, since all diagonal elements of

$$[g_{\alpha\beta} - n_\alpha n_\beta - \tilde{k}_\alpha \tilde{k}_\beta (1/\tilde{k}^2)] \leqslant 0.$$

The non-equal-time commutators of the fermion operators and the propagators are identical with those of the usual formulation. We may now perform a perturbation expansion to see the differences between this approach and the non-gauge-invariant approach.

The Schrödinger equation,

$$\mathrm{i} \frac{\partial \Phi^s}{\partial \tau} = H \Phi^s, \quad (37)$$

is transformed to the interaction representation:

$$\mathrm{i} \frac{\partial \Phi}{\partial \tau} = H_I(\tau) \Phi(\tau), \quad (38)$$

where

$$H_\mathrm{I} = e \int \mathrm{d}^3\tilde{x} : \bar{\psi}(\tilde{x}, \tau) \gamma^\alpha \psi(\tilde{x}, \tau) : A_\alpha^T(\tilde{x}, \tau)$$

$$+ \tfrac{1}{2} e^2 \int \mathrm{d}^3\tilde{x}\, \mathrm{d}^3\tilde{x}' : \bar{\psi}(\tilde{x}, \tau)\, n_\alpha \gamma^\alpha \psi(\tilde{\psi}, \tau)\, \bar{\psi}(\tilde{x}' \tau)\, n_\lambda \gamma^\lambda \psi(\tilde{\psi}', \tau)\, G(\tilde{x} - \tilde{x}').$$

Note that the interaction Hamiltonian, H_I, is manifestly gauge invariant; then any Feynman graph is gauge-invariant. The rules are different since the photon propagator is modified by replacing $g_{\alpha\beta}$ in the Feynman formulation by $g_{\alpha\beta} - n_\alpha n_\beta - \tilde{k}_\alpha \tilde{k}_\beta (1/\tilde{k}^2)$. It is important to realize that this differs from what has been used in the literature as the gauge-invariant photon propagator. The so-called Landau propagator replaces $g_{\mu\nu}$ by $g_{\mu\nu} - (k_\mu k_\nu)/k^2$ and introduces an additional singularity since, on the mass shell, we have not yet taken into account the effect of the pure Coulomb term. We can write the contribution from this term in the form

$$\int_{-\infty}^{\infty} \mathrm{d}\tau\, H_C(\tau) = \tfrac{1}{2} e^2 \int \mathrm{d}^4x\, \mathrm{d}^4x' : n \cdot j(x)\, n \cdot j(x') : \delta(\tau - \tau')\, G(\tilde{x} - \tilde{x}'). \quad (39)$$

Since we can write

$$\delta(\tau)\, G(\tilde{x}) = -(2\pi)^4 \int \mathrm{d}^4k\, \mathrm{e}^{-ik\cdot x}\, (1/\tilde{k}^2), \tag{40}$$

we can combine this term with the photon propagator to obtain an effective photon propagator:

$$\{g_{\alpha\beta} - [k_\alpha k_\beta - n\cdot k\,(k_\alpha n_\beta + k_\beta n_\alpha)]\,(1/\tilde{k}^2)\}\,(1/k^2). \tag{41}$$

This can be used in all orders of perturbation theory, except for graphs with an electron propagator and a photon propagator between the same two vertices. In this case the Coulomb term does not contribute because there are no equal-time contractions when it is normally ordered. This means that the Coulomb term does not contribute to electron self-energy diagrams.

5 GENERAL RELATIVITY

The foregoing indicates how a theory containing a gauge group may be quantized without sacrificing any of the invariance properties. It is appropriate to ask why this procedure has not been applied to the general theory of relativity. The major obstacle to the first part of the procedure is the fact that the secondary constraint equations have the form

$$\Pi^{s}_{;s} = 0. \tag{42}$$

The momentum canonically congugate to g_{st} is not transverse and we cannot solve the constraint equation except in the linearized theory. Thus we may someday find a simplifying transformation which will enable us to solve Eq. (42), but this will still not enable us to proceed as in electrodynamics.

We may also note that Eq. (42) which is derived from the Dirac formulation[10] is not manifestly covariant. However, the procedure of explicitly introducing the coordinates of the observer employed for electrodynamics may be used to obtain a manifestly covariant formulation. We will then be able to find the matrix ε_{mn} and the observables, but we cannot use the results to calculate Feynman graphs. This is due to the fact that the Hamiltonian is a constraint. Then all observables commute with the Hamiltonian and are constants of the motion. Since the observables are constant, the non-equal-time commutators are the equal-time commutators, and there is no Schrödinger equation because $H = 0$. Thus an entirely new procedure must be developed for quantum calculations. Classically this problem finds its analogue in the Cauchy problem.

We are still trying to apply the Bergmann procedure to relativity, but we realize that finding the observables and their commutators will only be a first step.

REFERENCES

1. P. G. Bergmann and I. Goldberg, *Phys. Rev.*, **98**, 531 (1955).
2. First-class constraints have zero Poisson bracket with all other constraints.
3. I. Goldberg, *Phys. Rev.*, **112**, 1361 (1958).
4. Greek letters run from 0 to 3; Roman letters from 1 to 3.
5. P. A. M. Dirac, *Proc. Royal Soc.*, **235**, 138 (1950).
6. I. Goldberg, *Phys. Rev.*, **139B**, 1665 (1955).
7. I. Goldberg and E. Marx, *Nuovo Cimento*, **57B**, 485 (1968).
8. J. M. Jauch and F. Rohrlich, *The Theory of Photons and Electrons*, Reading, Massachusetts, 1955.
9. See [7], p. 499.
10. P. A. M. Dirac, *Proc. Royal Soc.*, **246A**, 333 (1958).

PAPER 20

Equations of motion in general relativity

JOSHUA N. GOLDBERG
Syracuse University, New York, U.S.A.

Ten years ago at Royaumont I presented a general discussion of the equations of motion without being able to suggest, even in principle, a specific method of applying the formalism in the absence of an approximation method. Today, because of the work of various people[1-5] on gravitational radiation fields, a new and more understandable approach can be made. In fact, Newman has already carried out extensive work on this problem[6]. I shall report on my attempt to understand his preliminary description which appeared in *Phys. Rev. Letters*.

The basis for the motion of particles being determined by the Einstein equations is in the covariance with respect to general coordinate transformations. Although the ten field equations are linearly independent, they are not wholly independent in that they satisfy differential identities, the Bianchi identities. With the help of approximation methods these have been shown to determine the motion of point particles and even of spinning particles with appropriate restrictions[7,8].

Because of the Bianchi identities, one can write

$$-2\xi^{\mu}\sqrt{-g}\,G^{\nu}_{\mu} = U^{\nu_{\nu}\sigma}_{,\sigma} - t^{\nu} \tag{1}$$

where $U^{\nu_{\nu}\sigma}$ is an appropriate superpotential and t^{ν} the corresponding pseudotensor or tensor, and ξ^{μ} a suitable set of weighting functions or descriptors. It follows that when the field equations are satisfied, $G^{\nu}_{\mu} = 0$,

$$\oint U^{n_{\nu}0}_{,0}\, n_n\, \mathrm{d}S = \oint t^{n} n_n\, \mathrm{d}S, \tag{2}$$

The left-hand side represents the time derivative of certain quantities while the right-hand side is the flux through the surface of integration. Once a solution has been obtained this integral relationship is trivally satisfied and therefore gives no information. Nonetheless, the integral separation given in Eq. (2) may provide an interpretation. However, one would like to find a way to apply these conditions prior to satisfying all of the field equations. Furthermore, one wants to show that the resulting integrals are independent of the surface S and therefore can be identified with particle properties rather than field properties.

Consider a world tube in whose external neighborhood the empty space Einstein equations are satisfied. There may be several such isolated regions containing matter, but we focus on the empty space surrounding one such region. Topologically the world tube is an $S_2 \times R_1$ and we want to think of it as a one parameter family of closed two surfaces $S_0(u)$. Following Tamburino and Winicour[5], we construct a local null coordinate system by taking surfaces $x^0 = u =$ constant to be generated by the outgoing null geodesics normal to $S_0(u)$. In the usual way one can arrive at a coordinate system such that[1-5]

$$ds^2 = 2\phi\, du^2 + 2e^{2b}\, du\, dr - r^2 h_{AB}\, (dx^A - U^A\, du)\, (dx^B - U^B\, du) \tag{3}$$

where x^A ($A = 2, 3$) may be chosen to be angles on the sphere (θ, φ) and $x^1 = r$ the luminosity distance so that $|h_{AB}| = \sin^2 \theta$. With this form for the mectric the field equations break up into four groups[5].

i) $G_{1\mu} = 0$ contain no time derivatives, hence are hypersurface equations;

ii) $G_{AB} - \frac{1}{2} g_{AB} g^{CD} G_{CD}$ contain time derivatives and are the propagation equations;

iii) G_{0A} and G_{00} contain time derivatives but are not wholly independent of the previous equations;

iv) $g^{CD} G_{CD}$ is trivally satisfied once groups (i) and (ii) are solved.

There are only six unknown functions in the metric (3) which in principle can be determined by (i) and (ii). These solutions contain a number of arbitrary functions which are then related by the supplementary equations (iii). We shall argue that these supplementary conditions, taken close to the matter, not at infinity, are just the sought for equations of motions.

From the Bianchi identities it follows that having satisfied groups (i) and (ii) that

$$\left(\sqrt{-g}\, \xi^A G^1_A\right)_{,1} = 0 \tag{4}$$

for weighting functions independent of r. Hence from (2) we see that the relationship

$$\oint U^{10}[\xi^A]_{,0}\, d\Omega = \oint t^1[\xi^A]\, d\Omega \tag{5}$$

is independent of the surface S in u = constant. Next we find

$$\left(\xi^0 \sqrt{-g}\, G_0^1\right)_{,1} = 0$$

and

$$\oint U^{10}[\xi^0]_{,0}\, \mathrm{d}\Omega = \oint t^1[\xi^0]\, \mathrm{d}\Omega \tag{6}$$

is likewise surface-independent. From the work of Bondi[1], Sachs[2], and Goldberg[3] we know that (5) gives angular momentum restrictions while (6) gives energy-momentum restrictions and moments thereof.

The one remaining problem is to determine the world tube or rather the one-parameter family of two surfaces $S_0(u)$. Unfortunately, I have no general prescription to offer. Instead, I shall find properties one would like to have and show that these are just the conditions chosen by Newman and Posadas.

One question which has perplexed people is how to recognize that a point singularity is spherically symmetric or how to endow the singularity with well defined multipole structure. The definition given by Janis and Newman[10] for the asymptotic field is also meaningful for the situation being considered here. The conformal tensor can be expressed in terms of scalars[3] Ψ_A $(A = 0 \cdots 4)$ with the asymptotic behavior of $O(r^{A-5})$. Thus Ψ_4 represents the radiative part while Ψ_0, Ψ_1 and Ψ_2 are present in static or stationary fields. Assume that for our problem near matter

$$\Psi_2 = \Sigma \frac{\Psi_2^n}{r^{3+n}} \tag{7a}$$

$$\Psi_1 = \Sigma \frac{\Psi_1^n}{r^{4+n}} \tag{7b}$$

$$\Psi_0 = \Sigma \frac{\Psi_0^n}{r^{5+n}} \tag{7c}$$

then

$$M = \frac{1}{4\pi} \oint \Psi_2^0\, \mathrm{d}\Omega \tag{8a}$$

$$N_m = \frac{1}{4\pi} \oint \Psi_1^0\, {}_1\bar{Y}_{1m}\, \mathrm{d}\Omega \tag{8b}$$

$$Q_{lm} = \frac{1}{4\pi} \oint \Psi_0^{(l-2)}\, {}_2\bar{Y}_{lm}\, \mathrm{d}\Omega . \tag{8c}$$

From the field equations one finds that $\Psi_0 \neq 0$ implies that the null rays of u = constant have shear $\sigma \neq 0$. In particular, Ψ_0 singular implies σ singular. Null rays which emanate from one point in flat space are shear free. For the Schwarzschild solution those rays which appear to come from $r = 0$ are also shear free.

Therefore, to restrict attention to point particles without structure, assume that in terms of a tetrad related to the surfaces u = constant, needed to define the tube $S_0(u)$, only Ψ_2 becomes singular for $r \to 0$. One might like the shear to be zero on u = constant, but that is asking too much. In linearized gravitational theory we can superpose two solutions each of which has only $\overset{(1,2)}{\Psi}_2 \neq 0$ with respect to its own characteristic tetrad. This solution represents two particles moving with straight line motion. The Ψ_A calculated with respect to the tetrad characteristic of particle (1) will exhibit in the neighborhood of the first particle behavior such that

$$\Psi_0 = O(1), \quad \Psi_1 = O(1). \tag{9}$$

This behavior implies $\sigma = O(r)$ is a correct choice. That is, σ should not be forced to be zero, but should be non-singular for $r \to 0$.

These considerations should extend to spinning particles where Ψ_1 and Ψ_2 may be singular, but not Ψ_0. In that case $\sigma = O(r)$ is still a suitable condition to require. Spinning particles may move on arbitrary world lines unless a suitable center of mass can be defined. In this case $\operatorname{Re} \Psi_1^0 = 0$ gives a suitable condition. The extension to particles with quadrupole moment appears difficult. With σ being singular, the description of a good world tube $S_0(u)$ becomes questionable. Perhaps it can be done by requiring $\sigma = O(1/r^5)$ but I have not investigated this question.

Newman and Posadas require that the surface u = constant contain rays whose divergence is characteristic of spherical expansion $l^\mu_{;\mu} = 2/r + O(r)$, the shear is non-singular, $\sigma = O(r)$ and finally that the induced metric on the two-surfaces S be non-singular as $r \to 0$. These latter two conditions appear to be important in order to be sure that the quantities calculated by (5) and (6) be characteristic of the particles.

Finally, I want to emphasize that in this brief report I have only attempted to understand the paper of Newman and Posadas[6]. In this investigation I have benefited from earlier discussions with Newman on the meaning of the multipole definitions and the structure of the metric when the origin is taken on an arbitrary time-like curve in Minkowski space.

REFERENCES

1. H. BONDI, M. G. J. VAN DER BURG and A. W. K. METZNER, *Proc. Roy. Soc.*, **A269**, 21 (1962).
2. R. K. SACHS, *Proc. Roy. Soc.*, **A270**, 103 (1962).
3. E. T. NEWMAN and R. PENROSE, *J. Math. Phys.*, **3**, 566 (1962).
4. E. T. NEWMAN and T. W. J. UNTI, *J. Math. Phys.*, **3**, 891 (1962).
5. L. A. TAMBURINO and J. H. WINICOUR, *Phys. Rev.*, **150**, 1939 (1966).
6. E. T. NEWMAN and R. POSADAS, *Phys. Rev. Letters*, **22**, 1196 (1969).
7. L. INFELD and J. PLEBANSKI, *Motion and Relativity*, PWN, Warsaw, 1960.
8. J. N. GOLDBERG, *Equation of Motion in Gravitation*, in: L. WITTEN (ed.). *Gravitation: An Introduction to Current Research*, Wiley, New York, N.Y., 1962.
9. J. N. GOLDBERG, *Phys. Rev.*, **131**, 1367 (1963).
10. A. J. JANIS and E. T. NEWMAN, *J. Math. Phys.*, **6**, 902 (1965).

PAPER 21

Modification of the classical gravitational field equations due to a virtual quantized matter field

L. HALPERN
University of Windsor, Ontario, Canada

The modifications of the gravitational field equations due to a quantized matter field have certain experimental support in the well known attraction of conducting planes in the vacuum predicted by Casimir[1]. One can conclude that, due to these attractive forces, a closed vessel with conducting walls is subject to a pressure which contributes to the total energy, and thus increases the mass beyond that of the material of the walls. This additional mass has to be taken into account in the field equations of general relativity; being a quantum effect it may manifest itself in a consistent quantized theory of the gravitational field. As we still lack such a theory we proceed to incorporate it phenomenologically as an additional term on the right hand side of the gravitational field equations. Formally the term is taken into account by writing

$$R_i^k - \tfrac{1}{2}\delta_i^k R = -\varkappa\,(T_i^k + \langle\tau_i^k\rangle), \tag{1}$$

where

$$\langle\tau_i^k\rangle = \frac{(\mathrm{O}_{\text{out}}\,|\tau_{\mathrm{H}i}^k(x)|\,\mathrm{O}_{\text{in}})}{(\mathrm{O}_{\text{out}}\mid \mathrm{O}_{\text{in}})}. \tag{1a}$$

Here, $\tau_{\mathrm{H}i}^k(x)$ is the symmetric energy momentum tensor of the virtual matter field expressed in terms of the field operators in the Heisenberg picture.

Calculated in lowest order of perturbation theory, $\langle\tau_i^k\rangle$ diverges. The divergences were first removed by DeWitt, and DeWitt and Utiyama[2]. There occur divergences of fourth order and second order which may be removed by requiring

$$\langle \tau_i^k \rangle_{;k} = 0, \tag{2}$$

(i.e., covariance) in analogy to the case of the current in electrodynamics. The above authors also ascribed the subtracted terms to renormalization of the cosmological constant and of the gravitational coupling constant. I believe it is not the cosmological constant that is renormalized but rather the rest mass of the graviton. A cosmological constant would not result in terms that are non-covariant.

After the subtraction of the non-gauge invariant terms, we are left with a term of the form

$$\int \mathrm{d}^4 p / p^2 \, (p + k)^2, \tag{3}$$

that diverges logarithmically. This term can be made finite by subtracting a logarithmically divergent expression. This procedure corresponds in first order to a renormalization of nonlinear terms in the Lagrangian of the gravitational field. In the particular case of a virtual electromagnetic field these nonlinear terms are of the form:

$$C(-g)^{1/2} \, (-R^2 + 3R_{ik}R^{ik}). \tag{4}$$

This is the case to which we have specialized here. However, we were surprised to find that the divergent expression, if subtracted at $k = 0$ like the other two divergences, again gives rise to a logarithmic divergence—a kind of infrared divergence[3]. This infrared divergence cannot be removed in the conventional way. As all other attempts have failed, we now suggest that the divergent term be subtracted at a value $k \neq 0$. This means that in first approximation the constant C of the non-linear term cannot be renormalized to zero in the domain of long wavelengths (large distances from the source). If we proceed in the indicated way we are left with an undetermined constant b^2, the square of the wavelength at which we subtract. We believe that this constant shows up only in a contact term which is of no physical significance. The result of the subtraction is an expression of the form:

$$\frac{\varkappa}{15\,(2\pi)^4} \, (k^2)^2 \ln \left| \frac{k^2}{b^2} \right|. \tag{5}$$

This means, crudely speaking, that we have to replace the propagator $1/-k^2$ of the linearized theory by the propagator:

$$\frac{1}{-k^2 + \dfrac{\varkappa^2}{15\,(2\pi)^4}(k^2)^2 \ln\left|\dfrac{k^2}{b^2}\right|} \tag{6}$$

The Fourier transform of this propagator has, in the static case, the form:

$$\frac{4\pi}{r}\int_0^\infty \frac{\sin kr}{k\left[1 + \dfrac{\varkappa k}{15\,(2\pi)^4}\ln\left|\dfrac{k^2}{b^2}\right|\right]}\,\mathrm{d}k, \tag{6a}$$

where k is the absolute value of the 3-vector $\mathbf{k}$. We have up till now not succeeded in finding the expression in $\mathbf{x}$-space. We know of course that the Fourier transform of $1/k^2$ is $4\pi/r$, and we expect the Fourier transform of the above expression to be of the form

$$4\pi\left(\frac{1}{r} + \frac{\varkappa}{15\,(2\pi)^4}\frac{1}{r^n} + b^3\,\delta(r)\right). \tag{6b}$$

Assuming this to be the propagator of the linearized equations, one can obtain solutions of the non-linear equations by replacing $1/k^2$ by this propagator. This has not been done. One may be tempted to expect that with this propagator $g^{00} = 1/(1 - \alpha/r)$ may be replaced by

$$g^{00} = \frac{1}{1 - \dfrac{\alpha}{r} + \dfrac{\beta}{r^n}}$$

so that the Schwarzschild singularity disappears at least for certain values of α.

We acknowledge suggestions and discussions by B. Jouvet, *Collège de France, Paris*.

DISCUSSION

Deser

This problem is very intricate and contributions to it come from many different quarters. I would like to mention that Schwinger and one of his graduate students (A. Radkowski) have analyzed the particular case of a closed graviton loop. They managed after much work to get rid of the infrared divergence and remained with a term proportional to $1/r^3$.

Halpern

The term $1/r^3$ is probably the same as occurs here. There is however a fundamental difference in the infrared divergences, in the case of closed graviton loops, and the present case of closed photon loops. Weinberg[4] has shown that the conventional gravitational infrared divergences cancel in all orders irrespective of the spin and mass of the particles—thus this covers also the closed graviton loop, as the graviton is a particle of spin two. There is however definitely no such remedy in the case of the closed photon loop, so that we find here an undetermined constant. We may separate this term which multiplies a delta function as a contact interaction which is apparently without physical significance. The meaning of it is not known to us.

Weber

The attraction between the walls in Casimir's work may alternatively be described by the well known Van der Waals forces instead of by the zero point energy, and the zero point energy which is infinite can be removed by altering the order of the operators. I wonder therefore whether it contributes at all to the gravitational mass.

Halpern

The levels of the zero point energy are altered by the gravitational field and this difference is physically real and cannot be removed.

REFERENCES

1. H. B. G. Casimir and D. Polder, *Phys. Rev.*, **73**, 360 (1948).
2. B. S. de Witt, Thesis, Harvard University, 1952. B. S. de Witt and R. Utiyama, *J. Math. Phys.*, **3**, 608 (1962).
3. L. Halpern, *Arkiv. f. Fysik*, **34**, 539 (1967).
4. S. Weinberg, *Phys. Rev.*, **140**, B516 (1965).

PAPER 22

Neutrinos in Rainich geometry*

A. INOMATA

State University of New York at Albany, New York, U.S.A.

INTRODUCTION

The Rainich geometry is characterized by the Ricci tensor $R_{\mu\nu}$ satisfying the following algebraic conditions,

$$R = 0; \quad R_{\lambda\mu}R^{\lambda\nu} = \varrho^2\delta^\nu_\mu; \quad R_{00} \geqslant 0, \tag{1}$$

where

$$\varrho^2 = \tfrac{1}{4}R_{\varrho\sigma}R^{\varrho\sigma} \geqslant 0. \tag{2}$$

The geometry, if it is non-null, i.e., if $\varrho^2 > 0$, reproduces, in the sense of Rainich, Misner and Wheeler[1], the whole content of the coupled theory of Einstein's general relativity and Maxwell's electrodynamics under the differential condition,

$$\partial_\mu\alpha_\nu - \partial_\nu\alpha_\mu = 0, \tag{3}$$

where

$$\alpha_\mu = \tfrac{1}{4}e_{\mu\nu\varrho\sigma}R^\sigma_\lambda\nabla^\varrho R^{\nu\lambda}/\varrho^2. \tag{4}$$

In the case where $\varrho^2 = 0$, despite the failure of the differential condition (3), the geometry provides a field which is identifiable with the null electromagnetic field[2]. The null field so identified is, however, not unique. By a non-constant duality rotation, another null field can be found satisfying the Maxwell equations. An alternative interpretation seems to be admissible. The null geometry, subjected to the differential condition,

$$\nabla_\lambda R_{\mu\nu} = 0, \tag{5}$$

* Work supported in part by the Research Foundation of the *State University of New York*.

may describe a class of two-component neutrinos coupled with gravitation[3]. Ambiguity, thus, remains in understanding the implication for the null geometry.

The purpose of this paper is to point out that the Rainich–Misner–Wheeler scheme of the non-null geometry may also accommodate a class of four-component neutrinos.

RESTRICTED CLASS OF NEUTRINOS

What we refer to as the neutrino field is a c-number field ψ which obeys the Dirac equation defined in curved space-time[4],

$$\gamma^{\mu}\nabla_{\mu}\psi = 0. \tag{6}$$

The stress-energy tensor to serve as the source to the Einstein equation is

$$T_{\mu\nu} = \tfrac{1}{4}(\bar{\psi}\gamma_{\mu}\nabla_{\nu}\psi - \nabla_{\nu}\bar{\psi}\gamma_{\mu}\psi + \bar{\psi}\gamma_{\nu}\nabla_{\mu}\psi - \nabla_{\mu}\bar{\psi}\gamma_{\nu}\psi). \tag{7}$$

If use is made of the Pauli–Kofink identity,

$$(\bar{\psi}Q\gamma_{\lambda}\psi)\,\gamma^{\lambda}\psi = (\bar{\psi}Q\psi)\,\psi - (\bar{\psi}Q\gamma_5\psi)\,\gamma_5\psi, \tag{8}$$

where $Q = I, \gamma_{\mu}, \gamma_5, \gamma_{\mu}\gamma_5$, and $\gamma_{\mu\nu} = \frac{1}{2}(\gamma_{\mu}\gamma_{\nu} - \gamma_{\nu}\gamma_{\mu})$, then we can easily verify that the field ψ constrained by the nonlinear equation

$$\nabla_{\mu}\psi = (\bar{\psi}\gamma_{\lambda\mu}\psi)\,\gamma^{\lambda}\psi + (\bar{\psi}\gamma_{\lambda\mu}\gamma_5\psi)\,\gamma^{\lambda}\gamma_5\psi \tag{9}$$

is a solution of Eq. (6). Therefore, Eq. (9) defines a special class of neutrinos. Under the two-component condition $\psi = \gamma_5\psi$, the righthand side of Eq. (9) vanishes and the constrained field loses its physical significance. Thus, the neutrino field in question is necessarily four-component.

ALGEBRAIC CONDITIONS

Substitution of Eq. (9) reduces the stress-energy tensor (7) to the form

$$T_{\mu\nu} = \xi_{\lambda\mu}\xi^{\lambda}_{\nu} + {}^{*}\xi_{\lambda\mu}{}^{*}\xi^{\lambda}_{\nu} \tag{10}$$

where

$$\xi_{\mu\nu} = \mathrm{i}\,(\bar{\psi}\gamma_{\mu\nu}\psi); \quad {}^{*}\xi_{\mu\nu} = (\bar{\psi}\gamma_{\mu\nu}\gamma_5\psi). \tag{11}$$

By definition,

$$\mathrm{i}\,\gamma_{\mu\nu}\gamma_5 = \tfrac{1}{2}e_{\mu\nu\varrho\sigma}\gamma^{\varrho\sigma} \tag{12}$$

where $e_{\mu\nu\varrho\sigma}$ is the permutation tensor. Hence, it is seen that $*\xi_{\mu\nu}$ is the tensor dual to $\xi_{\mu\nu}$; namely,

$$*\xi_{\mu\nu} = \tfrac{1}{2}e_{\mu\nu\varrho\sigma}\xi^{\varrho\sigma}. \tag{13}$$

Since the form (10) of the stress-energy tensor coincides with that of the electromagnetic field, it is obvious that, through the Einstein equation,

$$R_{\mu\nu} - \tfrac{1}{2}g_{\mu\nu}R = \varkappa T_{\mu\nu}, \tag{14}$$

the first two conditions of (1) are fulfilled. In this case, we have

$$\varrho^2 = \varkappa^2 \, [(\bar{\psi}\gamma_\lambda\psi)\,(\bar{\psi}\gamma^\lambda\psi)]^2 \tag{15}$$

which is non-zero unless the field is two-component. The bilinear vector $(\bar{\psi}\gamma_\mu\psi)$ becomes null when the field ψ reduces to a two-component field. Whether or not the geometry is positive-definite, as is required by the third condition of (1), is not immediately clear. The lack of positive-definiteness of the energy is, in fact, a feature of all c-number fields with half-integral spin. In the reduced form (10), the energy component T_{00} may still assume values negative as well as positive. There is, nonetheless, a way of assuring positive-definiteness of R_{00}. Since the stress-energy tensor of this neutrino source is traceless, the Einstein equation (14) becomes

$$R_{\mu\nu} = \varkappa T_{\mu\nu}. \tag{16}$$

Therefore, geometry remains invariant under the change of the sign of T_{00} only if the sign of the coupling constant $\varkappa$ is covariant. The same assertion has been adopted in geometrization of two-component neutrinos[3]. Some generality of this assertion has also been discussed[5]. Positive-definiteness of R_{00} is indeed a necessary consequence of the integrability requirement of Eq. (9), detailed calculations of which will be given elsewhere.

DIFFERENTIAL CONDITIONS

The neutrino field constrained by Eq. (9) also satisfies the differential conditions (3). To see this, we make use of some of the following relations obtainable from Eq. (9);

$$\nabla_\mu\,(\bar{\psi}\psi) = 2\,(\bar{\psi}\gamma_5\psi)\,(\bar{\psi}\gamma_\mu\gamma_5\psi) \tag{17}$$

$$\nabla_\mu\,(\bar{\psi}\gamma_5\psi) = 2\,(\bar{\psi}\psi)\,(\bar{\psi}\gamma_\mu\gamma_5\psi) \tag{18}$$

$$\nabla_\mu\,(\bar{\psi}\gamma_\nu\psi) = 2\,(\bar{\psi}\psi)\,(\bar{\psi}\gamma_{\nu\mu}\psi) + 2\,(\bar{\psi}\gamma_5\psi)\,(\bar{\psi}\gamma_{\nu\mu}\gamma_5\psi) \tag{19}$$

$$\nabla_\mu\,(\bar{\psi}\gamma_\nu\gamma_5\psi) = 0. \tag{20}$$

After straightforward calculations, we obtain

$$e_{\mu\nu\varrho\sigma}T^{\sigma}_{\lambda}\nabla^{\varrho}T^{\nu\lambda} = 8\,[(\bar{\psi}\gamma_{\lambda}\psi)\,(\bar{\psi}\gamma^{\lambda}\psi)]^2 \cdot (\bar{\psi}\gamma_{\mu}\gamma_5\psi). \tag{21}$$

Because of Eq. (16), all R's in Eq. (4) can directly be replaced by the stress-energy tensor. Substituting, then, Eq. (15) and (21), together with Eq. (16), into Eq. (4), we arrive at the simple result,

$$\alpha_{\mu} = 2\mathrm{i}\,(\bar{\psi}\gamma_{\mu}\gamma_5\psi). \tag{22}$$

Now, Eq. (20) guarantees that the curl of the Rainich vector (4) vanishes

$$\partial_{\mu}\alpha_{\nu} - \partial_{\nu}\alpha_{\mu} = 0. \tag{23}$$

DUALITY COMPLEXION

The stress-energy (10) is invariant under the duality rotation,

$$\mathrm{e}^{*\alpha}\xi_{\mu\nu} = \xi_{\mu\nu}\cos\alpha + {}^*\xi_{\mu\nu}\sin\alpha \tag{24a}$$

$$\mathrm{e}^{-*\alpha}\xi_{\mu\nu} = -\xi_{\mu\nu}\sin\alpha + {}^*\xi_{\mu\nu}\cos\alpha, \tag{24b}$$

where α is a real function of space-time. If the antisymmetric field $\xi_{\mu\nu}$ satisfies the Maxwell equations, $\nabla^{\nu}\xi_{\mu\nu} = 0$, then one can define the concept of the duality complexion by identifying the gradient of α with the Rainich vector (22). From Eq. (9), however, follows equation for $\xi_{\mu\nu}$,

$$\nabla^{\nu}\xi_{\mu\nu} + 2\mathrm{i}\,{}^*\xi_{\mu\nu}\,(\bar{\psi}\gamma^{\nu}\gamma_5\psi) = 0. \tag{25}$$

Inasmuch as the field ψ is four-component, the second term of Eq. (25) remains nonvanishing. Such identification of α is, therefore, not apparent.

While the algebraic conditions (1) work to confine the stress-energy tensor to the form (10), the role of the differential condition (3) is to endow the source field with the character of the Maxwell field. In truth, the condition (3) requires only the vanishing of the divergence of the stress-energy tensor,

$$\nabla^{\nu}T_{\mu\nu} = 0, \tag{26}$$

and, in fact, this is satisfied by the field $\xi_{\mu\nu}$. As the stress-energy tensor (10) is invariant under the duality rotation (24), so is Eq. (26). Therefore there is a way to transform the second term away from Eq. (25).

Let us introduce an antisymmetric tensor field $f_{\mu\nu}$ by a duality rotation,

$$f_{\mu\nu} = e^{*\alpha}\xi_{\mu\nu}. \tag{27}$$

Then, we have

$$e^{-*\alpha}\,\nabla^{\nu} f_{\mu\nu} = \nabla^{\nu}\xi_{\mu\nu} + {}^{*}\xi_{\mu\nu}\partial^{\nu}\alpha. \tag{28}$$

Obviously the field $f_{\mu\nu}$ obeys the Maxwell equations,

$$\nabla_{\nu} f^{\mu\nu} = 0, \tag{29}$$

if the duality rotation angle α is so chosen that

$$\partial_{\mu}\alpha = 2\mathrm{i}\,(\bar{\psi}\gamma_{\mu}\gamma_{5}\psi) = \alpha_{\mu}. \tag{30}$$

This implies that the angle α is the duality complexion of $f_{\mu\nu}$ with reference to $\xi_{\mu\nu}$. In other words, the field $\xi_{\nu\nu}$ plays a role of an extremal field for the Maxwell field. On the other hand, the Maxwell field may be chosen as an extremal field for the neutrino field. The relative complexion is determined by

$$\alpha = \int \alpha_{\mu}\,\mathrm{d}x + \alpha_{0} \tag{31}$$

up to an additive constant.

CHIRALITY COMPLEXION

The duality rotation (24) is actually generated by the chirality transformation,

$$\psi' = \mathrm{e}^{i\beta\gamma_{5}}\psi, \tag{32}$$

with

$$\alpha = 2\beta. \tag{33}$$

The gauge function β determined by Eq. (31) and (33) defines the relative chirality complexion.

From Eq. (32), we obtain

$$e^{-\beta\gamma_{5}}\,\nabla_{\mu}\psi' = \nabla_{\mu}\psi + \mathrm{i}\,\partial_{\mu}\beta\gamma_{5}\psi, \tag{34}$$

which is equivalent to the expression (28). Thus, if the neutrino field is constrained by Eq. (9), the Dirac field corresponding to the Maxwell field (27) must be subjected to the equation,

$$\nabla_{\mu}\psi = (\bar{\psi}\gamma_{\lambda\mu}\psi)\,\gamma^{\lambda}\psi + (\bar{\psi}\gamma_{\lambda\mu}\gamma_{5}\psi)\,\gamma^{\lambda}\gamma_{5}\psi - (\bar{\psi}\gamma_{\mu}\gamma_{5}\psi)\,\gamma_{5}\psi, \tag{35}$$

and hence it is a c-number solution of the Heisenberg equation,

$$\gamma^{\mu}\nabla_{\mu}\psi + (\bar{\psi}\gamma^{\mu}\gamma_{5}\psi)\,\gamma_{\mu}\gamma_{5}\psi = 0. \tag{36}$$

As is mentioned earlier, the neutrino field obeying Eq. (9) must necessarily be four-component in order to retain its physical significance. In contrast, in the two-component limit, the Heisenberg equation (36) in the c-number theory reduces to the neutrino equation (6), and the constraint (35) yields the restricted neutrino equation previously proposed for the null geometry[6].

REFERENCES

1. G. Y. Rainich, *Trans. Am. Math. Soc.*, **27**, 106 (1925); C. W. Misner and J. A. Wheeler, *Ann. Phys.* (N.Y.), **2**, 525 (1957).
2. A. Peres, *Phys. Rev.*, **118**, 1105 (1960); *Ann. Phys.* (N.Y.), **14**, 419 (1961).
3. A. Inomata and W. A. McKinley, *Phys. Rev.*, **140**, B1467 (1965).
4. The same notations as those in ref. 3 are used here.
5. A. Inomata and D. Peak, *Nuovo Cimento* (to be published).
6. Eq. (24) in ref. [3].

PAPER 23

Approximate solutions of Einstein's equations

RICHARD KERNER
Institut Henri Poincaré, Paris, France

I

It is well known that any Riemannian manifold of dimension n can be imbedded locally and isometrically into a Euclidean space of dimension $\frac{1}{2}n\,(n+1)$. This dimension can however be somewhat smaller if the Riemannian manifold in question has some particular symmetry. For example, the De Sitter space can be imbedded, even globally, into a 5-dimensional Euclidean space (instead of dimension 10), the Schwarzschild solution can be imbedded into a 6-dimensional Euclidean space, etc.

In particular any solution of the Einstein equations (being a 4-dimensional Riemannian manifold) can be regarded as a hypersurface of dimension 4, imbedded into an appropriate Euclidean space. All the information concerning the internal geometry of this hypersurface is then contained in the imbedding functions z_α, $\alpha = 1, 2, \ldots, N$, N being the dimension of the Euclidean space, $N \leqslant 10$. The internal metric of the imbedded manifold is given by the formula

$$g_{ij} = \eta^{\alpha\beta}\partial_i z_\alpha \partial_j z_\beta, \tag{1}$$

where

$$\eta^{\alpha\beta} = \mathrm{diag}\,(p+, q-), \quad p+q=N,\ p \geqslant 1, \quad q \geqslant 3 \tag{2}$$

is the metric of the Euclidean space of the imbedding.

Also all other geometrical quantities can be expressed by means of the derivatives of the imbedding functions $z_\alpha(x^i)$. In particular, the Einstein equations in vacuo can be expressed symbolically as

$$R_{jk}\,(\partial_i z_\alpha) = 0. \tag{3}$$

It is quite natural to suppose that a slight and appropriately smooth deformation of the imbedded manifold should not affect its internal properties very much. Thus in the case when the imbedded manifold is an Einstein space, i.e. when its internal metric satisfies Eq. (3), it is reasonable to assume that after a deformation of the imbedding functions characterized by a small parameter ε the equations $R_{jk} = 0(\varepsilon)$ will be fulfilled.

Henceforth we shall investigate the deformations of the imbedded manifold given by the variations of the imbedding functions instead of producing small disturbances of the metric itself in a form, say,

$$g_{ij} \to g_{ij} + \varepsilon \overset{(1)}{g}_{ij} + \varepsilon^2 \overset{(2)}{g}_{ij} + \cdots \tag{4}$$

expecting that such an approach should give us some more insight into the problem.

II

From now on we assume that an imbedding of some Riemannian manifold is given, satisfying the Einstein equations:

$$z_\alpha = z_\alpha(x^i), \quad \alpha = 1, 2, \ldots, N; \quad i, j = 0, 1, 2, 3. \tag{5}$$

Here z_α are the Cartesian coordinates in E^N, so that $\mathrm{d}s^2 = \eta^{\alpha\beta}\, \mathrm{d}z_\alpha\, \mathrm{d}z_\beta$, where $\eta^{\alpha\beta}$ is given by (2). On the other hand $\mathrm{d}s^2 = \eta^{\alpha\beta}\partial_j z_\alpha \partial_k z_\beta\, \mathrm{d}x^j\, \mathrm{d}x^k$, which gives us the expression (1) for g_{jk}. According to our assumptions, this g_{jk} satisfies Eq. (3): $R_{jk}(z_\alpha) = 0$.

Now let us produce a slight deformation of the imbedding:

$$z_\alpha(x^i) \to w_\alpha(x^i) = z_\alpha(x^i) + \varepsilon v_\alpha(x^i). \tag{6}$$

We can write symbolically

$$R_{jk}(w_\alpha) = R_{jk}(z_\alpha) + \varepsilon \overset{(1)}{R}_{jk}(z_\alpha, v_\beta) + \varepsilon^2 \overset{(2)}{R}_{jk}(z_\alpha, v_\beta) + \cdots \tag{7}$$

If the functions v_α with their first and second derivatives, satisfy the Lipschitz condition with a constant $K \ll \varepsilon^{-1}$, then we can write

$$R_{jk}(w_\alpha) = R_{jk}(z_\alpha) + O(\varepsilon) + O(\varepsilon^2) + \cdots \tag{8}$$

and our new manifold given by the imbedding functions w_α can be called an approximate solution of the Einstein equations.

So far everything has been trivial; the first non-trivial step is to require that

$$\overset{(1)}{R}_{jk}(z_\alpha, v_\beta) = 0. \tag{9}$$

Then the approximation will hold up to the second order in ε.

The existence of the solutions of (9) depends naturally on the choice of the functions z_α. As always in variational problems of this kind, we face the following three possibilities:

a) The functions $z_\alpha(x^i)$ are such that Eq. (9) cannot be solved in $v_\alpha(x^i)$ (except the most trivial solution which is always good, v_α = const, which corresponds to a simple translation in E^N). In this case we shall call the corresponding Einstein space an "isolated point" in the manifold of all Einstein spaces.

b) The functions $z_\alpha(x^i)$ are such that Eq. (9) are fulfilled identically for any set of functions $v_\alpha(x^i)$. In this case we shall call the corresponding Einstein space an "extremal point" in the manifold of all Einstein spaces.

c) When a) and b) are not true, there can still exist some explicit solutions of (9). In this case we shall say that there exist "extremal directions" in the space of functions $v_\alpha(x^i)$, which make the left hand side of (9) vanish.

It is also worthwhile investigating the problem of the convergence of the series (7) in the functional space to which the v_α functions belong. At least theoretically this might indicate a possibility of constructing some new exact solutions of the Einstein equations, starting from known ones.

III

Let us assume an imbedding of an Einstein space given by the set of functions $z_\alpha = z_\alpha(x^i)$. We define

$$\overset{(0)}{g}_{jk} = \eta^{\alpha\beta}\partial_j z_\alpha \partial_k z_\beta. \tag{1}$$

Now if we pass to the deformed manifold with the imbedding functions

$$w_\alpha(x^i) = z_\alpha(x^i) + \varepsilon v_\alpha(x^i), \tag{6}$$

then the corresponding metric is equal to

$$g_{jk} = \overset{(0)}{g}_{jk} + \varepsilon \overset{(1)}{g}_{jk} + \varepsilon^2 \overset{(2)}{g}_{jk}, \tag{10}$$

where

$$\overset{(1)}{g}_{jk} = \eta^{\alpha\beta}\,(\partial_j z_\alpha \partial_k v_\beta + \partial_j v_\alpha \partial_k z_\beta), \tag{10a}$$

$$\overset{(2)}{g}_{jk} = \eta^{\alpha\beta}\,(\partial_j v_\alpha \partial_k v_\beta). \tag{10b}$$

The contravariant metric tensor is also affected by the deformation; assuming the $\overset{(0)}{g}{}^{jk}$ known, we put

$$g^{jk} = \overset{(0)}{g}{}^{jk} + \varepsilon \overset{(1)}{g}{}^{jk} + \varepsilon^2 \overset{(2)}{g}{}^{jk}, \tag{11}$$

where

$$\overset{(1)}{g}{}^{il} = -\overset{(0)}{g}{}^{im}\overset{(0)}{g}{}^{ln}\overset{(1)}{g}_{mn}, \tag{11a}$$

and

$$\overset{(2)}{g}{}^{il} = \overset{(0)}{g}{}^{im}\overset{(0)}{g}{}^{ln}\overset{(0)}{g}{}^{jk}\overset{(1)}{g}_{jn}\overset{(1)}{g}_{km} - \overset{(0)}{g}{}^{im}\overset{(0)}{g}{}^{ln}\overset{(2)}{g}_{mn}. \tag{11b}$$

Then we have

$$g_{ij} g^{jk} = \delta_i^k + O(\varepsilon^3), \tag{12}$$

which is good enough for our purpose.

Now, developing the Christoffel symbols in the powers of the parameter ε,

$$\Gamma^i_{jk} = \overset{(0)}{\Gamma}{}^i_{jk} + \varepsilon \overset{(1)}{\Gamma}{}^i_{jk} + \varepsilon^2 \overset{(2)}{\Gamma}{}^i_{jk} + \cdots, \tag{13}$$

we obtain the following results:

$$\overset{(1)}{\Gamma}{}^i_{jk} = \eta^{\alpha\beta}\overset{(0)}{g}{}^{im}\partial_m z_\alpha \nabla_j(\partial_k v_\beta) + \eta^{\alpha\beta}\overset{(0)}{g}{}^{im}\partial_m v_\alpha \nabla_j(\partial_k z_\beta) \tag{14}$$

$$\overset{(2)}{\Gamma}{}^i_{jk} = \eta^{\alpha\beta}\overset{(0)}{g}{}^{im}\,(\partial_m v_\alpha \nabla_j(\partial_k v_\beta) - \overset{(0)}{g}{}^{ln}\overset{(1)}{g}_{mn}\,(\partial_l z_\alpha \partial^2_{jk} v_\beta + \partial_l v_\alpha \partial^2_{jk} z_\beta), \tag{15}$$

where ∇_j stands for the covariant differentiation with respect to the non-perturbed connection $\overset{(0)}{\Gamma}{}^i_{jk}$.

For the moment we restrict ourselves only to the first approximation, given by (9); now we can write this equation as

$$\begin{aligned}\overset{(1)}{R}_{jk} = {} & \eta^{\alpha\beta}\overset{(0)}{g}{}^{im}\,[(\nabla_i\nabla_m z_\alpha)(\nabla_j\nabla_k v_\beta) - (\nabla_j\nabla_m z_\alpha)(\nabla_i\nabla_k v_\beta)] \\ & + \eta^{\alpha\beta}\overset{(0)}{g}{}^{im}\,[(\nabla_i\nabla_m v_\alpha)(\nabla_j\nabla_k z_\beta) - (\nabla_j\nabla_m v_\alpha)(\nabla_i\nabla_k z_\beta)] \\ & + 2\eta^{\alpha\beta}\overset{(0)}{g}{}^{im}\overset{(0)}{R}{}^l_{ijk}\nabla_l v_\alpha \nabla_m z_\beta = 0.\end{aligned} \tag{16}$$

Here we have made use of the symmetry properties of the $\overset{(0)}{R}{}^l_{ijk}$ and replaced $\partial_j z_\alpha$ and $\partial_j v_\alpha$ by the covariant derivatives, since for any value of the index α the functions z_α and v_α can be regarded as scalars with respect to the coordinate transformations of the x^i.

Locally the derivatives of z and v-functions can be always replaced by some vector fields with vanishing rotation:

$$\nabla_j z_\alpha = Y_j^\alpha, \qquad \nabla_j v_\beta = X_j^\beta, \tag{17}$$

$$\nabla_j Y_k^\alpha - \nabla_k Y_j^\alpha = 0, \quad \nabla_j X_k^\alpha - \nabla_k X_j^\alpha = 0. \tag{18}$$

Then our problem can be formulated as follows: given N rotationless vector fields corresponding to the derivatives of the imbedding functions we want to find another set of N rotationless vector fields X_i^α satisfying the equations

$$\eta_{\alpha\beta} \overset{(0)}{g}{}^{im} [\nabla_i Y_m^\alpha \nabla_j X_k^\beta - \nabla_j Y_m^\alpha \nabla_i X_k^\beta + \nabla_i X_m^\alpha \nabla_j Y_k^\beta - \nabla_j X_m^\alpha \nabla_i Y_k^\alpha]$$
$$+ 2\eta_{\alpha\beta} \overset{(0)}{g}{}^{im} \overset{(0)}{R}{}^n_{ijk} Y_m^\alpha X_n^\beta = 0. \tag{19}$$

Having all these formulae at our disposal, we can now investigate the simplest possible case, when the original Einstein space is the Minkowskian space-time.

IV

It must be emphasized that, because of their tensorial character Eq. (9) do not depend on the choice of the coordinate system x_i. The sum on the right hand side of (7) is a tensor, and so is the first term; hence the sum of the remaining terms must also be a tensor. As the parameter ε is arbitrary, this can be possible only in the case when all the terms are tensors.

Keeping this fact in mind we can choose for the imbedding of the Minkowskian space-time the simplest coordinates possible:

$$z_1 = t, \quad z_2 = x, \quad z_3 = y, \quad z_4 = z, \quad z_5 = \ldots = z_N = 0. \tag{20}$$

It is easy to see then that $\overset{(1)}{R}_{jk} = 0$ for any set of functions v_α; in other words, following our classification, the Minkowskian space-time is an extremal one.

Thus we see that the interesting things begin here with the next approximation given by the equation

$$\overset{(2)}{R}_{jk} = 0. \tag{21}$$

It is easy to show in our coordinate system that this equation reduces to the following one:

$$\eta_{\alpha\beta} \overset{(0)}{g}{}^{im} (\nabla_i X_m^\alpha \nabla_j X_k^\beta - \nabla_j X_m^\alpha \nabla_i X_k^\beta) = 0 \tag{22}$$

where the summation over the indices α and β is only from 5 to N.

The following observations concerning this equation have to be pointed out: (a) we make use of the fact that $\overset{0}{R}{}^i_{jkl} = 0$ for the Minkowskian space-time, (b) the deformations can be divided into two groups: the ones tangent to the hyperplane $z_5 = \cdots = z_N = 0$, i.e. such that only v_1, v_2, v_3, v_4 are not zero, and the ones perpendicular to our hyperplane, i.e. such that only v_5 to v_N are not zero. But the first class of deformations keeps the hyperplane unchanged, and therefore just corresponds to the change of the x^i coordinate system; as such it is of no interest. We therefore restrict ourselves to the perpendicular deformations only, thereby obtaining Eq. (22). This situation is more general: from purely geometrical considerations we can see that for any hypersurface the only interesting deformations will be those perpendicular to the hypersurface, while the tangent ones will just correspond to a change of coordinates or to internal symmetries of the hypersurface.

If we study the deformations in one direction only, we get the equation for only one vector field:

$$\overset{(0)}{g}{}^{im}(\nabla_i X_m \nabla_j X_k - \nabla_j X_m \nabla_i X_k) = 0, \tag{23}$$

with the subsidiary condition

$$\nabla_j X_k - \nabla_k X_j = 0. \tag{18}$$

One particular solution to this is obtained by putting $X_j = \partial_j v$ and $v(x^i) = f(k_i x^i)$, where f is any real function and k_i any constant vector. The corresponding metric has the form

$$g_{jk} = \overset{(0)}{g}_{jk} + \varepsilon^2 \dot{f}^2 k_j k_k. \tag{24}$$

We note that the vector k_j need not be null; this corresponds to gravitational waves propagating with any speed. The condition $k_i k^i = 0$ can be imposed if we want the dual tensor to satisfy the exact equation (24) $g_{ij} g^{jk} = \delta^k_i$ instead of the approximate equation (12).

The same reasoning is valid for any value of the index α from 5 to N; so that we can generalize our plane-wave solution a bit, by putting $v_\alpha(x^i) = f_\alpha(k^{(\alpha)}_m x^m)$, $5 \leqslant \alpha \leqslant N$. The corresponding metric tensor will then be of the form:

$$g_{jk} = \overset{(0)}{g}_{jk} + \sum_{\alpha=5}^{N} \varepsilon^2 \dot{f}^2_\alpha(k^{(\alpha)}_m x^m)\, k^{(\alpha)}_j k^{(\alpha)}_k. \tag{25}$$

V

As a concluding remark we should like to say that this method of investigating approximate solutions of the Einstein equations is theoretical rather than practical. The most interesting results will be those giving the information about the structure of the manifold of all the solutions of Einstein's equations. It is quite easy to show that the Minkowskian space-time is the extremal one i.e. that it is a good background for the linear approximations in agreement with known results. It will be interesting to classify the known exact solutions in this way and to find the "isolated points".

The possibility of introducing a Banach or even Hilbert structure to the space of deforming functions v_α suggests that the manifold of all exact solutions of the Einstein equations could be given the structure of an infinitely-dimensional locally Banachian (resp. Hilbertian) differential manifold. Then the very strong theorems of modern functional analysis could be applied.

Finally, this approach is close to the tetrad approach, giving it a clear geometrical meaning: it is sufficient to regard the components of the tetrad as the derivatives of the imbedding and deforming functions; then the main difference is just that of dimension (N tetrads instead of 4).

We have given here only an outline of this method, illustrating it by the simplest application possible—the Minkowskian case. Investigations of the approximate solutions of the Einstein equations in the vicinity of other solutions (the Schwarzschild solution, the Weyl solution etc.) are in progress.

Acknowledgements

The author is highly indebted to Professors Mme Y. Choquet-Bruhat, A. Lichnerowicz, M. Flato and A. Papapetrou for many useful discussions concerning this paper.

PAPER 24

Third principle of relativity

G. KNAPECZ
Budapest, Hungary

ABSTRACT

The concept of the geometric object ("coplet") is generalized. The new objects obey the transitivity and the identity functional equations for infinitesimal general coordinate transformations. By the solution of these functional equations the existence of the new geometric objects is proved. It is shown that the spinors of the Poincaré group also belong to these objects. The new geometric objects may be applied in the description of nature if one requires the expression of natural laws to be covariant under infinitesimal general coordinate transformations.

1 INTRODUCTION

Under the third principle of relativity we understand the requirement that the description of nature and its phenomena should be covariant under infinitesimal general coordinate transformations. More precisely, this principle consists of two requirements: (I) the entities in terms of which the natural phenomena are to be described should be objects of the group of infinitesimal general coordinate transformations (for brevity the "Einstein group"), and (II) the expressions of natural laws should be concomitants of these objects.

The third principle is the weakened form of Einstein's general principle of relativity. It is not a new one, since it has often been applied already[1,2], but it is not exhausted sufficiently so far.

The aim of this paper is to show that the third principle involves some new, algebraic possibilities which may be useful in the general relativistic physics. These possibilities are related to the Einstein group.

2 A DISTINGUISHED PROPERTY OF THE EINSTEIN GROUP

It is a known fact that the Poincaré subgroup of the groupoid of all general coordinate transformations (for brevity "Einstein groupoid") has some representations which are not representations of the Einstein groupoid itself[3-7]. For example, the spinor representation breaks down at the transition from the Poincaré group to the Einstein groupoid. This "violation rule" appears to be valid also in the case of other subgroups and subgroupoids of the Einstein groupoid.

The violation rule has the consequence that the Einstein groupoid does not have some representations (e.g. the spinor representation) which would be needed in general relativistic physics. This deficiency of the Einstein groupoid thereby diminishes the applicability of the general principle of relativity.

In this situation it seems reasonable to investigate whether the violation rule is also valid at the transitions from the subgroups of the Einstein group to the Einstein group itself? Namely, if the violation does not occur then the third principle may be very useful.

As will be shown below the violation law is not valid in the case of the Einstein group and its subgroups: the Einstein group has the important property that *it has all representations of all its subgroups*. Particularly, since the infinitesimal Poincaré group has a spinor representation the Einstein group has it also.

3 THE ALGEBRAIC "CONSERVATION RULE"

Before giving the proof of the above statements it is necessary to mention that Physics is interested both in the representations of some transformation groups (groupoids), and in the vectors of the abstract carrier spaces of these representations, which sometimes are called "covariant multiplets", or briefly "*coplets*" of the group (groupoid) in question. For example, the theory of hadrons concetrates its attention on the representations of the unitary groups SU3, etc., while relativistic physics concentrates its attention on the coplets $g_{ik}(x)$, $\Gamma^i_{kl}(x)$, $A_i(x)$, etc. Since we are dealing with general relativity theory, we will concentrate our attention on the coplets of the Einstein group.

Definition A coplet of the Einstein group is a system of entities $\psi_A(x^k)$ $(A = 1, 2, \ldots, M)$ which in the case of infinitesimal general coordinate transformations

$$\bar{x}^k = x^k + \varepsilon^k(x^l), \quad \varepsilon^k(x) \ll 1, \tag{1}$$

or

$$x^k = \bar{x}^k - \bar{\varepsilon}^k(\bar{x}^l)), \quad \bar{\varepsilon}^k(\bar{x}) = \varepsilon^k(x), \tag{2}$$

has an *explicite* transformation law

$$\bar{\psi}_A(\bar{x}^k) = \Gamma_A\left[\psi_B\,(x(\bar{x})), \bar{x}^k, x^l(\bar{x}^k), \frac{\partial x^l}{\partial \bar{x}^k}, \ldots\right], \tag{3}$$

or

$$\bar{\psi}_A(\bar{x}^k) = L_A\left[\psi_B\,(\bar{x} - \bar{\varepsilon}), \bar{x}^k, \bar{\varepsilon}^l(\bar{x}^k), \frac{\partial \bar{\varepsilon}^l}{\partial \bar{x}^k}, \ldots\right], \tag{4}$$

which obey the infinitesimal transitivity functional equation

$$\begin{aligned}
&\Gamma_A\left\{\psi_C\left[\bar{\bar{x}}^k - \left(\bar{\varepsilon}^k + \bar{\bar{\varepsilon}}^k\right)\right], \bar{\bar{x}}^k, \bar{\bar{x}}^k - \left(\bar{\varepsilon}^k + \bar{\bar{\varepsilon}}^k\right), \delta^k_l - \frac{\partial\left(\bar{\varepsilon}^k + \bar{\bar{\varepsilon}}^k\right)}{\partial \bar{\bar{x}}^l}, \ldots\right\} \\
&= \Gamma_A\left\{\Gamma_B\left[\psi_C\,(\bar{x}^k - \bar{\varepsilon}^k), \bar{x}^k, \bar{x}^k - \bar{\varepsilon}^k, \delta^k_l - \frac{\partial \bar{\varepsilon}^k}{\partial \bar{x}^l}, \ldots\right],\right. \\
&\qquad \left.\bar{\bar{x}}^k, \bar{\bar{x}}^k - \bar{\bar{\varepsilon}}^k, \delta^k_l - \frac{\partial \bar{\bar{\varepsilon}}^k}{\partial \bar{\bar{x}}^l}, \ldots\right\},
\end{aligned} \tag{5}$$

as well as the identity functional equation

$$\Gamma_A\,[\psi_B(x^k), x^k, x^k, \delta^k_l, 0, \ldots] = \psi_A(x^k). \tag{6}$$

As will be shown below the transformation formula of spinors obeys Eq. (5) and (6). In addition, the transformation formulas of any coplet of the Einstein groupoid also obey these equations.

Proposition One particular solution of (5) and (6) is the transformation formula

$$\bar{\psi}_A(\bar{x}^k) = \psi_A\,(x^l(\bar{x}^k)) + K^{Ss}_{At}\,\frac{\partial \bar{\varepsilon}^t\,(\bar{x}^k)}{\partial \bar{x}^s}\,\psi_S\,(x^l(\bar{x}^k)), \tag{7}$$

where K^{Bb}_{Aa} are *arbitrary* constants. (The summation convention of labels is understood.)

Proof On the one hand one has

$$\bar{\bar{\psi}}_A \equiv \Gamma_A \left\{\psi_C \left(\bar{\bar{x}} - \bar{\varepsilon} - \bar{\bar{\varepsilon}}\right), \bar{\bar{x}}, \bar{\bar{x}} - \bar{\varepsilon} - \bar{\bar{\varepsilon}}, \frac{\partial \bar{\bar{x}} - \bar{\varepsilon} - \bar{\bar{\varepsilon}}}{\partial \bar{\bar{x}}}\right\}$$

$$= \psi_A + K_{At}^{Ss} \frac{\partial \bar{\varepsilon}^t + \bar{\bar{\varepsilon}}^t}{\partial \bar{\bar{x}}^s} \psi_S, \tag{8}$$

and on the other one

$$\bar{\bar{\psi}}_A \equiv \Gamma_A \left\{\Gamma_B \left[\psi_C (\bar{x} - \bar{\varepsilon}), \bar{x}, \bar{x} - \bar{\varepsilon}, \frac{\partial \bar{x} - \bar{\varepsilon}}{\partial \bar{x}}, \ldots\right], \bar{\bar{x}}, \bar{\bar{x}} - \bar{\bar{\varepsilon}}, \frac{\partial \bar{\bar{x}} - \bar{\bar{\varepsilon}}}{\partial \bar{\bar{x}}}, \ldots\right\}$$

$$= \bar{\psi}_A + K_{At}^{Ss} \frac{\partial \bar{\bar{\varepsilon}}^t}{\partial \bar{\bar{x}}^s} \bar{\psi}_S$$

$$= \psi_A + K_{Av}^{Tu} \frac{\partial \bar{\varepsilon}^v}{\partial \bar{x}^u} \psi_T + K_{At}^{Ss} \frac{\partial \bar{\bar{\varepsilon}}^t}{\partial \bar{\bar{x}}^s} \left(\psi_S + K_{Sz}^{Vw} \frac{\partial \bar{\varepsilon}^z}{\partial \bar{x}^w} \psi_V\right)$$

$$= \psi_A + K_{At}^{Ss} \left(\frac{\partial \bar{\varepsilon}^t}{\partial \bar{x}^s} + \frac{\partial \bar{\bar{\varepsilon}}^t}{\partial \bar{\bar{x}}^s}\right) \psi_S. \tag{9}$$

The formulas (8) and (9) coincide because

$$\frac{\partial \bar{\varepsilon}}{\partial \bar{x}} \cdot \frac{\partial \bar{\bar{\varepsilon}}}{\partial \bar{\bar{x}}} = 0 \quad \text{and} \quad \frac{\partial \bar{\varepsilon}}{\partial \bar{x}} = \frac{\partial \bar{\varepsilon}}{\partial \bar{\bar{x}}} \tag{10}$$

up to the necessary order. QED.

Considering that in the case of the Einstein group the constants K_{Bb}^{Aa} are arbitrary, while in the case of its subgroups the corresponding constants have either special, or at most arbitrary values, the conservation rule is also proved.

We see that the Einstein group does not suffer from the deficiencies characterizing the Einstein groupoid. For example, at the transition, e.g., from the infinitesimal Poincaré group to the infinitesimal Einstein group, the coplets, as well as the representations, are conserved. This fact has the important consequence that there is no split between infinitesimal special relativity on the one hand, and infinitesimal general relativity theory on the other one.

4 EXAMPLES OF COPLETS OF THE EINSTEIN GROUP

Any coplet of the finite Einstein groupoid is a coplet of the infinitesimal Einstein group too (but the inverse is not true in general). For example, the infinitesimal transformation formula of a covector

$$\bar{V}_k(\bar{x}) = V_k(x) - \bar{\varepsilon}^s_{,k} V_s(x) \tag{11}$$

obeys Eqs. (5) and (6), with the transformation constants

$$K^{Bb}_{Aa} = -\delta^b_A \delta^B_a . \tag{12}$$

The transformation rule of a density of weight one

$$\bar{d}(\bar{x}) = d(x) - \bar{\varepsilon}^s_{,s}\, d(x) \tag{13}$$

also obeys (5) and (6), with the constants

$$K^{Bb}_{Aa} = -\delta^b_a \delta^B_A . \tag{14}$$

True coplets of the infinitesimal group are, e.g., the "deformators", the "rotators", and many other coplets. The transformation formula of a deformator is

$$\bar{D}_k(\bar{x}) = D_k(x) + \tfrac{1}{2}\,(\bar{\varepsilon}^s_{,k} + \bar{\varepsilon}^k_{,s})\, D_s(x), \quad k, s = 0, 1, 2, 3, \tag{15}$$

where the summation in s is understood, irrespective of where the label stands. This formula obeys (5) and (6). The transformation formula of a rotator reads

$$\bar{R}_k(\bar{x}) = R_k(x) - \tfrac{1}{2}\,(\bar{\varepsilon}^s_{,k} - \bar{\varepsilon}^k_{,s})\, R_s(x) \tag{16}$$

which also obeys (5) and (6).

The transformation laws (15) and (16) cannot be *explicitly* generalized to the Einstein *groupoid.*

5 WEYL SPINORS

As the last example, we show that the spinors are coplets of the Einstein group. Namely, the constants K^{Aa}_{Bb} can be fitted to the corresponding constants of the infinitesimal Poincaré group.

The transformation formula of Weyl spinors under an infinitesimal Poincaré transformation reads[8,9]

$$\bar{\psi}_\gamma(\bar{x}) = \exp\left\{-\frac{\mathrm{i}}{2}\,(\alpha_\varkappa + \mathrm{i}\beta_\varkappa)\,\sigma_{\varkappa\gamma\sigma}\right\}\psi_\sigma(x) \tag{17}$$

where $\sigma_\varkappa = (\sigma_1, \sigma_2, \sigma_3)$ are the Pauli matrices, $\gamma, \sigma = 1, 2$, and $\alpha_\varkappa$ and $\beta_\varkappa$ denote the infinitesimal rotation angles and dimensionless velocities of an infinitesimal Poincaré transformation, respectively.

This formula, as is well known, cannot be *explicitly* generalized to the *finite* general coordinate transformations, but it can be generalized to infinitesimal general coordinate transformations if the constants K_{Bb}^{Aa} are suitably fitted. A possible (but not the unique) generalization of (17) to the Einstein group is

$$\bar{\psi}_\gamma(\bar{x}) = \psi_\gamma(x) + \tfrac{1}{4}\{[\varepsilon^{0}_{,1} + \varepsilon^{1}_{,0} - \mathrm{i}\,(\varepsilon^{2}_{,3} + \varepsilon^{3}_{,2})]\,\sigma_{1\gamma\sigma}$$

$$+ [\varepsilon^{0}_{,2} + \varepsilon^{2}_{,0} - \mathrm{i}\,(\varepsilon^{3}_{,1} + \varepsilon^{1}_{,3})]\,\sigma_{2\gamma\sigma}$$

$$+ [\varepsilon^{0}_{,3} + \varepsilon^{3}_{,0} - \mathrm{i}\,(\varepsilon^{1}_{,2} + \varepsilon^{2}_{,1})]\,\sigma_{3\gamma\sigma}\}\,\psi_\sigma(x). \qquad (18)$$

We repeat that the formula (18) is a possible generalization. Whether it is the best one will not be discussed here.

The formula (18) obeys, Eq. (5) and (6), because the constants of the formula (18) are a special case of the constants of the general formula (7). Thus the Weyl spinors are simultaneously coplets of the infinitesimal Poincaré group and of the Einstein group, which is a subgroup of the Einstein groupoid.

6 DISCUSSION

As seen from (18) and (7), spinors and other infinitesimal coplets can be defined without any relation to the structure of space time. The use of tetrad fields, or the introduction of the metric field is not unavoidable. The spinors may be treated and applied as algebraic (group theoretic) entities, which are connected with the Einstein group, and not with any other group. Thus according to the present definition, the spinors are not gauge-, or bein-quantities.[10,11]

If, according to the third principle, one requires the expressions of natural laws to be invariant under infinitesimal general coordinate transformations, then the transformation law of infinitesimal coplets (7) can be applied, to give us additional tools for the description of natural phenomena, in general relativistic physics.

REFERENCES

1. P. G. Bergmann, *Phys. Rev.*, **75**, 680 (1949).
2. N. Rosen, *Ann. Phys.*, **38**, 170 (1966).
3. A. Nijenhuis, *Theory of Geometric Objects*, University of Amsterdam, Amsterdam, 1952.
4. J. Schouten, *Ricci-calculus*, Springer, Berlin, 1954.
5. J. Aczél and S. Golab, *Funktionalgleichungen der Theorie der geometrischen Objekte*, Warszawa, Panstwowe Wydawnictwo Naukowo (1960).
6. M. Kucharzewski and M. Kuczma, *Basic Concepts of the Theory of Geometric Objects*, Warszawa, Panstwowe Wydawnictwo Naukowe (1964).
7. G. Knapecz, *Acta Phys. Hung.*, **24**, 97 (1968).
8. F. Gursey, *Relativity, Groups, and Topology*, Gordon and Breach, New York, N.Y., 1964.
9. W. Heisenberg, *Introduction to the Unified Field Theory of Elementary Particles*, Interscience, New York, N.Y., 1966.
10. A. Peres, Suppl. to *Nuovo Cim.*, **24**, 389 (1962).
11. B. De Witt, *Relativity, Groups, and Topology*, Gordon and Breach, New York, N.Y., 1964.

PAPER 25

Cosmological models with non-zero pressure

J. KULHANEK and G. SZAMOSI
University of Windsor, Ontario, Canada

The aim of this note is to construct relativistic cosmological models which have great similarity to the Friedman model and which allow non-zero pressure.

The validity of Einstein's field equations is assumed. It is assumed furthermore that the three-space is isotropic and homogeneous i.e. the line element is, in co-moving coordinates, of the Robertson–Walker type:

$$ds^2 = (dx^0)^2 - \exp[\varphi(x^0) + f(r)] (dr^2 + r^2 (d\theta^2 + \sin^2\theta \, d\varphi^2)).$$

The homogeneously and isotropically distributed matter is considered as a mixture of a perfect gas and radiation (which is not necessarily the black-body type) called cosmological fluid.

The cosmological fluid is assumed to have the following properties:

1) It is a perfect fluid. The energy-momentum tensor is written as

$$T_{ab} = (e + p) u_a u_b - p g_{ab}$$

with $u^a u_a = 1$.

2) Its pressure is the function of the rest temperature (T) and the proper material density (ϱ) in the following form:

$$p = \tfrac{3}{2}\alpha\varrho T - \tfrac{1}{2}\varrho + e_0 \tag{1}$$

where α and e_0 are constants the meaning of which is given below. Eq. (1) is referred to as the thermal equation of state.

3) The proper internal energy (ε) is also a function of the proper material density and the rest temperature,

$$\varepsilon = \varepsilon(\varrho, T).$$

This function (sometimes called the caloric equation of state) is assumed to have the following form

$$\varepsilon = \frac{3}{2}\left(\frac{2e_0}{\varrho} + \alpha T - 1\right). \tag{2}$$

We note that ε is related to the proper energy density e and to the proper material density via the relation[1,2]

$$(T_{00} =)\, e = \varrho\,(1 + \varepsilon). \tag{3}$$

Simple algebra shows then that p can be written in the form

$$p = e_g \alpha T + \tfrac{1}{3} e_r, \tag{4}$$

where

$$e_g = \frac{e}{1+\varepsilon}; \quad e_r = \frac{e}{1+\varepsilon}\,\varepsilon = e - \varrho.$$

From here one obtains the physical meaning of α and e_0:

$$\alpha = \frac{1}{3T_{\text{cr}}}, \quad e_0 = p_{\text{cr}}.$$

For the expression of the change of the proper entropy we obtain

$$\mathrm{d}s = \frac{\alpha}{T}\,\mathrm{d}T + \frac{p + 3e_0}{T}\,\mathrm{d}\left(\frac{1}{\varrho}\right).$$

This expression is, in general, integrable if we assume an extra equation, preferably between p and e:

$$p = p(e). \tag{5}$$

This is the condition which will make the hydrodynamical problem determined in the sense that it characterizes the finite change of state of the system[3]. Eq. (5) is usually referred to as the equation of state. It can be given arbitrarily, as any form would be compatible with Eq. (1) and (2).

Using Einstein's equation and denoting the scale factor function by $U(x_0) = \mathrm{e}^{\varphi(x_0)/2}$ we obtain, after a routine calculation,

$$\left(\frac{\mathrm{d}U}{\mathrm{d}x^0}\right)^2 = \frac{B}{2} + \frac{1}{3}\,e_0 U^2 - \frac{1}{6}\,A U^{-4}, \tag{6}$$

where A is a constant which determines the critical volume of the cosmological fluid,

$$V_{\text{cr}}^2 = -\frac{A}{4e_0}.$$

This is meaningful if $A < 0$. The constant B is the curvature of the three space $B \gtreqless 0$. Eq. (6) is generally integrable in terms of Weierstrass elliptic functions, which shows that this model has Friedmanian nature[4].

From Einstein's equation we obtain the conservation law in the form:

$$\frac{\mathrm{d}}{\mathrm{d}t} U^3 (e - 3p - 4e_0)^{1/2} = 0. \tag{7}$$

Integrating and using Eq. (2) and (3) we obtain

$$\varrho (1 - 3\alpha T) = 4e_0 + AU^{-6}. \tag{8}$$

The physical meaning of the conserved quantity may be given after the relation (5) or other condition is given. For example, for the condition s = constant (isentropic change), (8) expresses the conservation of proper material density. In fact, it can be easily shown[2] that the condition s = constant is an example of a condition which make our problem determinate without using the explicit form of Eq. (5).

Combining Eq. (8) with Eqs. (3) and (7) one verifies that

$$e = e_0 - \tfrac{1}{2}AU^{-2}(t); \quad p = -e_0 - \tfrac{1}{2}AU^{-6}(t), \tag{9}$$

where $U(t)$ is a solution of Eq. (6). Once we know $U(t)$ as an explicit function of the time then, from Eq. (9) we have e and p as functions of the time. However, it follows from (8) that in order to know ϱ and T and hence e_r, e_g, p_r and p_g and S as functions of the time, Eq. (5) or some other condition is also necessary.

Another, perhaps a bit fanciful, property of the solution of Eq. (6) is obtained if one considers the wave equation in a homogenous isotropic universe. Schrödinger considered this problem[5] and solved the familiar wave equation

$$\frac{1}{\sqrt{-g}} \frac{\partial}{\partial x^a} \left(g^{ab} \sqrt{-g} \frac{\partial \psi}{\partial x^b} \right) + \varkappa^2 \psi = 0, \tag{10}$$

for the Robertson–Walker line element. For the time-dependent part $\chi(t)$ of the wave function Schrödinger obtained the equation

$$\frac{\mathrm{d}^2\chi}{\mathrm{d}\tau^2} + [n(n + 2)\, U^4 + \varkappa^2 U^6]\, \chi = 0, \tag{11}$$

where n is a non-negative integer and $n(n + 2)$ is a constant of separation. Here

$$\mathrm{d}\tau = U^{-3}\, \mathrm{d}t,$$

and Eq. (6) is re-written, using τ, as:

$$\frac{d^2U}{d\tau^2} + \left[\frac{1}{6} A - \frac{3}{2} BU^4 - \frac{4}{3} e_0 U^6\right] U = 0. \tag{12}$$

Eq. (12) and (6) are equivalent if

$$B = -\tfrac{2}{3}n(n+2), \quad A = 0, \quad e_0 = -\tfrac{3}{4}\varkappa^2.$$

From Eq. (11) and (12), one obtains

$$\frac{d\chi}{d\tau} U - \frac{dU}{d\tau} \chi = \text{const.}$$

Since the relation

$$\chi_1 \frac{d\chi_2}{d\tau} - \chi_2 \frac{d\chi_1}{d\tau} = \text{const.}$$

is valid for any two solutions of Eq. (11), it follows that U is also a solution of Eq. (11). Thus the time-dependent part of the wave function, with the above restrictions, gives us directly the scale factor function.

REFERENCES

1. A. Taub, *Ann. Math.*, **13**, 472 (1957).
2. A. Lichnerowicz, *Relativistic Hydrodynamics and Magnetohydrodynamics*, W. A. Benjamin, Inc., New York–Amsterdam, 1967.
3. J. L. Synge, *Proc. Lond. Math. Soc., Ser. II*, **43**, 376 (1937).
4. O. Heckmann and E. Schucking, *Relativistic Cosmology*, in: L. Witten (ed.), *Gravitation: An Introduction to Current Research*, Wiley, New York, N.Y., 1962.
5. E. Schrödinger, *Physica*, **6**, 899 (1939).

PAPER 26

Analogues of the Landau–Lifshitz pseudotensor

ELIHU LUBKIN*

University of Wisconsin-Milwaukee, U.S.A.

ABSTRACT

The identities of Landau and Lifshitz which express the ten mechanical conservation laws in Einstein's theory of gravitation are reviewed in slightly generalized form. The expressions obtained are used to construct ten conservation laws related to de-Sitter synmetry in the same way that the usual ten laws are related to Poincaré symmetry. The pattern of conservation laws associated with symmetry of a tail zone, or of a symmetric comparison background, is stated as a conjecture. Another illustration is given, in which an antisymmetric form is compared to the canonical antisymmetric form.

1 INTRODUCTION

In Sections 2–5, the expression of the ten mechanical conservation laws in the Einstein theory of gravitation given by the Landau–Lifshitz pseudotensor is reviewed, inasmuch as this used subsequently and also to note that slightly greater generality[1,2] is possible than may appear from a direct reading of the usual textual presentation[3]. The usual argument is applicable to expressions (4a), derived from any four-index symbol (1) antisymmetric in a pair of indices and symmetric in the exchange of that pair with the other pair. For example, one readily obtains ten divergenceless expressions from $F^{ij}F^{kl}$, where $F^{ij} = -F^{ji}$ is the electromagnetic field or any other antisymmetric symbol. The identities are not necessarily limited to energy or to general relativity.

* Supported in part by Graduate School research funds.

The identities themselves are presented in Section 2, their generality is emphasized in Section 3, then in Section 4, reduction of $(n-1)$-volume to $(n-2)$-bounding-surface integrals is done, free of the usual but inessential 3-index antisymmetry assumption (3). In Section 5, a remark of Wheeler's on the degeneration of totals to zero for examples with empty bounding surface, is called into question.

In Sections 6, 7 a conjecture on the relationship between symmetry of a homogeneous background and conservation laws is formulated in terms of a definition of "mediator". A mediator maps a contravariant vector field associated with symmetry to a conserved current density: $j = \tau\xi$ combines an infinitesimal automorphism[4,5] ξ of a homogeneous background G-structure[6] with a "mediator" τ which contains the elements of structure of an inhomogeneous G-structure, to produce a divergenceless current density j, which vanishes where the two structures agree. Section 8 rewords Section 2 in this mediator language; the background is flat Minkowski space. In Section 9, the L & L expressions are used again to construct a new example of a mediator for the case of a background with de-Sitter symmetry. Since the Poincaré group is not the symmetry group of the natural background space in a de-Sitter universe, it is more natural there to replace the usual ten mechanical conservation laws by ten conservation laws associated with the infinitesimal algebra of de-Sitter symmetry, and the new example of a mediator, constructed with the aid of the old one in a constructed space of one higher "radial" dimension, provides such a replacement. Appendix 2 draws attention to the interesting character of an exception to this work.

In Section 10, a third example of a mediator is given, constructed from the exterior derivative of an antisymmetric form, in which the symmetry is that of the canonical antisymmetric form (exterior derivative zero), this example being unrelated to the L & L expressions, except through the notion of mediator.

Unfortunately, the idea of a mediator for a general G-structure with homogeneous background does not itself generate the examples; it is therefore to be taken only as a suggestion for a unifying idea. The other main theme here is direct manipulation of the L & L expressions.

2 L & L IDENTITIES

Let h^{ijkm} have the symmetry properties

$$h^{ijkm} + h^{ijmk} = 0 \quad \text{and} \quad h^{ijkm} = h^{kmij}; \tag{1a,b}$$

these imply

$$h^{ijkm} + h^{jikm} = 0. \tag{1c}$$

The usual L & L expressions

$$h^{ijkm} = h^{ik}h^{jm} - h^{im}h^{jk}, \quad \text{for} \quad h^{ik} = h^{ki}, \tag{2}$$

as well as the Riemann tensor R_{ijkm}, also have the further property that antisymmetrizing on 3 index positions (any 3), yields 0; equivalently, that

$$h^{ijkm} + h^{imjk} + h^{ikmj} = 0, \tag{3}$$

but (2, 3) will not be used.

Define

$$\tau^{ik} = h^{ijkm}_{\ ,j,m}, \tag{4a}$$

where the commas denote literal differentiation ($_{,j} = \partial/\partial x^j$) with respect to a list of coördinates x^j, and

$$M^{ijk} = x^i\tau^{jk} - x^j\tau^{ik}. \tag{4b}$$

Then

$$\tau^{ik}_{\ ,k} = 0 \tag{5a}$$

because $h^{ijkm}_{\ ,k,j,m}$ is at once symmetric and antisymmetric in k, m; also,

$$\tau^{ik} = \tau^{ki}; \tag{5b}$$

and in immediate consequence of (5a, b),

$$M^{ijk}_{\ ,k} = 0. \tag{5c}$$

If h^{ik} is the contravariant density associated with the signatured-Riemannian metric in physical 4-space, then (5a, c) represent the laws of conservation of energy, momentum, angular momentum, and center of mass in a null-momentum system, as given by L & L. Thus, τ^{ik} includes both the "source" energy-momentum-stress and the "gravitational" energy-momentum-stress. Separation of τ^{ik} into such parts plays no role here, except in the following remark. Although use of any symmetric 2-indexed symbol in (2) yields the same identities[7], the choice of the contravariant density by L & L was no doubt motivated by the desire that the gravitational part vanish at a center of geodesic coordinates[8].

3 GENERALITY OF THE IDENTITIES

It is easy to write 4-index symbols with symmetry properties (1), other than the example used by L & L. Thus, given any such h^{ijkm}, fh^{ijkm} is another, for any function f. The Riemann tensor provides another example. Any antisymmetric symbol F^{ij} produces yet another example, by $h^{ijkm} = F^{ij}F^{km}$. Thus, we have ten conservation laws for the Maxwell field, independent of Maxwell's equations; i.e., for any electric charges and magnetic monopoles subject to the identities of current conservation. These ten conserved quantities involve second derivatives of the F^{ij}, unlike the usual ten mechanical conserved quantities of the Maxwell theory, which involve the fields themselves.

Any conserved current $j^i, j^i_{,i} = 0$, is locally of form $j^i = F^{ij}_{,j}, F^{ij} + F^{ji} = 0$. However, F^{ij} is ambiguous up to an additive $F^{ijk}_{,k}$, with F^{ijk} totally antisymmetric. This ambiguity prevents $h^{ijkm} = F^{ij}F^{km}$ from leading to ten conserved currents *unambiguously* associated with any one conserved current; furthermore, the F^{ij} are deduced from j^i only by integration. For example, for the free Maxwell field, $j^i = 0$, so $F^{ij} = 0$ and $\tau^{ij} = 0$ is one possible choice; another free Maxwell field in fact gives nonzero answers for the ten currents.

4 SURFACE INTEGRALS

Since L & L involve (3) in their reduction of volume integrals to surface integrals, because (3) follows in the L & L example (2), but because (3) does not follow from (1) alone (e.g., $h^{ijkm} = F^{ij}F^{km}$, $F^{ij} + F^{ji} = 0$, is a counterexample), expressions (6a, b) for surface integrals are given here, valid independently of (3); and also, incidentally, for general dimension n, but with language which best fits $n = 4$. Instead of giving the totals $\int d^{n-1}x\, \tau^{i0}$ and $\int d^{n-1}x\, M^{ij0}$ over an $(n-1)$-volume, as $(n-2)$-surface integrals, it will be noted that the integrands are literal divergences "$\nabla\cdot$", not involving the 0 coordinate:

$$\tau^{i0} \equiv h^{ij0m}_{,j,m} = \nabla \cdot h^{ij0\rightarrow}_{,j}. \tag{6a}$$

The index 0 doesn't figure because of the antisymmetry of h in its last 2 indices, and the arrow signifies contraction with $\nabla\cdot$ and omission of the index value 0. By including the x^i, x^j in (4b) inside one (4a) differentiation, compensating therefor, then noting that the compensation terms $-h^{jmki}_{,m} + h^{imkj}_{,m}$ cancel for $m = 0$, one finds that

$$M^{ij0} = \nabla \cdot (x^i h^{jm0\rightarrow}_{,m} - x^j h^{im0\rightarrow}_{,m} - h^{j\rightarrow 0i} + h^{i\rightarrow 0j}). \tag{6b}$$

By using (1) and (3), the last two terms may be written more compactly, as $-h^{0\rightarrow ji}$; it is only in this detail that the use of (3) enters in L & L.

Method of squashing function

The fact that fh^{ijkm} also gives $\frac{1}{2}n\,(n-1)$ conserved quantities for free f allows employment of a "squashing" function f which is 1 over an $(n-1)$-volume at definite early x^0, which descends to 0 in a thin shell, is 0 outside the shell, is 0 where one monitors a flux, and which is 0 everywhere at some late x^0, to give 0 for the early total, and therefore to give minus the shell contribution for the unmodified volume contribution. This method of obtaining surface integrals by squashing is made unneccessary by the explicit expressions (6), but the squashing approach shows that if by introducing inner-boundary $(n-2)$-surfaces, topological complications can be cut away, and if by inserting false simple regions in place of the complications the manifold can be made amenable to converting the integral of an $(n-1)$-divergence into an $(n-2)$-surface integral over several pieces of surface, then the $(n-1)$-volume integrals of the expressions (6) may be turned into surface integrals.

5 "0 = 0"

Any of the $(n-2)$-surface integrals corresponding to (6) may be interpreted as a total quantity corresponding to an $(n-1)$-volume proceeding from either side of the surface. Change of the orientation of surface elements entails a minus sign between these two interpretations: "side of surface" refers to surface orientation rather than to the question of whether the surface really divides the volume into two disjoint pieces, which it may. In any case, the total on one side plus that on the other is zero.

It has been suggested by Wheeler[9] that essentially this argument reduces such conservation laws to the triviality "0 = 0", for all-universe totals. However, the weakness of the derivation in depending on the literal coördinates of one coördinate patch, a point which should be emphasized in order generally to obviate misunderstanding, should be noted here. Except for the 1-patch limitation, the "0-total" argument applies whether the $(n-1)$-space is compact or not, in particular, to topologically Euclidean 3-space (in which case a 2-sphere does separate an "inside" from an "outside"). In this case, and for ordinary gravitation, the balancing "outside" totals reside mainly on a "surface at infinity"; a truncation of a weak gravitational field to the Minkowski flat metric over a distant shell would produce balancing totals in that shell. These balancing totals, rather than being an embarrassment, are merely an expression of the squashing-function method of computing totals as surface integrals. In a compact case singularly covered by one coördinate patch, a *non*zero contribution to the zero total could be expected from an infinitesimal surface around the coördinate singularity.

6 RÔLE OF A SYMMETRIC BACKGROUND

It is of course puzzling to have conservation laws associated with the Poincaré group where the geometry is unrelated to the Poincaré group. Yet the Poincaré group has entered implicitly in the Cartesian operation of literal differentiation. The manipulations have blindly found the Poincaré symmetry of an implicit flat comparison metric. It is when (2) is built from a metric actually close in some" tail zone" to such a comparison metric that the physically interesting picture of nearly "vacuous" regions (small values of τ^{ij}), might supplement the formal use of a comparison metric to produce a kind of scattering theory, with conservation laws for totals which are mainly sums over vacuously separated zones of "matter".

Below, the rôle of a Poincaré-group-symmetric background metric is generalized. What is actually accomplished is the presentation of two other examples, but first a general pattern which fits all three examples is stated in the language of *G*-structures.

7 DEFINITION OF "MEDIATOR"

Consider two *G*-structures, *A* and *B*, with the same (n by n) real matrix group *G*, and over the same n-real-dimensional manifold *M*. These *G*-structures are to coincide except in one coördinate patch *P*, and they are to coincide in *P* in a designated subset, but may differ over the rest of *P*. The region of designated coincidence, including the complement of *P*, will be called "tail", or "region of asymptotic symmetry". *A*, the structure of interest, is to be compared with *B*, the background structure. *B* is given to be highly symmetrical; possibly a homogeneous space. Its group of infinitesimal automorphisms is to be nontrivial, and possibly to be locally transitive. A single coördinate system over *M* is to be used in common for *A* and *B*, and the coincidence of *A* and *B* in the tail is to be expressed by choice of common frames there, or common metrics, etc., in case it is not necessary to introduce frames explicitly. (It will usually be desirable to consider the structure of primary interest and B as being "close" rather than coincident in the tail, but conceptual clarity is attained here by replacing this structure of primary interest by a varied structure "A" which in fact coincides with B in a designated tail zone, and then to discuss separately the relaxation of this variation.)

Infinitesimal automorphisms are given by a real vector space of contravariant vector fields $\xi^i(x^1, \ldots, x^n)$; the field of scalars is the real constant scalar functions: The characterizing property of an automorphism is that its action as a motion on a field of frames for the structure B should produce a field of frames[4,5]; this is familiar as the notion of "Killing vector" in Riemannian geometry.

A conservation law will be simply a divergenceless contravariant vector density field, or "current".

A *mediator* is to be a mixed tensor density field $\tau^i_j(x)$ such that

$$j^i(x) = \tau^i_j(x)\,\xi^j(x), \quad \text{or} \quad j = \tau\xi, \tag{7}$$

is to be a conserved current, for each infinitesimal B-automorphism ξ, i.e.,

$$j^i_{,i} = 0 \quad \text{for all } B\text{-automorphisms } \xi, \tag{8}$$

and such that:

$$j \text{ is to vanish in the tail.} \tag{9}$$

The structure A enters only in that τ, the mediator, is to depend on A. The particular formulae given for this dependence in the three examples are not, however, deduced from a more general expression. Therefore the idea that there exists a natural notion of mediator to relate deviation from a symmetric background to currents in all cases appears in this paper as only an illustrated conjecture.

If τ is found so that (8) is satisfied but not (9), then the obvious renormalization

$$\tau'(A) = \tau(A) - \tau(B) \tag{10}$$

engenders a mediator which also satisfies (9).

In spite of the ease with which (9) may therefore be established, (9) is important, because it makes the conserved "charge" corresponding to a current into a sum of separate pieces if the tail's complement at a typical fixed "time" is a union of two or more separate "islands". Thus, a sum of subtotals is required to be constant, and one has a rudimentary scattering theory. Of course, here "time" is one of $x^1, \ldots, x^n$, say, x^1, and the total charge at time x^1 is $\int j^1 (x^1, \ldots, x^n)\, dx^2 \cdot \cdots \cdot dx^n$ = sum over separate non-tail "islands". The interest in choosing a time dimension here depends on the possibility of seeing, either exactly or approximately, such island structure for some G-structures "A".

The term "mediator" is chosen because τ mediates between tail-symmetry and conserved currents, and also between the inhomogeneous A and the (relatively) homogeneous B. The "pseudotensor" character of the L & L "mediator" is ascribed to a variety of ways one A can be regarded as being obtained by distortion of one homogeneous B.

8 THE L&L PSEUDOTENSOR AS A MEDIATOR

It will be convenient for the sequel to put

$$h^{ij} = \sigma^w g^{ij}, \tag{11a}$$

$$\sigma = (\det (g_{..}))^{1/2} \tag{11b}$$

(it will be immaterial whether σ or $|\sigma|$ appears); where w is an arbitrary weight[8,7], rather than 1. Here, A, B are real $O(n_+, n_-)$-structures, $g_{ij} = g_{ji}$ is the real covariant metric tensor of A, g^{ij} is the reciprocal contravariant tensor, and η_{ij}, a diagonal matrix of n_+ ones and n_- minus ones, is the metric tensor of B; the dimension of the manifold M is $n = n_+ + n_-$. When it becomes convenient, purely imaginary coördinates will be used for dimensions of negative signature, and η_{ij} will be replaced by δ_{ij}. The mediator here is

$$\tau^i_k = \tau^{ij}\eta_{jk},$$

or

$$j^i = (\tau\xi)^i = \tau^{ik}\xi_k, \quad \text{where} \quad \xi_k = \eta_{km}\xi^m. \tag{12}$$

In fact, introduction of $\xi^k = \delta^k_m$ (mth "translation") reproduces the energy and momentum densities τ^{im}, and the introduction of $\xi^k = x^m\eta^{pk} - x^p\eta^{mk}$ ((m, p)th "rotation") reproduces the angular momentum and center-of-mass-related quantities M^{imp}, Eq. (4b).

9 TAILS OF UNIFORM CURVATURE

B is here represented as the $(n - 1 = n_+ + n_- - 1)$-dimensional sub-metric-space

$$\eta_{ij}x^i x^j = R^2 \quad (\text{“}R\text{-sphere”}) \tag{13}$$

of (n_+, n_-)-flat metric space,

$$\mathrm{d}s^2 = \eta_{ij}\,\mathrm{d}x^i\,\mathrm{d}x^j; \tag{14}$$

or

$$\delta_{ij}x^i x^j = R^2, \tag{13a}$$

$$\mathrm{d}s^2 = \delta_{ij}\,\mathrm{d}x^i\,\mathrm{d}x^j \tag{14a}$$

with the last n_- coördinates pure imaginary. This "positive definite notation" will be utilized to cover implicitly the indefinite cases. A is an arbitrary metric on the manifold of the sphere (13), which in the case of nonempty tail must, for the sake of continuity, agree in signature with B.

The plan is to utilize the L & L pseudotensor in n-space in comparing extensions of A and B to n-space, in order to arrive at a mediator which in fact refers only to the given $(n - 1)$-dimensional problem. Satisfactory estensions are found, with the x^i essentially homogeneous coördinates. The metric $\mathrm{d}s^2 = g_{ij}(x^1, \ldots, x^n)\,\mathrm{d}x^i\,\mathrm{d}x^j$, with the functions g_{ij} homogeneous of degree Δ, i.e.,

$$g_{ij}(\lambda x^1, \ldots, \lambda x^n) = \lambda^\Delta g_{ij}(x^1, \ldots, x^n), \quad \text{all} \quad \lambda > 0, \tag{15}$$

restricts to an arbitrary metric on the R-sphere, and conversely is a uniquely defined extension of an arbitrary R-sphere metric if Δ is given. Condition (15) is imposed only along half-rays, $\lambda > 0$, to avoid trouble at the origin, and so that the conditions $g_{ij}(-x) = (-)^\Delta g_{ij}(x)$ which would restrict the form of the metric on the R-sphere, are not entailed.

Homogeneous degree Δ for the $g_{ij}(x)$, "H.D. $(g_{ij}) = \Delta$", obviously entails definite homogeneous degrees for the functions (11b, 11a, 2, 4a, 4b), namely,

$$\text{H.D.}\,(\sigma) = \tfrac{1}{2}n\Delta, \tag{16a}$$

$$\text{H.D.}\,(h^{ij} = \sigma^w g^{ij}) = (\tfrac{1}{2}wn - 1)\,\Delta, \tag{16b}$$

$$\text{H.D.}\,(h^{ijkm}) = (wn - 2)\,\Delta, \tag{16c}$$

$$\text{H.D.}\,(\tau^{ik}) = (wn - 2)\,\Delta - 2; \tag{16d}$$

$$\text{H.D.}\,(x^i\tau^{jk} - x^j\tau^{ik}) = (wn - 2)\,\Delta - 1. \tag{16e}$$

The L & L pseudotensor of course provides divergenceless currents in n-space: $\sum_{i=1}^{n} (\partial j^i/\partial x^i) = 0, j = \tau\xi$, for each of the n translations and $\frac{1}{2}n(n-1)$ rotations ξ. The plan is to obtain $(n-1)$-space divergenceless currents from the rotations, where the $(n-1)$-space is the R-sphere. If $\theta^1, \ldots, \theta^{n-1}$, "angles", are coördinates of half-rays $\{\lambda x; \lambda > 0\}$, and $R = \theta^n = (\delta_{ij}x^i x^j)^{1/2}$, and

$$J^i(\theta) = \det\left(\frac{\partial x^{\cdot}}{\partial \theta^{\cdot}}\right)\frac{\partial \theta^i}{\partial x^j}\, j^j(x) \tag{17}$$

give the components of the current in question, transformed as a contravariant vector density[8] to the "polar coördinates", then $\sum_1^n (\partial J^i/\partial \theta^i) = 0$, or

$$\sum_1^{n-1} \frac{\partial J^i}{\partial \theta^i} + \frac{\partial J^R}{\partial R} = 0, \tag{18}$$

where $J^R = J^n$. Equation (18) will become an R-sphere conservation law if

$$\frac{\partial J^R}{\partial R} = 0 \tag{19}$$

at radius R.

It is shown in Appendix 1 that condition (19) can be achieved for all rotational currents, by appropriate choice of Δ; also, though incidentally, for all R, if $w \neq 2/n$.

Since by (19), J^R does not figure, (7) is summed for only $i < n$. Since the rotational $\xi^{\cdot}$ lie tangent to spheres, the index $j = n$ does not figure in (7). Here (7) is understood as written in polar coördinates, being obtained from the L & L quantities in n-space Cartesian coördinates by transforming τ as a mixed density and ξ as a contravariant vector field, so as to validate (17), where $\tau\xi$ has in fact been transformed as a contravariant vector density (in spite of the fact that it looks like it should have weight $2w$, and other "pseudo" properties). Condition (9) is attained by the device (10), B having the uniform curvature appropriate to a particular value of R. Thus, a mediator τ' acting in the $(n-1)$-space and vanishing in R-uniformly-curved tail, has been given.

10 AN EXAMPLE INVOLVING ANTISYMMETRIC FORMS

A, B are real symplectic structures, with γ_{ij} the antisymmetric form of A, η_{ij} that of B; the η_{ij} constant, γ and η nonsingular; say η in the canonical form of matrices $\begin{pmatrix} 0 & 1 \\ -1 & 0 \end{pmatrix}$ repeated along successive (2 by 2) diagonal blocs, 0's elsewhere. The infinitesimal automorphisms of B are the canonical motions $\xi^i = \eta^{ij}H_{,j}$, H any smooth function, where $\eta_{ij}\eta^{kj} = \delta^k_i$. If a mediator τ^i_j is given, $\tau^i_j\xi^j = \tau^{ij}\xi_j$, where $\tau^{ij} = \tau^i_k\eta^{kj}$ and $\xi_j = \eta_{mj}\xi^m$; the infinitesimal B-automorphisms are given by $\xi_j = H_{,j}$.

The requirement (8) becomes $(\tau^{ij}H_{,j})_{,i} = 0$, all H, or

$$\tau^{ij}H_{,i,j} + \tau^{ij}_{,i}H_{,j} = 0, \quad \text{all } H.$$

A large class of τ^{ij} satisfying this is the class of antisymmetric contravariant divergenceless densities: $\tau^{ij} + \tau^{ji} = 0$, and $\tau^{ij}_{,i} = 0$. Such an object is given most generally locally as the divergence of a third-rank antisymmetric density, $\tau^{ij} = \tau^{ijk}_{,k}$, where a change of τ^{ijk} by adding a divergence of a fourth-rank antisymmetric density leaves τ^{ij} the same, and is the most general change of τ^{ijk} leaving τ^{ij} the same.

To have a mediator of this form, it is necessary to give a formula by which τ^{ij} (or τ^{ijk}) is given from γ^{ij}.

$$\gamma_{ijk} \equiv \gamma_{ij,k} + \gamma_{jk,i} + \gamma_{ki,j}$$

is a covariant antisymmetric tensor. Let the notation for $\gamma_{..}$'s reciprocal be given by $\gamma_{ik}\gamma^{jk} = \delta^j_i$. Let $\gamma^{ijk} \equiv \gamma^{ia}\gamma^{jb}\gamma^{kc}\gamma_{abc}$; this is a contravariant antisymmetric tensor. Let $\sigma \equiv [\det(\gamma_{..})]^{1/2}$; σ is a scalar density. $\tau^{ijk} \equiv \sigma\gamma^{ijk}$ is, then, a contravariant antisymmetric third-rank density dependent on A;

$$\tau^{ij} = (\sigma\gamma^{ijk})_{,k}$$

gives a mediator, $\tau^k_i = \tau^{ij}\eta_{kj}$. Since $\gamma_{ijk} = 0$ for B and tail, no step (10) of renormalization is needed here.

Aside from the somewhat vague notion of mediator shared in common by this symplectic example and the examples related to the L & L expressions, the only other common property of the τ^i_j's is that they involve, respectively, second derivatives of the γ_{ij} and g_{ij}.

APPENDIX 1 ATTAINING A VANISHING $\partial J^R/\partial R$

From (17), $J^R = \det(\partial x^{\cdot}/\partial\theta^{\cdot})\,(\partial R/\partial x^j)\,j^j(x)$. R-dependence is to be made explicit. $(\partial R/\partial x^j) = (x^j/R)$ is of H.D. 0. For $i < n$, H.D. $(R\,(\partial\theta^i/\partial x^j)) = 0$; i.e., these depend only on $\theta^1, \ldots, \theta^{n-1}$, not on R. Therefore,

$$\text{H.D.}\left(\det\begin{bmatrix} R\,\dfrac{\partial\theta^1}{\partial x^1} & \cdots & R\,\dfrac{\partial\theta^1}{\partial x^n} \\ \vdots & & \\ R\,\dfrac{\partial\theta^{n-1}}{\partial x^1} & \cdots & R\,\dfrac{\partial\theta^{n-1}}{\partial x^n} \\ \dfrac{\partial(R=\theta^n)}{\partial x^1} & \cdots & \dfrac{\partial(R=\theta^n)}{\partial x^n}\end{bmatrix}\right) = 0;$$

$$\text{H.D.}\,(\det(\partial\theta^{\cdot}/\partial x^{\cdot})) = 1 - n;$$

$$\text{H.D.}\,(\det(\partial x^{\cdot}/\partial\theta^{\cdot})) = n - 1;$$

$$\det(\partial x^{\cdot}/\partial\theta^{\cdot}) = R^{n-1}$$

times a function of

$$\theta^1, \ldots, \theta^{n-1} = R^{n-1} f(\theta^1, \ldots, \theta^{n-1}).$$

$$\partial\,(\det(\partial x^{\cdot}/\partial\theta^{\cdot}))/\partial R = (n-1)\,R^{n-2} f = (n-1)\,R^{-1}\det(\partial x^{\cdot}/\partial\theta^{\cdot}).$$

$$J^R = R^{-1}\det(\partial x^{\cdot}/\partial\theta^{\cdot})\,x^i j^i;$$

$$\frac{\partial J^R}{\partial R} = -R^{-2}\det\left(\frac{\partial x^{\cdot}}{\partial\theta^{\cdot}}\right)x^i j^i + R^{-1}(n-1)\,R^{-1}\det\left(\frac{\partial x^{\cdot}}{\partial\theta^{\cdot}}\right)x^i j^i$$
$$+ R^{-1}\det\left(\frac{\partial x^{\cdot}}{\partial\theta^{\cdot}}\right)\frac{\partial(x^i j^i)}{\partial R}.$$

The desideratum (19) is therefore

$$0 \overset{?}{=} (n-2)\,x^i j^i + R\,\partial\,(x^i j^i)/\partial R;$$

equivalently, that

$$\text{H.D.}\,(x^i j^i) \overset{?}{=} 2 - n.$$

The homogeneous degree of the rotational currents, (16e), was $(wn - 2)\Delta - 1$; therefore H.D. $(x^i j^i) = (wn - 2)\Delta$.

Hence, (19) is attained (for $w \neq 2/n$; see Appendix 2) by setting $2 - n = (wn - 2)\Delta$; i.e., by choosing Δ to be

$$\Delta = \frac{2 - n}{wn - 2}. \tag{20}$$

For the weight $w = 1$ of L & L, $\Delta = -1$.

APPENDIX 2 PARENTHETICAL NOTE ON WEIGHTS

$\det(\sigma^w g^{\cdot\cdot}) \equiv 1$ if and only if $w = 2/n$. For all other w, the values of the $\frac{1}{2}n(n + 1)$ distinct g_{ij}'s can be recovered from $h^{\cdot\cdot}$, but for $w = 2/n$, only their $\frac{1}{2}n(n + 1) - 1$ ratios can be recovered. Therefore, $h^{\cdot\cdot}$ for $w = 2/n$ is a convenient algebraic object for representing the conformal geometry corresponding to the signatured-metric geometry; in G-structure language, the group G is enlarged from "congruences" to "similarities": arbitrary uniform stretches of the old frames are admitted as frames. It is for this interesting case that the construction of Section 9 breaks down: the denominator of (20), in Appendix 1, is zero when $w = 2/n$.

Acknowledgments

That L & L's h^{ijkm} (2) satisfy (3) was taught me by Ivor Robinson. The 2-sided nature of surface conservation laws which relate conservation to Killing vector fields was used by Jeffrey M. Cohen in a conversation. I was made aware of the reference to J. N. Goldberg by Dr. Robinson and an anonymous referee. The references to Dorn and Schild and to McCrea and Synge are due to kind communications from Fred Cooperstock and J. L. Synge.

REFERENCES

1. J. McCrea and J. L. Synge, *Quart. Appl. Math.*, **24**, 355 (1967), Section 5, also present the more general form.
2. W. S. Dorn and A. Schild, *Quart. Appl. Math.*, **14**, 210 (1956). This is really a converse, showing that h^{ijkm} exist, given τ^{ij}, divergenceless and symmetric, but of course immediately suggests the easier direct theorem. h^{kmij} arrived at in this converse way are now usually called "superpotentials"; Dorn and Schild use the term "stress functions".

3. L. D. LANDAU and E. M. LIFSHITZ, *Classical Theory of Fields*, 2nd ed., Pergamon, Oxford, 1962, hereafter referred to as L & L.
4. E. A. RUH, *On the Automorphism Group of a G-Structure*, Thesis, Brown University, June 1964. Also refs. 6, 5.
5. E. LUBKIN, *Ann. Phys.*, (N.Y.) **32**, 218 (1965).
6. S. STERNBERG, *Lectures on Differential Geometry*, Prentice-Hall, Englewood Cliffs, 1964; A. LICHNEROWICZ, *Géometrie des Groupes de Transformations*, Dunod, Paris, 1958; S. KOBAYASHI and K. NOMIZU, *Foundations of Differential Geometry*, Interscience, New York, N.Y., 1963; also ref. 5.
7. J. N. GOLDBERG, *Phys. Rev.*, **111**, 315 (1958). GOLDBERG therein refers to R. SACHS.
8. O. VEBLEN, *Invariants of Quadratic Differential Forms*, Cambridge University Press, Cambridge, 1952.
9. J. A. WHEELER, *Geometrodynamics*, Academic Press, New York, N.Y., 1962, p. 64.

PAPER 27

Classification of space-time curvature tensor

R. S. MISHRA

Banaras Hindu University, India

1 INTRODUCTION

In this article I shall develop the classification theory of the curvature tensor $'\mathscr{K}$ of the space-time V_4, which is the operational space of the general theory of relativity. Some of the work in this direction has been done jointly by the late Professor Hlavatý and me and some by himself[1]. In the last paper, which he was working on before his death in January 1969, he left a note that in case of his death or incapacitation I would be willing to complete his unfinished manuscript. I am working on that unfinished paper and the purpose of this discussion is to give the technique used in that paper and the papers which preceded it.

Let $\mathscr{G}$ denote the gravitational metric tensor field of the relativistic Riemann space-time V_4 with the signature $(+ + + -)$, ${}^{-1}\mathscr{G}$ its inverse, $\mathscr{K}$ the corresponding curvature tensor,

$$'\mathscr{R}(X, Y) = (C_1^1\mathscr{K})(X, Y), \tag{1.1a}$$

$$\mathscr{G}(\mathscr{R}(X), Y) = {}'\mathscr{R}(X, Y), \tag{1.1b}$$

the Ricci tensor and

$$r = C_1^1\mathscr{R}, \tag{1.1c}$$

the scalar curvature of V_4. The Einstein field equations will be assumed in a certain general form[2]

$$\mathscr{R}(X) - \tfrac{1}{2}rX = mu(X)\,U + \mathscr{F}(X) + pX. \tag{1.2}$$

Maxwell's equations are

$$(\operatorname{div} k)(X) = -\eta X, \tag{1.3a}$$

$$(D'_X k)(Y, Z) + (D'_Y k)(Z, X) + (D'_Z k)(X, Y) = 0. \tag{1.3b}$$

Here $m > 0$ is the mass at rest of the test particle, U the time-like velocity unit vector:

$$-1 = \mathscr{G}(U, U) = u(U),$$

$\mathscr{F}$ the energy momentum tensor of the electromagnetic field, $'k(X, Y) = -'k(Y, X)$ (which is supposed not to be a null field), and $p > 0$ is the pressure. $\mathscr{F}$ may be put explicitly in the form

$$\mathscr{F}(X) = \frac{M^2}{2}\{-\underset{11}{U}\overset{11}{u}(X) - \underset{21}{U}\overset{21}{u}(X) + \underset{31}{U}\overset{31}{u}(X) - \underset{41}{U}\overset{41}{u}(X)\}, \tag{1.4}$$

where $\underset{11}{U}, \ldots, \underset{41}{U}$ are mutually perpendicular unit vectors ($\underset{41}{U}$ being time-like) and $\overset{11}{u}, \ldots, \overset{41}{u}$ their inverses. The vectors $\underset{11}{U}, \underset{21}{U}$ (the vectors $\underset{31}{U}, \underset{41}{U}$) are known up to ordinary (up to Lorentz) rotations. M^2 is a real function related to the eigenvalues of k (with respect to $\mathscr{G}$). Unless stated explicitly otherwise, we always assume $m \neq 0$.

Assuming $\mathscr{R}$ known, one has to express

$$\mu, p, \mathscr{F}, \eta \quad \text{etc.}$$

as algebraic concomitants of $\mathscr{R}$. In order to solve this problem one has to analyze every possible algebraic structure of $\mathscr{R}$ from the point of view of geometry as well as physics. It turns out that not every geometrically possible solution is admissible from the physical point of view and only four of algebraically different $\mathscr{R}$ admit solutions. If one requires the continuity equation to be satisfied, this number still diminishes.

If the Ricci tensor does (does not) satisfy Einstein equations, we call it admissible (inadmissible). Its Weierstrass characteristic Ch_2 (with regard to $\mathscr{G}$) will be termed admissible (inadmissible) Ch_2-characteristic. If the curvature tensor has the property that its Ricci tensor is admissible (inadmissible) it will be termed admissible (inadmissible). Its Weierstrass characteristic Ch_4 (with respect to a tensor ε proportional to the Levi–Civita tensor) will be termed admissible (inadmissible).

The problems to be solved are as follows:

1) To find all admissible Ricci tensors and all their admissible curvature tensors. This is equivalent to "To find all admissible Ch_2 and Ch_4 characteristics".

2) To express m, U, $\underset{31}{U}$, $\underset{41}{U}$, M^2, p as algebraic concomitants of $\mathscr{R}$.

The importance of the problem lies in geometrodynamics. As far as we know the first of the problem has been partly solved by Petrov[3], Newmann[4], Shell[5], Struik[6], Churchill[7] and Ruse[8].

2 PREREQUISITES

The continuity equation is

$$\operatorname{div}(mU) = 0. \tag{2.1a}$$

This is compatible with the field equations if the pressure p is constant along the trajectory of the test particle:

$$U \cdot p = 0. \tag{2.1b}$$

The orthogonal components of the covariant curvature tensor are given by

$$(('\mathscr{K})) = \left(\begin{pmatrix} P & Q \\ Q^t & R \end{pmatrix}\right). \tag{2.2a}$$

Here

$$((P)) = \left(\begin{pmatrix} 1414 & 1424 & 1434 \\ 2414 & 2424 & 2434 \\ 3414 & 3424 & 3434 \end{pmatrix}\right), \quad ((Q)) = \left(\begin{pmatrix} 1423 & 1431 & 1412 \\ 2423 & 2431 & 2412 \\ 3423 & 3431 & 3412 \end{pmatrix}\right),$$

$$((R)) = \left(\begin{pmatrix} 2323 & 2331 & 2312 \\ 3123 & 3131 & 3112 \\ 1223 & 1231 & 1212 \end{pmatrix}\right). \tag{2.2b}$$

(Q^t is the transpose of Q and $ijkl$ stands for $'\mathscr{K}(e_i, e_j, e_k, e_l)$.

The orthogonal components of the metric tensor $\mathscr{G}$ are given by

$$((G)) = \operatorname{Diag}(1, 1, 1, -1). \tag{2.3}$$

From (2.2) and (2.3) we get the orthogonal components of the Ricci tensor as

$$(('R)) = ((1bc1 + 2bc2 + 3bc3 - 4bc4)). \tag{2.4}$$

This shows that diagonal terms in $((Q))$ have no importance for $((\overset{1}{R}))$. Therefore the diagonal terms will be symbolized by asterisks.

The field Eqs. (1.2) are equivalent to

$$\mathscr{R}(X) - \varrho_1 X = \mathscr{T}(X) \tag{2.5a}$$

where

$$\mathscr{T}(X) = mu\,(X)\,U + M^2\,\{\overset{31}{u}(X)\,\underset{31}{U} - \overset{41}{u}(X)\,\underset{41}{U}\} \tag{2.5b}$$

and

$$\varrho_1 = \tfrac{1}{2}r - \tfrac{1}{2}M^2 + p. \tag{2.5c}$$

The tensors $\mathscr{R}$ and $\mathscr{T}$ have the same eigenvectors (with respect to I_4) and their eigenvalues ϱ_a and τ_a are related by

$$\tau_a = \varrho_a - \varrho_1. \tag{2.6}$$

Both tensors $\mathscr{R}$ and $\mathscr{T}$ have the same Ch_2-characteristics.

From (2.5), it is clear that ϱ_1 is an eigenvalue of $\mathscr{R}$. The rank of $\mathscr{T}$ is $\leqslant 3$. We say a Ch_2-characteristic of $\mathscr{R}$ is admissible if Eqs. (2.1) and (2.5) admit at least one solution

$$m,\ U,\ M^2,\ p,\ \overset{31}{u},\ \overset{41}{u}.$$

3 ADMISSIBLE Ch_2-CHARACTERISTICS

Let $r = 3$. Then the eigenvalue ϱ_1 of $\mathscr{R}$ is single. Therefore the following six characteristics are possible

$$[1111],\ [1\,(11)\,1],\ [1\,(111)],\ [112],\ [1\,(112)],\ [13].$$

We will first consider the characteristic $[1111]$. The eigenvalues of $\mathscr{R}$ then are $\varrho_1, \varrho_2, \varrho_3, \varrho_4$. If we calculate these, using (2.5b) and

$$u(X) = (\varkappa\,\underset{21}{U} + \beta\,\underset{31}{U} + \gamma\,\underset{41}{U})\,(X),$$

we find

$$M^2 = \tau_2 = \varrho_2 - \varrho_1, \quad m = \tau_2 - (\tau_3 + \tau_4) = \varrho_1 + \varrho_2 - (\varrho_3 + \varrho_4)$$

$$2p = -(\varrho_3 + \varrho_4), \quad \varkappa^2 = \frac{\tau_3\tau_4}{\tau_2\,(\tau_2 - \tau_3 - \tau_4)}. \tag{3.1}$$

M^2 and $\varkappa^2$ are positive. If we require mass and pressure also to be positive then

$$\tau_2 > 0, \quad \tau_2 > \tau_3 + \tau_4, \quad \tau_3\tau_4 > 0, \quad \varrho_3 + \varrho_4 < 0.$$

These inequalities are possible if we have either the arrangement,

$$\varrho_2 > \varrho_1 > \varrho_3 > \varrho_4; \tag{3.2a}$$

or the arrangement

$$\varrho_2 > \varrho_4 > \varrho_3 > \varrho_1, \tag{3.2b$_1$}$$

$$\varrho_3 + \varrho_4 < \varrho_1 + \varrho_2. \tag{3.2b$_2$}$$

Therefore *the necessary and sufficient conditions for the Ch_2-characteristic* [1111] *to hold are*

1) *The eigenvalues satisfy (3.2a) or (3.2b) while in both cases*

$$\tau_3\tau_4 > 0. \tag{3.3}$$

2) *The eigenvalue ϱ_4 leads to a time-like eigenvector.*

3) $\varrho_3 + \varrho_4 < 0$.

4) *The continuity equation holds.*

When these conditions are satisfied M^2, m, p etc. are given by (3.1). $\underset{31}{U}$, $\underset{41}{U}$ can also be calculated.

We can similarly prove that *the necessary and sufficient conditions that the Ch_2-characteristic* [112] *is admissible are*

1) *Assuming*

$$\tau_1 > 0, \tag{3.4a}$$

$$\tau_2 > 2\tau, \tag{3.4b}$$

$$\varrho < 0. \tag{3.4c}$$

2) *The continuity equation holds.*

When these conditions are satisfied

$$M^2 = \tau_2, \quad m = \tau_2 - 2\tau, \quad p = -\varrho. \tag{3.5}$$

U, $\underset{31}{U}$, $\underset{41}{U}$ can also be calculated.

It can be proved that all other characteristics are inadmissible when $r = 3$.

When $r = 2$, the only admissible characteristic is [(11) 11]. *Necessary and sufficient conditions for the Ch_2-characteristic* [(11) 11] *to be admissible are*

1) *The following inequalities hold*

$$\text{(a) } \varrho_3 > \varrho_1\ (= \varrho_2), \quad \text{(b) } \varrho_4 < \varrho_3, \quad \text{(c) } \varrho_1 + \varrho_4 < 0. \tag{3.6}$$

2) *The eigenvalue* ϱ_4 *leads to a time-like eigenvector.*

3) $$\underset{41}{V} \cdot (\varrho_1 + \varrho_4) = 0.$$

If these conditions are satisfied

$$M^2 = \tau_3, \quad m = (\varrho_3 - \varrho_4), \quad 2p = -(\varrho_1 + \varrho_4), \quad U = \underset{41}{V}.$$

We can also obtain $\underset{31}{U}$ and $\underset{41}{U}$.

When $r = 1$, the only admissible characteristic is [(111) 1]. *Necessary and sufficient conditions for the* Ch_2*-characteristic* [(111) 1] *to be admissible are*

1) *The single eigenvalue leads to a time-like eigenvector.*

2) *The following inequalities hold:*

$$\varrho_4 < \varrho_1 \, (= \varrho_2 = \varrho_3), \tag{3.7a$_1$}$$

$$\varrho_1 + \varrho_4 < 0. \tag{3.7a$_2$}$$

3) *The following condition holds:*

$$\underset{41}{V} \cdot (\varrho_1 + \varrho_4) = 0. \tag{3.7b}$$

When these conditions are satisfied

$$m = -\tau_4, \quad 2p = -(\varrho_1 + \varrho_4), \quad U = \underset{41}{V}. \tag{3.8}$$

The only admissible complex Ch_2-characteristic is $[11\bar{1}\bar{1}]$. This happens when $r = 3$.

Necessary and sufficient conditions for Ch_2*-characteristic* $[11\bar{1}\bar{1}]$ *to be admissible are*

1) *Either*

$$\varrho_2 > \varrho_1 > \varrho^*; \tag{3.9a}$$

or

$$\varrho_2 > \varrho^* > \varrho_1, \tag{3.9b$_1$}$$

$$\varrho_1 + \varrho_2 > 2\varrho^*. \tag{3.9b$_2$}$$

In both cases the following inequalities hold

$$\tau_3 \tau_4 > 0, \tag{3.9c}$$

$$\varrho^* < 0. \tag{3.10}$$

2) *The equation of continuity holds.*

When these conditions are satisfied, the physical quantities are given by (3.1).

4 Ch_4-CHARACTERISTICS

We enumerate in this Section all eleven possible types of Ch_4-characteristics

$$[q\,r\,s\,\underbrace{1\,\cdots\,1}_{p\text{ times}}]$$

of the curvature tensor $'\mathscr{K}$ of our relativistic space time V_4.

	N	q	r	s	p	Basic	Derived
I	3	2	2	2	0	2	2
II	2	2	2	0	2	4	11
III	2	3	2	0	1	1	4
IV	2	3	3	0	0	2	1
V	2	4	2	0	0	1	1
VI	1	2	0	0	4	3	15
VII	1	3	0	0	3	2	7
VIII	1	4	0	0	2	2	3
IX	1	5	0	0	1	1	1
X	1	6	0	0	0	1	0
XI	0	0	0	0	6	4	18
						23 (Total)	63

Here the first column denotes the type of Ch_4-characteristic, the column headed by N indicates the total number of numerals q, r, s which are different from 0 and 1, the column headed by q indicates the numeral q. Similar significance have the columns r, s and p. The last but one column indicates the total number of basic real and complex Ch_4-characteristics of the type under consideration. The last column indicates the total number of derived real and complex Ch_4-characteristics. At least one of the numbers q, r, s, p is zero and must be disregarded in the Ch_4-characteristic. We have

$$q + r + s + p = 6.$$

Thus, for instance, the Ch_4-characteristic of the type V will be written as [42].

The eigenvalues of the curvature tensor will be denoted by

$$\alpha, \beta, \gamma, \lambda, \mu, \nu.$$

If one of these eigenvalues, say α, is double, we write α instead of β and similarly for higher multiplicity. The order of the eigenvalues is the same as the order of the numerals in the corresponding Ch_4-characteristic. Thus for instance, if we have the characteristic [321], the eigenvalues will be written as

$$\alpha, \alpha, \alpha, \lambda, \lambda, \nu.$$

If an eigenvalue say β is complex, and γ is its complex conjugate, we write

$$\beta = \beta_1 + i\beta_2, \quad \gamma = \beta_1 - i\beta_2.$$

We will put

$$((\mathscr{W}(\varrho))) = ((\mathscr{R} - \varrho\mathscr{G})). \tag{4.1a}$$

Thus, in particular

$$((\mathscr{W}(\varrho_1))) = (('\mathscr{T})). \tag{4.1b}$$

5 CANONICAL FORM OF $'K$

For the classification of $'K$, we will first obtain the canonical form of $'K$. We will first consider the real characteristic of the type I, namely [222]. This will serve as a model for other characteristics. Since the characteristic is [222], the canonical matrix of $^*\mathscr{R}$ is given by

$$((^*\mathscr{R})) = \begin{pmatrix} \alpha & 1 & \cdot & \cdot & \cdot & \cdot \\ \cdot & \alpha & \cdot & \cdot & \cdot & \cdot \\ \cdot & \cdot & \gamma & 1 & \cdot & \cdot \\ \cdot & \cdot & \cdot & \gamma & \cdot & \cdot \\ \cdot & \cdot & \cdot & \cdot & \mu & 1 \\ \cdot & \cdot & \cdot & \cdot & \cdot & \mu \end{pmatrix}. \tag{5.1a}$$

Since $C_1^1 {}^*\mathscr{R} = 0$, we have

$$2(\alpha + \beta + \gamma) = 0. \tag{5.1b}$$

We now construct the canonical matrix $'e$.

We have

$$'^*\mathscr{R}(X, Y) = 'e(X, {}^*\mathscr{R}(Y)) = 'e({}^*\mathscr{R}(X), Y) = '^*\mathscr{R}(Y, X). \tag{5.2}$$

Using (5.1a) in this equation, we obtain, in view of the requirement of symmetrization

$$(('e)) = \begin{pmatrix} \cdot & \varepsilon_1 & \cdot & \cdot & \cdot & \cdot \\ \varepsilon_1 & \cdot & \cdot & \cdot & \cdot & \cdot \\ \cdot & \cdot & \cdot & \varepsilon_2 & \cdot & \cdot \\ \cdot & \cdot & \varepsilon_2 & \cdot & \cdot & \cdot \\ \cdot & \cdot & \cdot & \cdot & \cdot & \varepsilon_3 \\ \cdot & \cdot & \cdot & \cdot & \varepsilon_3 & \cdot \end{pmatrix}, \quad \varepsilon_1, \varepsilon_2, \varepsilon_3 = \pm 1. \tag{5.3}$$

Without loss of generality, we may assume $\varepsilon_1, \varepsilon_2, \varepsilon_3 = 1$.

Let us now define a P-tetrahedron the edges of which satisfy the conditions

$$'\mathscr{E}(X, Y) = \varepsilon(F(X), F(Y)). \tag{5.4}$$

$$(('\mathscr{E})) = \begin{pmatrix} \cdot & \cdot & \cdot & 1 & \cdot & \cdot \\ \cdot & \cdot & \cdot & \cdot & 1 & \cdot \\ \cdot & \cdot & \cdot & \cdot & \cdot & 1 \\ 1 & \cdot & \cdot & \cdot & \cdot & \cdot \\ \cdot & 1 & \cdot & \cdot & \cdot & \cdot \\ \cdot & \cdot & 1 & \cdot & \cdot & \cdot \end{pmatrix}. \tag{5.5}$$

Then this P-tetrahedron will be termed a projective frame.

If

$$'e(X, Y) \overset{\text{def}}{=} '\mathscr{E}(V(X), V(Y)), \tag{5.6}$$

we can construct the appropriate frame $\underset{x}{V}$.

From (5.3), (5.5) and (5.6), we at once get

$$\underset{\varkappa}{V}^A = \delta_x^A, \quad \varkappa, x = 1, \text{I};\ 2, \text{IV};\ 3, \text{II};\ 4, \text{V};\ 5, \text{III};\ 6, \text{VI}.$$

From (5.1), (5.2) and (5.3) we at once get

$$(('\mathscr{R})) = \begin{pmatrix} \cdot & \alpha & \cdot & \cdot & \cdot & \cdot \\ \alpha & 1 & \cdot & \cdot & \cdot & \cdot \\ \cdot & \cdot & \cdot & \gamma & \cdot & \cdot \\ \cdot & \cdot & \gamma & 1 & \cdot & \cdot \\ \cdot & \cdot & \cdot & \cdot & \cdot & -(\alpha+\gamma) \\ \cdot & \cdot & \cdot & \cdot & -(\alpha+\gamma) & 1 \end{pmatrix}. \tag{5.7}$$

Hence if

$$'\mathscr{R}(X, Y) = '{}^*\mathscr{R}(V(X), V(Y)),$$

we obtain

$$((P)) = ((0)), \quad ((Q)) = \text{Diag}(\alpha, \gamma, -(\alpha+\gamma)), \quad ((R)) = \text{Diag}(1, 1, 1).$$

We have given above the treatment for the basic characteristic [222]. But attached with this basic characteristic there are two derived characteristics [(22) 2] and [(222)]. For the derived characteristics it can be proved that

$$((P)) = ((0)), \quad ((Q)) = \text{Diag}(\alpha, \alpha, -2\alpha), \quad ((R)) = \text{Diag}(1, 1, 1),$$

$$((P)) = ((0)), \quad ((Q)) = ((0)), \quad ((R)) = \text{Diag}(1, 1, 1).$$

We have considered above a real basic characteristic [222]. Now we will consider a complex basic characteristic $[22\bar{2}]$. For this case

$$((^*\mathcal{R})) = \begin{pmatrix} \alpha & 1 & . & . & . & . \\ . & \alpha & . & . & . & . \\ . & . & \gamma & 1 & . & . \\ . & . & . & \gamma & . & . \\ . & . & . & . & \bar{\gamma} & 1 \\ . & . & . & . & . & \bar{\gamma} \end{pmatrix}, \tag{5.8a}$$

whence

$$\alpha + 2\gamma_1 = 0. \tag{5.8b}$$

Requirement of symmetrization of $^*\mathcal{R}$ yields

$$(('e)) = \begin{pmatrix} . & 1 & . & . & . & . \\ 1 & . & . & . & . & . \\ . & . & . & i & . & . \\ . & . & i & . & . & . \\ . & . & . & . & . & -i \\ . & . & . & . & -i & . \end{pmatrix}. \tag{5.9}$$

The frame will be given by

$$\underset{1}{V}^A = \delta^A_{\text{III}}, \quad \underset{2}{V}^A = \delta^A_{\text{IV}}, \quad \sqrt{2}\,\underset{3}{V}^A = \delta^A_{\text{I}} + i\delta^A_{\text{II}}, \quad \sqrt{2}\,\underset{4}{V}^A = i\delta^A_{\text{IV}} + \delta^A_{\text{V}},$$
$$\sqrt{2}\,\underset{5}{V}^A = \delta^A_{\text{I}} - i\delta^A_{\text{II}}, \quad \sqrt{2}\,\underset{6}{V}^A = \delta^A_{\text{V}} - i\delta^A_{\text{IV}}.$$

$(('R))$ will then be given by

$$(('\mathcal{R})) = \begin{pmatrix} . & \alpha & . & . & . & . \\ \alpha & 1 & . & . & . & . \\ . & . & . & i\gamma & . & . \\ . & . & i\gamma & i & . & . \\ . & . & . & . & . & -i\bar{\gamma} \\ . & . & . & . & -i\bar{\gamma} & -i \end{pmatrix}.$$

Consequently in this case

$$((P)) = ((0)), \quad ((Q)) = \begin{pmatrix} \gamma_1 & -\gamma_2 & . \\ \gamma_2 & \gamma_1 & . \\ . & . & -2\gamma_1 \end{pmatrix}, \quad ((R)) = \begin{pmatrix} . & 1 & . \\ 1 & . & . \\ . & . & 1 \end{pmatrix}. \tag{5.10}$$

I have given above the canonical form of $'\mathcal{K}$ when its characteristic is [222], or $[22\bar{2}]$. We can similarly obtain the canonical forms of $'\mathcal{K}$ for all other characteristics.

6 ADMISSIBILITY OF DIFFERENT CHARACTERISTICS

In this section, I will continue considering the Ch_4-characteristic [222]. This will serve as a model for other characteristics. Broadly speaking the method is the same for all characteristics, but in certain characteristics new situations arise, which it is not possible to enumerate for want of space and time.

Since

$$((\mathscr{W}(\varrho))) = (('\mathscr{R} - \varrho\mathscr{G})),$$

$$((\mathscr{W}(\varrho))) = \text{Diag}\,(-(2+\varrho),\ -(2+\varrho),\ -(2+\varrho),\ \varrho)$$

so that

$$\Delta_4 = -(2+\varrho)^3\varrho.$$

Consequently 0 is a single eigenvalue and -2 is a triple eigenvalue of $\mathscr{R}$. Hence $\mathscr{R}$ must have one of the following Ch_2-characteristics

$$[(111)\,1],\ [1\,(111)],\ [31],\ [13]. \tag{6.1}$$

But according to Section 3, only [(111) 1] is not inadmissible. We should, then, have $\varrho_1 = \varrho_2 = \varrho_3 = -2$, $\varrho_4 = 0$. Here $\varrho_4 > \varrho_1$ which contradicts (3.7a). Hence the characteristic [(111) 1] is also inadmissible. Consequently the characteristic [222] is inadmissible.

We will now demonstrate the case of an admissible real characteristic [21111]. In this case

$$((P)) = \text{Diag}\,(0, -\sigma, -\omega),\quad ((Q)) = \text{Diag}\,(***),\quad ((R)) = \text{Diag}\,(1, -\sigma, -\omega),$$

where

$$2\sigma = \lambda - \gamma,\quad 2\omega = \nu - \mu.$$

Consequently

$$((\mathscr{W}(\varrho)))$$
$$= \text{Diag}\,(\omega + \sigma - \varrho, \omega - \sigma - 1 - \varrho, -(\omega - \sigma + 1) - \varrho, \omega + \sigma + \varrho), \tag{6.2}$$

so that the eigenvalues are

$$\pm(\omega + \sigma),\quad \omega - \sigma - 1,\quad -(\omega - \sigma + 1).$$

Let us first assume that

$$(\omega^2 - \sigma^2)\,(4\omega^2 - 1)\,(4\sigma^2 - 1) \neq 0. \tag{6.3}$$

Then the Ch_2-characteristic for $\mathscr{R}$ is [1111]. Since $-(\omega + \sigma)$ leads to a time-like eigenvector, $\varrho_4 = -(\omega + \sigma)$ cannot be $\omega + \sigma$, because in that case

$\varrho_3 + \varrho_4 = 0$ which contradicts (3.3b). Consequently we have the following four possibilities for the remaining eigenvalues

$$
\begin{array}{llll}
 & \varrho_1 & \varrho_2 & \varrho_3 \\
\text{a)} & \omega - \sigma - 1 & \omega + \sigma & -(\omega - \sigma + 1) \\
\text{b)} & \omega + \sigma & \omega - \sigma - 1 & -(\omega - \sigma + 1) \\
\text{c)} & \omega + \sigma & -(\omega - \sigma + 1) & \omega - \sigma - 1 \\
\text{d)} & -(\omega - \sigma + 1) & \omega + \sigma & \omega - \sigma - 1
\end{array}
\tag{6.4}
$$

(6.4b, c) are excluded, because neither of the two arrangements (3.2a, b) hold. For (6.4a), the arrangements (3.2a, b) yield respectively $\omega > \sigma > \frac{1}{2}$; $\frac{1}{2} > \sigma > \omega > 0$, For (6.4d) the arrangements (3.2a, b) yield respectively $\sigma > \omega > \frac{1}{2}$; $\frac{1}{2} > \omega > \sigma > 0$. Hence *when (6.3) is satisfied then the necessary and sufficient conditions for the characteristic* [21111] *to be admissible are*

$$\omega > \sigma > \tfrac{1}{2} \quad \text{or} \quad \tfrac{1}{2} > \sigma > \omega > 0,$$

when

$$\varrho_1 = \omega - \sigma - 1, \quad \varrho_2 = -\varrho_4 = \omega + \sigma, \quad \varrho_3 = -\omega + \sigma - 1$$

or

$$\sigma > \omega > \tfrac{1}{2}, \quad \text{or} \quad \tfrac{1}{2} > \omega > \sigma > 0,$$

when

$$\varrho_1 = -\omega + \sigma - 1, \quad \varrho_2 = -\varrho_4 = \omega + \sigma, \quad \varrho_3 = \omega - \sigma - 1.$$

When these conditions are satisfied, the physical quantities are given by (3.1).

Let us now assume

$$\omega + \sigma = 0.$$

Then $\varrho = 0$ is a double eigenvalue or a triple eigenvalue. Therefore the only admissible characteristics with these symbols can be [112] for $r = 3$, [(11) 11] for $r = 2$, [(111) 1] for $r = 1$.

[112] is inadmissible because (3.4c) viz $\varrho < 0$ is not satisfied (here $\varrho = 0$).

[(11) 11] and [(111) 1] are inadmissible because none of the single eigenvalues leads to a time-like eigenvector.

Let us now assume

$$\sigma = \tfrac{1}{2}.$$

Then $-(\omega + \frac{1}{2})$ is a double or triple eigenvalue, and the corresponding possible Ch_2-characteristics are [(11) 11] and [(111) 1]. In neither of these two cases do the single eigenvalues lead to a time-like eigenvector. Hence when $\sigma = \frac{1}{2}$ the characteristic [21111] is inadmissible.

We now assume

$$\sigma = -\tfrac{1}{2}.$$

Then $\omega - \frac{1}{2}$ is a double or triple eigenvalue and the corresponding possible Ch_2-characteristics are [(11) 11] and [(111) 1]. When the Ch_2-characteristic is [(11)11],

$$\varrho_1 = \varrho_2 = \omega - \tfrac{1}{2}, \quad \varrho_3(\varrho_4) = -\omega - \tfrac{3}{2}, \quad \varrho_4(\varrho_3) = -\omega + \tfrac{1}{2}.$$

In neither case, will the conditions (3.6) be satisfied. When the Ch_2-characteristic is [(111)1],

$$\varrho_1 = \varrho_2 = \varrho_3 = -1, \quad \varrho_4 = 1.$$

In this case the conditions (3.7) are not satisfied.

Similarly it can be proved that when $\omega = \pm\frac{1}{2}$, the Ch_4-characteristic [21111] is inadmissible.

Finally we shall take a basic complex characteristic $[2\bar{2}1\bar{1}]$. In this case

$$((P)) = \text{Diag}\,(.\ .\ -\mu_2), \quad ((Q)) = \begin{pmatrix} \alpha_1 & -\alpha_2 & . \\ \alpha_2 & \alpha_1 & . \\ . & . & -2\alpha_1 \end{pmatrix},$$

$$((R)) = \begin{pmatrix} . & 1 & . \\ 1 & . & . \\ . & . & \mu_2 + 2\alpha_1 \end{pmatrix}.$$

Consequently

$$((\mathscr{W}(\varrho))) = \text{Diag}\,(-(\mu_2 + 1 + \varrho), -(\mu_2 + 1 + \varrho), \mu_2 - 2 - \varrho, \mu_2 + \varrho)$$

so that the eigenvalues are

$$-(\mu_2 + 1), \quad -(\mu_2 + 1), \quad \mu_2 - 2, \quad -\mu_2.$$

In this case the Ch_2-characteristics [(11) 11] and [(111) 1] are not *a priori* inadmissible. For the Ch_2-characteristic [(11)11], the following arrangements are possible

$$\varrho_1 = \varrho_2 = -(\mu_2 + 1), \quad \varrho_3 = -\mu_2, \quad \varrho_4 = \mu_2 - 2, \tag{6.5a}$$

$$\varrho_1 = \varrho_2 = -(\mu_2 + 1), \quad \varrho_3 = \mu_2 - 2, \quad \varrho_4 = -\mu_2. \tag{6.5b}$$

For (6.5a), ϱ_4 leads to a space-like eigenvector. For (6.5b), we have in view of (3.6)

$$\mu_2 > 1,$$

and the equation of continuity. Hence these *are necessary and sufficient conditions for the admissibility of* $[2\bar{2}1\bar{1}]$. *When these conditions are satisfied*

$$M^2 = 2\mu_2 - 1, \quad m = 2(\mu_2 - 1), \quad 2p = 2\mu_2 + 1 \quad \text{etc.}$$

The classification of all the cases (23 basic and 63 derived characteristics) can be carried out on the lines suggested above. After the classification has been carried out, the results can be applied to the corresponding physical situations.

REFERENCES

1. V. HLAVATÝ, *Ann. di Mat.*, **61**, 121 (1963). V. HLAVATÝ and R. S. MISHRA, *Tensor*, **16**, 138 (1963). V. HLAVATÝ and R. S. MISHRA, *Rend. Circ. Matem. di Palermo*, **11**, 319 (1962); **13**, 1 (1964). R. S. MISHRA, *A Course in Tensors with Applications to Riemannian Geometry*, Allahabad, 1965.
2. Here, and in what follows, the equations hold for arbitrary vector fields X, Y, $Z \in V_4$.
3. A. Z. PETROV, *Russian Scientific Notes of Kazan University*, **114**, 55 (1964).
4. E. NEWMAN, *J. Math. Phys.*, **2**, 324 (1961).
5. J. F. SHELL, *J. Math. Phys.*, **2**, 202 (1961).
6. D. J. STRUIK, *J. Math. Phys. MIT*, **7**, 193.
7. R. V. CHURCHILL, *Trans. Am. Math. Soc.*, **34**, 126 (1932).
8. H. S. RUSE, *Proc. London Math. Soc.*, **41**, 302 (1936).

PAPER 28

A new solution of the field equations with perfect fluid

R. M. MISRA and UDIT NARAIN
University of Gorakhpur, India

ABSTRACT

A new solution of Einstein's field equations has been presented. The universe represented by it contains perfect fluid, the density of which depends upon the position, even when the cosmological constant is taken to be zero. Under certain conditions this solution reduces to Einstein's static solution.

In this note a new solution of Einstein's field equations, representing a universe filled with perfect fluid has been obtained. The universe characterized by this solution is static, non-rotating, shearfree and expansionfree. If, however, a particular case of this general solution is considered one obtains the Einstein's static universe in which the density of matter is non-vanishing even when the cosmological constant is taken to be zero[1]. A brief description of the solution is as follows.

The line element with the above properties may be chosen in the form

$$\mathrm{d}s^2 = (\mathrm{d}x^1)^2 + (2X^2 + Y^2)(\mathrm{d}x^2)^2 + 2X\,\mathrm{d}x^2\,\mathrm{d}x^3 + (\mathrm{d}x^3)^2 - Z^2(\mathrm{d}x^4)^2, \quad (1)$$

where X, Y and Z are functions of x^1 only. The non-vanishing components of the Christoffel symbols of the second kind[2] are obtained as

$$\Gamma^1_{22} = -(2XX_1 + YY_1), \quad \Gamma^2_{13} = \frac{X_1}{2(X^2 + Y^2)}, \quad (2)$$

$$\Gamma^1_{23} = -\frac{1}{2} X_1, \qquad \Gamma^3_{12} = \frac{-2X_1^2X^2 - 2XYY_1 + Y^2X_1}{2(X^2+Y^2)},$$

$$\Gamma^1_{44} = ZZ_1, \qquad \Gamma^3_{13} = -\frac{XX_1}{2(X^2+Y^2)}, \tag{2}$$

$$\Gamma^2_{12} = \frac{3XX_1 + 2YY_1}{2(X^2+Y^2)}, \qquad \Gamma^4_{14} = \frac{Z_1}{Z},$$

where the subscripts denote differentiation with respect to variable x^1. With the help of the Christoffel symbols and the components of the metric tensor we easily obtain the components of the contracted curvature tensor as

$$R_{11} = \frac{Z_{11}}{Z} + \frac{3X_1^2 + 2XX_{11} + 2Y_1^2 + 2YY_{11}}{2(X^2+Y^2)} - \left(\frac{XX_1 + YY_1}{X^2+Y^2}\right)^2,$$

$$R_{22} = \frac{Z_1}{Z}(2XX_1 + YY_1) + 2XX_{11} + YY_{11}$$

$$+ \frac{-2X_1^2X^2 - 6XX_1YY_1 + 3X_1^2Y^2 + 2Y_1^2X^2}{2(X^2+Y^2)},$$

$$R_{23} = \frac{1}{2}\left(X_{11} + \frac{X_1Z_1}{Z} - \frac{2XX_1^2 + YY_1X_1}{X^2+Y^2}\right), \tag{3}$$

$$R_{33} = -\frac{X_1^2}{2(X^2+Y^2)},$$

$$R_{44} = -ZZ_{11} - ZZ_1\frac{XX_1 + YY_1}{X^2+Y^2},$$

$$R_{12} = R_{13} = R_{14} = R_{24} = R_{34} = 0.$$

The field equations corresponding to this problem are taken to be

$$R_{ij} - \tfrac{1}{2}g_{ij}R - \lambda g_{ij} = -k\{(\mu + p)u_iu_j + pg_{ij}\} \tag{4}$$

where μ and p are the density and pressure of the fluid respectively. λ is the cosmological constant and u^i is the unit velocity vector tangent to worldlines of the perfect fluid satisfying equation

$$u_iu^i = -1. \tag{5}$$

We solve the field Eq. (4) in co-moving frame characterized by $u_i = Z\delta_i^4$ and $u^i = Z\delta_4^i$. The equations which must be satisfied are

$$R_{11} - \tfrac{1}{2}R - \lambda = -kp, \tag{6}$$

$$R_{22} - \tfrac{1}{2}(2X^2 + Y^2)R - \lambda(2X^2 + Y^2) = -kp(2X^2 + Y^2), \tag{7}$$

$$R_{23} - \tfrac{1}{2}XR - \lambda X = -kpX, \tag{8}$$

$$R_{33} - \tfrac{1}{2}R - \lambda = -kp, \tag{9}$$

$$R_{44} + \tfrac{1}{2}Z^2R + \lambda Z^2 = -k\mu Z^2. \tag{10}$$

The remaining equations vanish identically. Eq. (6) to (10) give rise to the following relations:

$$R_{11} = R_{33}, \tag{11}$$

$$R_{23} - XR_{33} = 0, \tag{12}$$

$$(X^2 + Y^2)R_{11} - R_{22} + XR_{33} = 0, \tag{13}$$

$$Z^{-2}R_{44} + R_{33} = -k(\mu + p), \tag{14}$$

$$Z^{-2}R_{44} - R_{33} = 2(-kp + \lambda), \tag{15}$$

$$R = k(-\mu + 3p) - 4\lambda. \tag{16}$$

In view of Eq. (3) and (12) we get

$$\frac{X_{11}}{X_1} + \frac{Z_1}{Z} - \frac{XX_1 + YY_1}{X^2 + Y^2} = 0. \tag{17}$$

Eq. (3) and (11) give rise to

$$\frac{Z_{11}}{Z_1} - \frac{XX_1 + YY_1}{X^2 + Y^2} = 0. \tag{18}$$

These two equations yield on integration

$$X_1^2 Z^2 = A^2(X^2 + Y^2), \tag{19}$$

$$Z_1^2 = B^2(X^2 + Y^2), \tag{20}$$

where A and B are integration constants. Eq. (19) and (20) determine X in terms of Z through the following relation

$$X_1^2 = \left(\frac{A}{B}\right)^2 \left(\frac{Z_1}{Z}\right)^2, \tag{21}$$

which on integration yields

$$X = \pm \frac{A}{B} \ln Z + C \tag{22}$$

where C is another constant of integration. Eq. (11) and (19) in view of (3) yield

$$\frac{Z_{111}}{Z_1} + \frac{Z_{11}}{Z} + \frac{A^2}{Z^2} \equiv Q = 0. \tag{23}$$

This equation determines Z. The energy density μ and the pressure p of the fluid, as determined by Eq. (3), (14), (15) and (16) are

$$\mu = \frac{Z_{11}}{kZ} + \frac{3A^2}{4kZ^2} - \frac{\lambda}{k} \tag{24}$$

and

$$p = \frac{Z_{11}}{kZ} - \frac{A^2}{4kZ^2} + \frac{\lambda}{k}. \tag{25}$$

It is immediately observed from Eq. (24) that the density of the fluid depends upon the position and it is non-zero when the cosmological constant is taken to be zero. Further, we have five equations to determine six unknowns, namely X, Y, Z, μ, p and λ. This leaves one of them, say λ in this case, arbitrary. However, if one wishes, λ may be specified without any loss of generality as follows. On first integration of Eq. (23) one will obtain a constant of integration which may be identified with λ. Hence

$$\int Q \, \mathrm{d}x^1 = \lambda. \tag{26}$$

However, the case of constant Z (say $Z = 1$) is particularly interesting. In this case, in view of expressions (3), $R_{44} = 0$, and one obtains from relation (11) the following equation for X

$$X_{111} + A^2 X_1 = 0. \tag{27}$$

This equation is readily integrated and the result is

$$X = G \cos Ax^1 + H \sin Ax^1 + K, \tag{28}$$

where G, H and K are constants of integration. In view of relation (19) with $Z = 1$ the function Y is determined as

$$Y^2 = (H \cos Ax^1 - G \sin Ax^1)^2 - X^2. \tag{29}$$

Further, the density and pressure for this case as given by Eq. (24) and (25) are

$$\mu = \frac{3A^2}{4k} + \frac{\lambda}{k} \tag{30}$$

and

$$p = -\frac{A^2}{4k} + \frac{\lambda}{k}. \tag{31}$$

The solution given by (28), (29), (30) and (31) is the general form of Einstein's static solution. Because, if one chooses $G = A = 1$ and $H = K = 0$ one obtains the usual form of line element corresponding to Einstein's static universe[3]. It is easier to understand the properties of the Einstein universe in the light of the solution obtained here.

In order to find out the Petrov type of the solution, we calculate the components of the conform tensor using the relation

$$\begin{aligned} C_{ijkl} = R_{ijkl} &+ \tfrac{1}{2}(g_{ik}R_{jl} - g_{il}R_{jk}) \\ &- \tfrac{1}{2}(g_{jk}R_{il} - g_{jl}R_{ik}) \\ &+ \tfrac{1}{6}R(g_{il}g_{jk} - g_{ik}g_{jl}). \end{aligned} \tag{32}$$

The non-vanishing components of the mixed conform tensor are

$$\begin{aligned} C^{14}_{14} &= C^{23}_{23} = -\frac{1}{3}\frac{Z_1}{Z}\frac{XX_1 + YY_1}{X^2 + Y^2}, \\ C^{24}_{24} &= C^{13}_{13} = -\frac{1}{6}\frac{Z_1}{Z}\frac{5XX_1 + 2YY_1}{X^2 + Y^2}, \\ C^{34}_{34} &= C^{12}_{12} = \frac{1}{6}\frac{Z_1}{Z}\frac{7XX_1 + 4YY_1}{X^2 + Y^2}, \\ C^{12}_{13} &= -C^{24}_{34} = \frac{1}{2}\frac{Z_1}{Z}\frac{X_1}{X^2 + Y^2}, \\ C^{13}_{12} &= -C^{34}_{24} = \frac{1}{2}\frac{Z_1}{Z}\frac{Y^2X_1 - 2X_1X^2 - 2XYY_1}{X^2 + Y^2}. \end{aligned} \tag{33}$$

The symmetry properties of conform tensor allow us to write this tensor as a six-by-six matrix C_{AB} where A and B take the values

$$14 \to 1, \quad 24 \to 2, \quad 34 \to 3, \quad 23 \to 4, \quad 31 \to 5, \quad 12 \to 6.$$

The eigenvalues of this matrix are found by solving the equations

$$|C_A^B - \lambda I_A^B| = 0. \tag{34}$$

The sum of eigenvalues will be zero (from $C^i_{jil} = 0$).

Using equations (18), (19), (20) and (34), the eigenvalues are obtained as

$$\begin{aligned}
\lambda_1 &= \lambda_4 = -\frac{1}{3}\frac{Z_{11}}{Z}, \\
\lambda_2 &= \lambda_5 = \frac{1}{6}\frac{Z_{11}}{Z} + \frac{1}{2}\frac{Z_1}{Z}\left\{\left(\frac{Z_{11}}{Z}\right)^2 + \frac{A^2}{Z^2}\right\}^{1/2}, \\
\lambda_3 &= \lambda_6 = \frac{1}{6}\frac{Z_{11}}{Z} - \frac{1}{2}\frac{Z_1}{Z}\left\{\left(\frac{Z_{11}}{Z}\right)^2 + \frac{A^2}{Z^2}\right\}^{1/2}.
\end{aligned} \tag{35}$$

Therefore the solution is of Petrov type I class.

REFERENCES

1. R. C. Tolman, *Relativity, Thermodynamics and Cosmology*, Clarendon Press, Oxford, 1934, pp. 335–45.
2. L. P. Eisenhart, *Riemannian Geometry*, Princeton University Press, Princeton, N.J., 1925, Chapter I.
3. I. Ozsvath, *J. Math. Phys.*, **6**, 590–610 (1965).

PAPER 29

Cosmological implications of the microscopic CP violation

YUVAL NE'EMAN and YOAV ACHIMAN*
Tel-Aviv University, Israel

ABSTRACT

The conjecture linking together statistical and cosmological time arrows implies interpreting the contracting phase in an oscillating cosmological model as a time-inverted expansion. The Fitch–Cronin effect now requires a redefinition of what is matter and what is antimatter in this time-inverted picture. However, if CPT is also broken, then the T-violating effects yield a different set of laws altogether for such reactions.

MICROSCOPIC IRREVERSIBILITY

It is by now a well-established fact that the Fitch–Cronin effect[1] represents a violation of the combined CP symmetry of the known interactions at the microscopic level. Strong and electromagnetic interactions respect both C (generalized charge parity) and P (space parity). Weak interactions break C and P but do respect the product operation CP. Using available experimental data one can deduce that the new effect may originate in either a P-conserving C-violating new "milli-strong" interaction (of strength similar to electromagnetic but uncoupled to photons), or in a "milli-weak" small component of the weak interactions, or alternatively in a "super-weak" new force of order $10^{-9}\, G_{\mathrm{Fermi}}$.

* Research sponsored by the Air Force Office of Scientific Research, Office of Aerospace Research, U.S. Air Force, under AFOSR grant number EOOAR-68-0010, through the European Office of Aerospace Research.

It would seem at first sight that one might salvage T invariance, i.e. symmetry under microscopic time-reversal, by abandoning CPT invariance. However, the situation is such that T is certainly violated[2,3]. Whether or not CPT holds only affects the size of this violation. With past experience having gradually forced us to abandon first P, then CP, we shall consider here both CPT invariant and CPT non-invariant situations.

Two recent studies[4,5] have independently dwelt upon problems arising from the possible links between the new microscopic "arrow of time" and the statistical and cosmological ones. In a general way, the new effect brings about an additional irreversibility, on top of the statistical one. It is difficult to conceive of situations in which thermodynamical irreversibility can be made to vanish, so as to bring out the direct effects of microscopic irreversibility. However, we would like to show in this paper how the usual assumption about the inter-relations between statistical and cosmological time-arrows brings about just such a situation.

A common impression[6] among astrophysicists and cosmologists has been that the two arrows are linked together. An expanding universe is then the only conceivable one; roughly, in a contracting universe entropy would have to increase, the universe changing from a disordered spread-out (large phase-space) state to a concentrated, ordered, (minute phase-space) state. This is of course an oversimplified statement, since models can be constructed in which an oscillating universe would go from one concentrated state to the next one with a permanent increase of entropy, for example. The simplicity of the first view has however led to its adoption as a plausible conjecture, fitting in nicely with the ideas of Steady-State theory, and adaptable to other models.

Consider now an oscillating model. The above view implies a reversal of phase-space considerations in the contracting phase, so as to make it appear as an expansion, from the statistics. Suppose the only available matter in the universe were a beam of K^0 (or $\overline{K}^0$) mesons. These would decay into π mesons; the fractional number of K^0 and $\overline{K}^0$ mesons remaining in a K^0 (or $\overline{K}^0$) beam is at time t (e.g. Rosen's preferred frame[7])

$$\begin{aligned} R^{\pm}(t) = \{\tfrac{1}{2}\,|1 - \varepsilon^2 + \delta^2|^{-2}\}\,\{&(1 + |\varepsilon|^2 + |\delta|^2 + 2\,\mathrm{Re}\,\varepsilon\delta^*)\,|1 \mp \varepsilon \pm \delta|^2\,\mathrm{e}^{-\gamma_S t} \\ &+ (1 + |\varepsilon|^2 + |\delta|^2 - 2\,\mathrm{Re}\,\varepsilon\delta^*)\,|1 \mp \varepsilon \mp \delta|^2\,\mathrm{e}^{-\gamma_L t} \\ &\pm (2\,\mathrm{Re}\,\varepsilon + 2\mathrm{i}\,\mathrm{Im}\,\delta)\,(1 \mp \varepsilon^* \pm \delta^*)\,(1 \mp \varepsilon \mp \delta)\,\times \\ &\times \mathrm{e}^{-(1/2)(\gamma_S + \gamma_L)t + \mathrm{i}\Delta m t} \pm (2\,\mathrm{Re}\,\varepsilon + 2\mathrm{i}\,\mathrm{Im}\,\delta)\,\times \\ &\times (1 \mp \varepsilon \pm \delta)(1 \mp \varepsilon^* \mp \delta^*)\,\mathrm{e}^{-(1/2)(\gamma_S + \gamma_L)t - \mathrm{i}\Delta m t}\}. \end{aligned}$$

ε is the CP violation parameter, δ the CPT violation. Both are complex numbers with the upper sign for K^0 and the lower one for $\bar{K}^0$. To first order in ε and δ (both are small) we may write,

$$R^{\pm}(t) = \{\tfrac{1}{2} \mp \mathrm{Re}\,(\varepsilon - \delta)\, e^{-\gamma_S t} + \{\tfrac{1}{2} \mp \mathrm{Re}\,(\varepsilon + \delta)\}\, e^{-\gamma_L t}$$
$$\pm \{\mathrm{Re}\,\varepsilon - \mathrm{i}\,\mathrm{Im}\,\delta\}\, e^{-(1/2)(\gamma_S + \gamma_L)t + i\Delta m t}$$
$$\pm \{\mathrm{Re}\,\varepsilon + \mathrm{i}\,\mathrm{Im}\,\delta\}\, e^{-(1/2)(\gamma_S + \gamma_L)t - i\Delta m t}. \quad (1)$$

THE CPT CONSERVING CASE

Suppose we now go over to a time $\tau - t$, with τ the oscillation period of the universe; in in fact to $-t$, which should look the same as $\tau - t$. Our pions are now being squeezed back to remake the original K^0 or $\bar{K}^0$ beam. Take the case where this was a K^0 beam; to describe this production process, we have to time-invert our formula. If CPT is conserved, a theorem[8] states that the time-reversed amplitude is the same as the CP-inverted one, to first order in the CP violation:

$$\langle B|\, H_{\mathrm{viol}}\, |A\rangle^* \equiv \langle \mathrm{T}B|\, \mathrm{T}H_{\mathrm{viol}}\mathrm{T}^{-1}\, |\mathrm{T}A\rangle = \langle \mathrm{T}B|\, \mathrm{C}^{-1}\mathrm{P}^{-1}H_{\mathrm{viol}}\mathrm{PC}\, |\mathrm{T}A\rangle$$
$$= \langle \bar{B}|\, H^{+}_{\mathrm{viol}}\, |\bar{A}\rangle - (\bar{B}|\, H^{-}_{\mathrm{viol}}\, |\bar{A}\rangle$$

for A, B spinless states, and H^{+}_{viol}, H^{-}_{viol} denoting the even and odd parity parts of H_{viol}. The two amplitudes are orthogonal to this order, so that

$$|\langle \mathrm{T}B|\, \mathrm{T}H_{\mathrm{viol}}\mathrm{T}^{-1}\, |\mathrm{T}A\rangle|^2 = |\langle \bar{B}|\, H_{\mathrm{viol}}\, |\bar{A}\rangle|^2. \quad (2)$$

The implication is that we have to use the $\bar{K}^0$ decay formula to describe K^0 production at $-t$. If we now consider the contraction phase between $\tau/2$ and τ as an expansion from the concentrated state at τ, we invert $t \to -t$ in the decay formula. This will then turn the production of K^0 into the decay of $\bar{K}^0$:

$$P^{+}(-t) = R^{-}(t). \quad (3)$$

Cosmologically, we learn that in the entropy-symmetric description, a contracting matter universe is the same as an expanding antimatter universe. Both descriptions use the same laws of nature, provided we replace matter by antimatter and vice versa. We note that this is now no trivial change, since the two do behave differently. Within one universe, they can now be distinguished in the relative sense. For instance,

$$K^0_L \to \pi^{\mp} + e^{\pm} + \overset{(-)}{\nu}_e \quad \text{or} \quad \to \pi^{\mp} + \mu^{\pm} + \overset{(-)}{\nu}_\mu$$

are CP violating decays[1], with the measured asymmetry ratio between rates

$$r = \frac{\Gamma_+ - \Gamma_-}{\Gamma_+ + \Gamma_-} \sim 2.10^{-3}$$

(the sign corresponds to the charge of the resulting electron or muon).

Communicating with physicists in a distant galaxy, we simply ask them to make K^0 or $\bar{K}^0$ mesons, to use the long-lived component K_L^0 and observe the asymmetry. We can tell them that when r is positive, the more numerous leptons are positrons and should be considered as antimatter by our conventions.

Returning to the cosmological situation we note that with CPT invariance, all textbooks will look the same in the time-inverted expanding universe except for a matter-antimatter replacement. What if CPT itself fails?

THE CPT VIOLATING CASE

The Lee–Oehme–Yang theorem[8] (2) does not hold. To explore this situation we shall study the behavior of ε and δ, the two complex parameters in (1) under the two operations we used:

a) a matter-antimatter replacement (i.e. CP)

b) overall time inversion (actually T since we do neutralize the effects of phase-space in the cosmological picture).

Under CP, both ε and δ change sign. In the 2-dimensional $\psi\,(K^0, \bar{K}^0)$ space, with

$$i\frac{d}{dt}\psi = (M - i\Gamma)\psi = \Lambda\psi \tag{4}$$

where M and Γ are 2×2 hermitean matrices, Λ a general 2×2 one, $\lambda_{S,L}$ the two eigenvalues (widths and masses) of Λ, short and long lived respectively

$$\varepsilon \simeq \frac{\Lambda_{12} - \Lambda_{21}}{\lambda_S - \lambda_L}. \tag{5}$$

Using the Wigner–Weisskopf method,

$$\begin{aligned}
\Lambda_{12} &\simeq \langle K^0|\, H_W\,|\bar{K}^0\rangle + \sum_f \frac{1}{M_{K^0} - M_f + i\varepsilon} \langle K^0|\, H_W\,|f\rangle \langle f|\, H_W\,|\bar{K}^0\rangle \\
\Lambda_{21} &\simeq \langle \bar{K}^0|\, H_W\,|K^0\rangle + \sum_f \frac{1}{M_{\bar{K}^0} - M_f + i\varepsilon} \langle \bar{K}^0|\, H_W\,|f\rangle \langle f|\, H_W\,|K^0\rangle.
\end{aligned} \tag{6}$$

Since CP exchanges K^0 and $\bar{K}^0$, $\Lambda_{12} \leftrightarrow \Lambda_{21}$ and thus the CP violation parameter changes sign,

$$\varepsilon \to -\varepsilon. \tag{7}$$

For the CPT violation parameter ($\delta = 0$ if CPT is conserved) we have

$$\delta \simeq \frac{\Lambda_{11} - \Lambda_{22}}{\lambda_S - \lambda_L} \tag{8}$$

$$\Lambda_{11} \simeq M_{K^0} + \langle K^0 | H_W | K^0 \rangle + \sum_f \frac{1}{M_{K^0} - M_f + i\varepsilon} \langle K^0 | H_W | f \rangle \langle f | H_W | K^0 \rangle$$

$$\Lambda_{22} \simeq M_{\bar{K}^0} + \langle \bar{K}^0 | H_W | \bar{K}^0 \rangle + \sum_f \frac{1}{M_{\bar{K}^0} - M_f + i\varepsilon} \langle \bar{K}^0 | H_W | f \rangle \langle f | H_W | \bar{K}^0 \rangle \tag{9}$$

so that

$$\delta \to -\delta. \tag{10}$$

We now try T. Wigner time reversal acts so that

$$\psi(x, t) \to \psi^*(x, -t)$$

to preserve the Schrödinger equation (4)

$$i \frac{d}{dt} \psi_T^* = (M^* - i\Gamma^*) \psi_T^* \tag{11}$$

because $M_T = M$, but $\Gamma_T = -\Gamma$. Since both M and Γ are hermitean, $M_{12}^* = M_{21}$, $\Gamma_{12}^* = \Gamma_{21}$ and we see in (5) that

$$\varepsilon_T = -\varepsilon \tag{12}$$

but from (8)

$$\delta_T = \delta. \tag{13}$$

This then answers our question. The new $P^+(-t)$ will look like $R^-(t)$ as far as the ε contribution is concerned, but will stay as in $R^+(t)$ for the δ terms. The most general case will then consist of *a Universe where replacing contraction by a time-inverted expansion implies that the resulting universe will have different laws of nature*! No matter-antimatter redefinition can settle this change.

The formulae (1) thus correspond to "true" decays. In the time-inverted expanding universe they will become,

$$R_{\mathrm{T}}^{\pm}(t) = \{\tfrac{1}{2} \pm \mathrm{Re}\,(\varepsilon + \delta)\}\,\mathrm{e}^{-\gamma_{\mathrm{S}}t} + \{\tfrac{1}{2} \pm \mathrm{Re}\,(\varepsilon - \delta)\}\,\mathrm{e}^{-\gamma_{\mathrm{L}}t}$$
$$\mp \{\mathrm{Re}\,\varepsilon + \mathrm{i}\,\mathrm{Im}\,\delta\}\,\mathrm{e}^{-(1/2)(\gamma_{\mathrm{S}}+\gamma_{\mathrm{L}})t + \mathrm{i}\Delta mt}$$
$$\mp \{\mathrm{Re}\,\varepsilon - \mathrm{i}\,\mathrm{Im}\,\delta\}\,\mathrm{e}^{-(1/2)(\gamma_{\mathrm{S}}+\gamma_{\mathrm{L}})t - \mathrm{i}\Delta mt}\,. \qquad (14)$$

The coefficients of the two diagonal states decay curves have been inverted, besides changing from K^0 to $\overline{\mathrm{K}}^0$. Another inversion occurs in the coefficients of the mixed oscillating terms.

REFERENCES

1. J.H. Christenson, J.W. Cronin, V.L. Fitch and R. Turlay, *Phys. Rev. Letters*, **13**, 138 (1964). For a recent review of the experimental situation see J. Steinberger in: *Proc. of the Topical Conf. on Weak Interactions*, Geneva (1969).
2. R.S. Casella, *Phys. Rev. Letters*, **21**, 1128 (1968); **22**, 554 (1969).
3. Y. Achiman, to be published (available as TAUP-66-68).
4. G. Zweig, paper presented at the *Conf. on Decays of K Mesons*, Princeton-Pennsylvania Accelerator, November 1967, unpublished.
5. Y. Ne'eman, paper presented at the March 1968 session of the *Israel Academy of Sciences*; published in *Proc. of the Israel Academy of Sciences and Humanities, Section of Sciences*, **13**, Jerusalem, 1969. See also *Int. J. Th. P.*, **3**, 1 (1970).
6. See for example T. Gold's article in *Recent Developments in General Relativity*, Pergamon-Macmillan, New York–Warsaw, 1962, p. 225; M. Gell-Mann, comments in Proc. of the Temple University panel on Elementary Particles and Relativistic Astrophysics (1967); E. Salpeter, lecture at the Tel-Aviv Seminar on Astrophysics (1968).
7. N. Rosen, *Proc. of the Israel Academy of Sciences and Humanities, Section of Sciences*, **12**, Jerusalem, 1968.
8. T.D. Lee, R. Oehme and C.N. Yang, *Phys. Rev.*, **106**, 340 (1957).

PAPER 30

Some notes on cosmology

J. PACHNER

*Theoretical Physics Institute, University of Alberta, Edmonton, Canada**

For many centuries the questions of the origin and development of the World belonged exclusively to the domain of religion, theology, and philosophy. Only after the foundations of the general relativity theory had been laid, the first (though not yet definite) answers to these questions, elaborated by scientific methods, were given. It seems to be worthwhile putting together some principles here on which scientific cosmology should be based in its further development.

The starting point of the present considerations is the assumption that the decisive phenomenon in cosmic evolution is gravitation and, possibly, some other cosmic field. The quantum phenomena might play an important role only in the period of the maximum contraction of the Universe when the mass density surpasses the value 10^{93} g/cm^3.

If we interpret the observed red shift of distant galaxies as a Doppler effect and make some plausible assumptions of the evolution of galaxies, then observational cosmology testifies (far more reliably than at the time of Einstein's first cosmological paper[1]) that the Universe from the global point of view is in a uniform and isotropic expansion. If we accept the very convincing arguments of Bondi[2] that the geometry of the cosmic space is Riemannian, its geometrical properties are described at the present epoch of cosmic evolution by the well-known Robertson–Walker line element[3,4] expressing the cosmological principle

$$ds^2 = [S^2(t)/(1 + kr^2/S_0^2)^2]\,(dx^2 + dy^2 + dz^2) - dt^2 \qquad (1)$$

* Present address: Department of Physics, University of Saskatchewan, Regina Campus, Regina, Sask., Canada.

with

$$k = \pm 1, 0, \quad S_0 = \text{const.}, \quad r^2 = x^2 + y^2 + z^2.$$

The "particles of the cosmic dust" that follow this cosmic expansion are generally accepted to be galaxies or clusters of galaxies, but we prefer to call them "vacuoles" in order to emphasize their internal structure.

The concept of a vacuole, i.e. of a spherical region in the Friedman universe inside which all matter is concentrated to its centre, was introduced into the relativistic cosmology by Einstein and Straus in 1945[5]. It turns out that the motion of test particles in such a vacuole is not influenced by the cosmic expansion. On the basis of Einstein field equations Schücking[6] deduced for its radius R_V a formula which may be easily reduced to the form[7]

$$R_V = (GM/qH^2)^{1/3}, \tag{2}$$

M being the total mass inside the vacuole, G the Newtonian constant of gravitation, H the Hubble factor of the cosmic expansion and q its deceleration parameter. Since H and q increase with decreasing curvature radius of the space, the radius of the vacuole simultaneously diminishes, as a consequence of which from a certain moment of the contraction of the Universe the larger and larger parts of the outer region of the vacuole begin to participate in the general cosmic contraction.

We now generalize the concept of the vacuole, defining it as a region (not necessarily of a spherical shape) inside which the celestial bodies and interstellar matter move without participating in the general cosmic expansion or contraction. The dimensions of this vacuole will depend on the spatial distribution of the matter inside it and on the assumed theory of gravitation. Instead of Eq. (2) we may merely suppose that its dimensions will decrease during the contraction of space. While the general cosmic expansion and contraction are uniform and isotropic, the characteristic feature of the motion inside the vacuoles is rotation.

Till now we have made no assumption on the gravitational law and on the existence of other cosmic fields, and thus, on the form of the function $S(t)$ in Eq. (1). If we assume the validity of the Einstein field equations without the cosmological term, Raychaudhuri's formula[8], reduced to the form[14]

$$\ddot{\varrho}/\varrho = 4\pi\varrho + \tfrac{4}{3}(\dot{\varrho}/\varrho)^2 + \Phi^2 - 2\,|\Omega|^2, \tag{3}$$

proves the inevitability of a singular state with an infinite mass density in any world model filled with a cosmic dust of arbitrary distribution whose motion has no rotational component (i.e. the square of the angular velocity $|\Omega|^2 = 0$).

Any mathematical singularity means a break-down of the physical theory. If we refuse to admit the existence of certain limits of our present knowledge, we must either modify the theory, or investigate whether the occurrence of a singularity is not a consequence of some improper assumptions made in order to facilitate the mathematical treatment of a correct physical theory.

From the mathematical point of view the modification of Einstein general relativity by introducing some cosmic field (e.g., the cosmological Λ-term of Einstein[1], the C-field of Hoyle[9], or of Hoyle and Narlikar[10], the negative pressure of McCrea[11], Pachner[12], and Rosen[13]) is certainly an easier way to remove the occurrence of singularities, but it is connected with the very difficult task of justifying the assumed modification from a physical point of view and to prove that just this modification represents the true natural law of cosmic evolution. Since the Einstein theory is modified in order to save the validity of the Robertson–Walker metric (1) for any stage of the cosmic evolution, the cosmic field must stop the contraction and revert to a new expansion before the internal motion of the vacuoles is disturbed by the contraction.

Einstein general relativity without the cosmological term represents the most simple and physically best-established gravitational theory. The author has therefore considered is worthwhile to investigate whether the occurrence of singularities is already incorporated in the Einstein field equations[14]. This approach implies that from a certain moment of the cosmic contraction the Robertson–Walker metric (1) loses its validity and must be replaced by a new one taking into account the local rotational motion of earlier vacuoles. Any world model that does not take into consideration this local rotational motion represents a potential world model, compatible with the Einstein field equations, but such a model does not describe the actual Universe with its characteristic local rotational motion.

The situation is mathematically not so hopeless as it seems at first sight, for Lichnerowicz has shown[15] that we may construct the global cosmic field by particular local fields. In the given case these local fields are created by the matter in a rotational motion with shear and contraction (or expansion) and may be uniquely extended to the empty space of the outer region of the earlier vacuoles[15]. Investigating the occurrence of singularities in relativistic cosmology, we may thus restrict ourselves to a local region occupied by a rotating ideal fluid.

The exact solution of Einstein equations without the cosmological term found by Maitra[16] proves that the rotation is really able to stop the contraction of space and convert it to a new expansion without the occurrence of a singularity. Recently it has been shown[14] that the rotation can create such a curvature of space-time which will avoid the contraction of incoherent matter also along the vorticity vector. A necessary, but not sufficient, condition for the existence of a maximum mass density (i.e. for the non-existence of a singularity) is that $\ddot{\varrho}$ shall become negative. The Raychaudhuri equation (3) shows that the fulfilment of this condition depends on the degree of the anisotropy of the motion (i.e. on the value of Φ^2). A generalization of Raychaudhuri's equation (3) on an ideal fluid shows[14] that an extremely high pressure (so high that the relativistic limit of the equation of state must be applied) creates an additional attraction and diminishes simultaneously the influence of rotation, and is thus responsible for a gravitational collapse in this case even if rotation is present.

If further investigation proves that the conditions under which the rotation is able to avoid the singularities correspond to the observational data in the vacuoles of the actual Universe, then in accordance with Ockham's razor "*Frustra fit per plura, quod fieri potest per pauciora*"—we ought to prefer the Einstein general relativity to any modification.

REFERENCES

1. A. Einstein, *S.-B. Preus. Akad. Wiss.*, **142** (1917).
2. H. Bondi, in: *Recent Developments in General Relativity*, Warszawa, Oxford, 1962, p. 47ff.
3. H. P. Robertson, *Astrophys. J.*, **82**, 284 (1935); **83**, 187, 257 (1936).
4. A. G. Walker, *Proc. London Math. Soc.*, **42**, 90 (1936).
5. A. Einstein and E. G. Straus, *Rev. Mod. Phys.*, **17**, 120 (1945); **18**, 148 (1946).
6. E. Schucking, *Z. Physik*, **137**, 595 (1954).
7. J. Pachner, *Phys. Rev.*, **137**, B1379 (1965).
8. A. Raychaudhuri, *Phys. Rev.*, **98**, 1123 (1955).
9. F. Hoyle, *Monthly Not. Roy. Astron. Soc.*, **108**, 372 (1948); **109**, 365 (1949); **120**, 256 (1960).
10. F. Hoyle and J. V. Narlikar, *Proc. Roy. Soc. (London)* **A278**, 465 (1964).
11. W. H. McCrea, *Proc. Roy. Soc. (London)* **A206**, 562 (1951).
12. J. Pachner, *Monthly Not. Roy. Astron. Soc.*, **131**, 173 (1965); *Bull. Astron. Inst. Czechol.* **16**, 321 (1965).
13. N. Rosen, *Int. J. Theor. Phys.*, **2**, 189 (1969).
14. J. Pachner, *Canad. J. Phys.*, **48**, 970 (1970); Erratum in press. *GRG* issue 3 (in press).
15. A. Lichnerowicz, *Theories Relativistes de la Gravitation et de l'Electromagnetisme*, Paris, 1955, Chapter 3.
16. S. C. Maitra, *J. Math. Phys.*, **7**, 1025 (1966).

PAPER 31

Invariant evolution of gravitational field

ASHER PERES

Technion-Israel Institute of Technology, Haifa

ABSTRACT

The initial values of the gravitational field variables, at some time $t = 0$, carry the imprint of their entire evolution, past and future. They can be used to write "equations of motion for absolute invariants", and to compute dynamical properties (such as collision cross-sections) without ever leaving the hypersurface $t = 0$.

מה-שהיה הוא שיהיה
ומה-שנעשה הוא שיעשה
ואין כל חדש תחת השמש
קהלת א, ט

Einstein's theory of gravitation, like other classical field theories, is based on the idea that our *physical observations* can be conveniently described by a system of *field variables* subject to some partial differential equations. While this method has been in general highly successful, it leads to some conceptual difficulties because the number of field variables often exceeds that of the physical degrees of freedom, i.e. there is no unambiguous way of prescribing the values of the field variables pertaining to a given physical situation. In the mathematical structure of the theory, this difficulty will appear as the existence of a *gauge group*, allowing transformations of the field variables while the physical situation is unchanged. This gauge group can always be eliminated by reducing the number of variables, but only at the expense of locality: the new dynamical equations become integro-differential, rather than partial differential equations.

The remarkable feature of Einstein's theory of gravitation, which sets it quite apart from other field theories, is that its gauge group consists of arbitrary distortions of the space-time coordinates, and thus cannot be easily

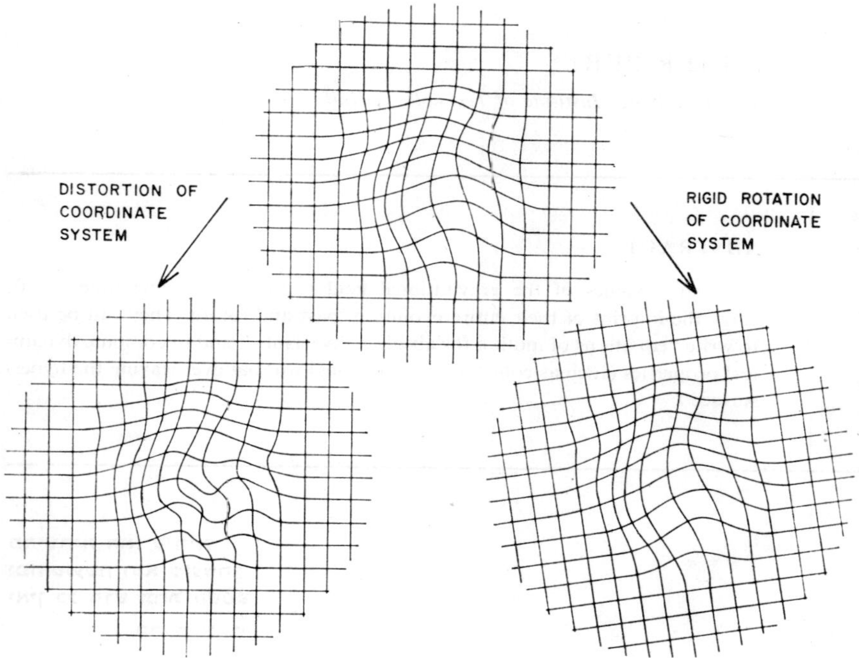

Figure 1

disentangled from the structure of space-time itself (see Figure 1). In particular, the time evolution of a physical system is locally indistinguishable from a gauge transformation—namely from a local distortion of the time coordinate (see Figure 2). It follows that the equations of motion of the gravitational field, $R_{mn} = 0$ (which give the "accelerations" in terms of the initial values and first time derivatives of the field), merely express an automorphism generated by the gauge group. Thus, *these "equations of motion" actually convey no relevant physical information and they are devoid of experimental interest.*

On the other hand, any meaningful question on the outcome of a real experiment (e.g. the scattering of a gravitational wave-packet by a static gravitational field) should be answered by considering the four remaining Einstein equations,

$$R^{00} - \tfrac{1}{2}g^{00}R = 0,$$

and

$$R_k^0 = 0.$$

These equations do not involve second time derivatives of the gravitational field; they are constraints imposed on the initial values of the field variables (and their conjugate momenta) at some initial time, say $t = 0$. *All gravitational physics is contained in these four constraints: the initial values of the field variables (and their conjugate momenta) already have the imprint of their entire time evolution, past and future*[1].

To show this explicitly, we must devise some formalism in which the four "dynamical constraints" are actually used to compute physical quantities, such as scattering cross-sections[2]. The first step in such calculations is to eliminate the gauge freedom, thus losing the nice local properties of the field equations, but gaining physical clarity. Let us express the six dynamical vari-

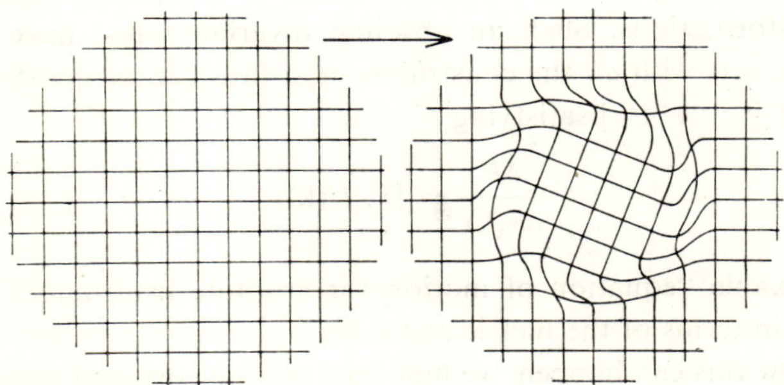

Figure 2

ables $g_{mn}(\mathbf{x})$ in terms of three "coordinate-like" scalar fields $y^A(\mathbf{x})$, and three invariant variables $\gamma^\alpha(\mathbf{k})$. For example, if space is asymptotically flat, the $y^A(\mathbf{x})$ are conveniently chosen as the harmonic coordinates, and the $\gamma^\alpha(\mathbf{k})$ then correspond to two transverse degrees of freedom with circular polarization $\gamma^{\pm}(\mathbf{k})$, and to one dilatational degree of freedom $\gamma^{0}(\mathbf{k})$, related to the Fourier transform of the energy density. (Fourier transforms, taken with respect to the harmonic coordinates, can be used to eliminate the $\mathbf{x}$ dependence of the γ^{α}.)

It then turns out that the three constraints $R_k^0 = 0$ imply the vanishing of the momenta canonically conjugate to the $y^A(\mathbf{x})$, and that the only remaining constraint can be expressed invariantly in terms of the $\gamma^{\alpha}(\mathbf{k})$ and their conjugate momenta $\pi_{\alpha}(\mathbf{k})$. At this stage, we carry the above reduction procedure one step further, by defining a "time-like" dynamical variable $\sigma(\mathbf{k})$, which is an invariant *functional of the metric at* $t = 0$. (Remember that we never have to leave the initial hypersurface $t = 0$, which must contain the imprint of the whole time evolution!) Physically, $-\sigma(\mathbf{k})$ is the dynamical variable which is canonically conjugate to the Fourier transform of the energy density $E(\mathbf{k})$.

The fourth constraint can then be rearranged in the form

$$E(\mathbf{k}) = H\,[\mathbf{k}, \gamma^{\pm}(\mathbf{1}), \pi_{\pm}(\mathbf{1}), \sigma(\mathbf{1})],$$

where H, which plays the role of an "effective Hamiltonian", is a function of $\mathbf{k}$, and a functional of $\gamma^{\pm}$, $\pi_{\pm}$, and σ. This equation can be appropriately called the "Hamiltonian constraint", because its Poisson brackets with *physical quantities* give their *invariant evolution* as functions of $\sigma(\mathbf{k})$.

To show this explicity, let us consider a *four-dimensional invariant*, i.e. a functional of the dynamical variables which is completely unaffected by gauge transformations. Such an absolute invariant must have vanishing Poisson brackets with *all* the constraints, and thus can be constructed as a functional $F\,[\gamma^{\pm}, \pi_{\pm}, \sigma]$ satisfying

$$\frac{\delta F}{\delta\sigma(\mathbf{k})} = [F, H(\mathbf{k})].$$

This remarkable "equation of motion for absolute invariants" is entirely formulated in terms of the metric at $t = 0$.

To see how this can happen, we first note that our original variables $\gamma^{\alpha}(\mathbf{k})$ and $\pi_{\alpha}(\mathbf{k})$ form a redundant set of initial data, because of the fourth constraint. A complete and nonredundant set of variables may be taken as the $\gamma^{\pm}(\mathbf{k})$, $\pi_{\pm}(\mathbf{k})$ and $\sigma(\mathbf{k})$. The fourth constraint then expresses $E(\mathbf{k})$ in terms of the above variables.

The point is, that by a suitable choice of variables[2] the redundant data on the initial hypersurface can be put in one-to-one correspondence with the evolution of the system outside the initial hypersurface[1]. The initial hypersurface therefore contains all the information on the evolution of the physical system.

Acknowledgements

I am indebted to A. Katz, C. W. Misner and M. Peshkin for many valuable comments.

REFERENCES

1. ECCLESIASTES, I, 9.
2. A. PERES, *Phys. Rev.*, **171**, 1335 (1968).

PAPER 32

A cosmological theory of gravitation

S. J. PROKHOVNIK
*Institute of Theoretical Astronomy, Cambridge, England**

ABSTRACT

A cosmological model, based on a uniformly expanding universe, can be represented as an acceleration field whose properties may provide the basis for the phenomenon of gravitation. The gravitational "constant" emerges as a parametric attribute of the model. The equivalence of gravitational and inertial mass is seen to be due to their common dependence on the field associated with a body.

1 BASIS OF THE FIELD

In previous communications[1,2] a model of a uniformly expanding universe was described in terms of a family of mutually-receding galaxies and associated fundamental observers. It was assumed that the distance $R(t)$ between any pair of fundamental observers is related to their mutual recession velocity w and their measure of cosmic time, t by the Hubble law

$$R(t) = wt, \tag{1}$$

where $R(t)$ and w are the estimates of these quantities by each fundamental observer from his light-intensity and Doppler redshift measurements, and where the present value of t is given by the reciprocal of the Hubble constant.

It was further assumed following McCrea[3], that this system of mutually-receding galaxies (the fundamental particles of our model) defines a basic reference frame or substratum for the propagation of light and indeed of all

* On leave from the University of New South Wales, Sydney, Australia.

forms of energy. This assumption leads to the result that if a light-signal passes successive fundamental observers F_0 and F at epochs (of cosmic time) t_0 and t respectively, then the distance r travelled by the signal relative to its source is given by

$$r = ct \log (t/t_0) \tag{2}$$

$r(t)$ being the space-interval (considered as a luminosity-distance) separating F_0 and F at epoch t. It has been shown elsewhere[1,4,5] that the resulting cosmological model is consistent with the operation of Einstein's principles and of relativistic phenomena in our universe, that it has a number of interesting astronomical implications, and that it may be represented as a hyperbolic velocity space. It also follows that the model manifests a cosmological acceleration field resulting from the time-dependence of its basic substratum.

For consider the view point of any fundamental observer F. In accordance with the Hubble law (1), any point in space, distant r from F and at cosmic time t, is associated with a recession velocity w, given by

$$w = \frac{r}{t}. \tag{3}$$

Thus w varies with respect to both r and t, such that, for instance a fixed space-interval is associated with decreasing recession velocities. These properties of the substratum, considered relative to F, are given exact expression by

$$\frac{\mathrm{d}w}{\mathrm{d}t} = \frac{-r}{t^2} + \frac{\dot{r}}{t} \tag{4}$$

which follows immediately from (3).

As our first additional assumption we will postulate that the acceleration factor described by (4) affects the motion, relative to F, of any material particle or of radiation at $P(r, t)$. In other words we are assuming that relative to any fundamental observer such as F, there exists a cosmological acceleration field

$$\ddot{r} = \frac{-r}{t^2} + \frac{\dot{r}}{t} \tag{5}$$

which applies equally to material bodies and radiation at any $P(r, t)$.

2 PROPERTIES OF THE ACCELERATION FIELD

The solution of (4) for a light-signal transmitted by F at $t = t_0$ is

$$r = ct \log (t/t_0)$$

since $r = 0$ and $\dot{r} = c$ when $t = t_0$. This result agrees with (2), so that our assumption of a cosmological acceleration effect on radiation is equivalent to assuming McCrea's hypothesis for a uniformly expanding universe. The observed behaviour of light passing in the vicinity of the sun suggests that a (gravitational) acceleration field affects equally the motion of both radiation and material bodies, so that our first new assumption constitutes only a small and justifiable step forward from the basic assumption underlying our model.

It is of interest to solve (5) for a body leaving F with uniform velocity u at $t = t_0$. We obtain

$$r = ut \log (t/t_0)$$

and

$$\dot{r} = u + u \log (t/t_0) = u + (r/t) = u + w.$$

Thus it follows from our assumptions that a body moving with a given velocity past a fundamental observer will proceed to pass every fundamental observer in its path with the same velocity; that is, it will, like radiation and in the absence of any other effects, move with constant velocity relative to the cosmological substratum. It is seen that the existence of a time-dependent substratum, as described by our model, would require a subtle modification of Newton's First Law of Motion. The Law would, of course, continue to hold precisely for bodies, such as fundamental observers or particles, stationary in the substratum. Applying (5) to determine the effect on the fundamental observer F, we have $r = \dot{r} = 0$ so that $\ddot{r} = 0$; and for any other fundamental observer or particle distant $r \neq 0$ from F, we have $\dot{r} = w = r/t$ so that $\ddot{r} = 0$ again.

The acceleration effect described by (5) consists of two parts. The second part, $\dot{r}/t$, is the larger and is directionally oriented. It may be interpreted as an effect due to a particle's (or light-signal's) recession from F so that r is increasing and so is the corresponding velocity-vector. $\dot{r}/t$ is the acceleration which enables the particle (or light-signal) to pass every fundamental observer with the same velocity. It is interesting that Bastin[6] obtained a similar result from his cosmological approach to gravitation.

The first part, $-r/t^2$, though apparently smaller, remembering that at present $t = T$ (of the order of 10^{10} years), is nevertheless the more interesting. We can consider it as due to the diminution of the velocity-vector with time for a fixed distance r from F; that is

$$\frac{\partial w}{\partial t} = (\ddot{r})_{r\,\text{constant}} = \frac{-r}{t^2}. \tag{6}$$

This is a very small effect and applies equally for all directions in space; hence the net effect on any body or light-signal will generally be zero. The effect would appear to be associated with the density of matter in our model universe, noting that for $t < T$ the density of the universe was greater and so was the value of $\partial w/\partial t$. Hence we will propose as our second new assumption that this effect depends on the density of matter (both its material and energy forms) in our universe and that local variations in the density are associated with corresponding variations of the effect in that locality. Combining this assumption with (6) and ignoring in the first place any possible effects of relative motion we have

$$(\ddot{r}) = \frac{-r}{t^2}\frac{\varrho}{\varrho_0} \tag{7}$$

where $(\ddot{r})$ is the gravitational acceleration of a particle (or light-signal) due to the presence of matter whose centre of mass at C is at a distance r from the particle; ϱ_0 is the average density of matter in our model universe at cosmic time t, and ϱ is the average density of matter in the spherical region which has its centre at C and the particle on its periphery so that the radius of the region is r.

In terms of the usual meaning for the density we have

$$\varrho = \frac{3M}{4\pi r^3};$$

and writing T for the present value of t, (7) becomes

$$(\ddot{r}) = \frac{-GM}{r^2} \tag{8}$$

where

$$G = \frac{3}{4\pi\, T^2 \varrho_0}. \tag{9}$$

The gravitational constant G has been related to many different combinations of the other natural constants and a relationship similar to (9) has been proposed by Sciama[7] and others from quite different considerations. However, in one other important respect, our approach does agree closely with that of Sciama. He contends[7] that gravitation and inertia are universal phenomena which must be related to the quantity and distribution of matter in the universe in accordance with Mach's Principle. It is seen that our model satisfies this requirement in terms of a velocity space and associated acceleration field which manifest both the large-scale recession of galaxies and the smaller-scale (intragalactic) operation of gravitational acceleration fields. Indeed it is suggested that the gravitational field phenomenon may be a property pertaining uniquely to a uniformly expanding universe.

Our approach leads to a theoretical value of G which is in close agreement with its observed value. Estimates of ϱ_0 range from 10^{-31} to 10^{-28} gm/cm^3. However if the quantity of intergalactic matter is as great as would appear from recent astronomical observations, then, following Bondi (1961), a figure such as 2×10^{-29} in the upper part of this range has claims to being a fair estimate. The most recent estimates of T are of the order of 1.3×10^{10} years or 4×10^{17} seconds. Employing these values for ϱ_0 and T in (9) yields

$$G = 7 \times 10^{-8} \text{ cm}^3 \text{gm}^{-1} \text{sec}^{-2}$$

which is of the same order as the observed value of 6.7×10^{-8}. Remembering that the estimates employed for ϱ_0 and T involve only one significant figure, the agreement between the two values is as good as could be expected. Note that in the context of our model, G is not an absolute constant but rather a parameter which depends on ϱ_0 and t such that its value increases directly with the measure of t. This may have relevance to problems associated with the evolution of the universe; the condensation and size of the galaxies, and the formation and size of stars may be linked with the values of G at certain epochs of the cosmological time-scale.

3 FURTHER IMPLICATIONS, THE EQUIVALENCE OF GRAVITATIONAL AND INERTIAL MASS

The result (8) describes the field associated with a body stationary in the substratum. We require one more assumption to deal with the problem of a body moving relative to the fundamental reference frame.

Our cosmological gravitational field is associated with the uniform expansion of our model universe and we have assumed that this field is intensified in the presence of a material body. It can therefore be considered as a form of energy deriving from the uniform expansion effect and associated with the presence of material bodies. A body stationary in the substratum will determine the strength* of a symmetric field in its vicinity. Now if the body moves relative to the substratum, in what manner will the field be affected? We will propose as our third new assumption that modifications of the original field, due to change in the body's position, travel through the substratum with the velocity of light, that is, movement of the body generates a flow of energy to modify the field. Thus at a point distant r km from a body, the field will remain unaltered for a period of r/c sec after the body has commenced moving but will then be affected by an energy wave which will modify the field.

Thus we are postulating the existence of gravitational waves travelling with the same velocity as light and associated with any change in the mass-energy of a body or of its position relative to the substratum. Recent reports suggest the observation of such waves. However it should be noted that the net effect of a non-impulsive movement of a body through the substratum would be a smoothly-continuous modification of the surrounding gravitational field; the effect on a particle stationary in the field would be equivalent to a relative velocity effect, that is, as if the body were stationary and the particle moving in the substratum. The Principle of Relativity applies here also, notwithstanding the existence of the substratum.

The mathematical description of a field affected by gravitational waves of finite velocity requires the employment of retarded potentials. This has already been attempted by Surdin[8] and more recently by Coster and Shepanski.[9] Surdin has shown that this approach is sufficient to explain the precession in the perihelion of Mercury and the observed deflection of light passing near the sun. Coster and Shepanski, on the other hand, have employed the approach to deduce that the gravitational field, due to a body of rest-mass m_0 and moving with velocity u, is given by

$$\ddot{r} = \frac{-Gm_0\,(1 - u^2/c^2)}{r^2\,(1 - u^2/\sin^2\theta/c^2)^{3/2}} \tag{10}$$

where θ is the angle between the direction of the body's motion and the radius vector of magnitude r. This result is, of course, analogous to a similar

* This strength defines the body's "gravitational mass".

law for a moving electric charge. Coster and Shepanski have also deduced from (10) and other considerations, that, in the circumstances of the field generated, both the effective gravitational mass and the effective inertial mass are given by

$$m = \frac{m_0 (1 - u^2/c^2)}{(1 - u^2 \sin^2 \theta/c^2)^{3/2}}. \tag{11}$$

All these results are valid and intelligible in terms of our cosmological acceleration field. The result (10) describes how the gravitational field associated with a body is modified by its movement in the substratum. Moving a body affects not only the symmetry of the field relative to the body, it also affects the energy of the field; indeed energy is required to alter the field. In our context this requirement is the source of a body's resistance to change of motion, that is of its property of inertia. In this way a body's gravitational mass and inertial mass are seen to be related to the same property of the body—the strength of the gravitational field associated with it. This strength depends on the strength, m_0, of the zero-motion field as well as on the velocity u relative to the substratum in accordance with (10) and (11). Such an interpretation of the equivalence of gravitational and inertial mass lends new meaning to the variation of mass formula previously derived from purely relativistic considerations.

4 FINAL OBSERVATIONS AND A CONJECTURE

The gravitational theory developed above resembles Einstein's rather than Newton's in so far as it is a field theory. However unlike Einstein's approach, it rests firmly on cosmological considerations which provide a fundamental reference frame and an acceleration field, centrally-directed with respect to every point in the universe.

This reference frame or basic substratum is observationally distinguishable from all other inertial and relativistically—equivalent reference frames; for the Doppler redshift in radiation from distant galaxies appears isotropic only to the fundamental observers, that is in respect to the basic substratum. It is not isotropic in respect to any other reference frame, for instance (as is well known), to one based on the earth or the sun.

The apparent existence of a uniform 3°K background radiation in our universe suggests a second criterion for defining the cosmological substratum. Only with respect to the fundamental reference frame should the frequency of this radiation appear isotropic. Observations by Conklin[10] suggest that this may indeed be the case.

These observational anisotropies and the implication that the background radiation also partakes in the expansion of the universe provide strong support for McCrea's light-propagation hypothesis and its far-reaching consequences; they strongly suggest that it is the propagation of radiation itself which is not isotropic relative to our terrestrial reference frame—even though, as a result of the operation of relativistic phenomena, the velocity of this propagation appears isotropic according to our measurements—and hence that our expanding universe does indeed constitute a basic substratum for energy propagation and for the operation of gravitational and other energy fields.

The gravitational field has, however, a cosmological significance which distinguishes it from the electromagnetic and other shorter-range fields. The former has its basis in the existence of a residual universal energy field—the cosmological acceleration field described by (5) and (6). Whereas the other fields can be interpreted in terms of transfers of energy which assumes the form of transverse waves during propagation but of discrete quanta (photons, mesons, etc.) on interaction with matter. The nature of this transfer has long been a matter of conjecture, and it is suggested that the residual energy field, deriving from the uniform expansion of the universe, may be the vehicle for the transfer of other forms of energy and so provide the basis for the manifestation of transverse energy waves. In this way Faraday's conjectures[11] "On Ray Vibrations" are seen to assume a new and cosmological significance.

REFERENCES

1. S. J. Prokhovnik, *Proc. Camb. Phil. Soc.*, **60**, 265 (1964).
2. S. J. Prokhovnik, *The Logic of Special Relativity*, Cambridge University Press, London, 1967.
3. W. H. McCrea, *Proc. Math. Soc. Univ. S'ton.*, **5**, 15 (1962).
4. S. J. Prokhovnik, *Int. J. Theor. Phys.*, **1**, 101 (1968).
5. S. J. Prokhovnik, *Proc. Camb. Phil. Soc.*, in print.
6. J. A. Bastin, *Proc. Camb. Phil. Soc.*, **56**, 401 (1960).
7. D. W. Sciama, *The Unity of the Universe*, Faber, 1959.
8. M. Surdin, *Proc. Camb. Phil. Soc.*, **58**, 550 (1962).
9. H. G. L. Coster and J. R. Shepanski, *J. Phys., A (Ser. 2)*, **2**, 22 (1969).
10. E. K. Conklin, *Nature*, **222**, 971 (1969).
11. M. Faraday, *Phil. Mag.*, **28**, 345 (1846).

PAPER 33

The recent renaissance of observational cosmology

D. W. SCIAMA*
University of Cambridge, England

It was just 51 years ago, in 1917, that Einstein inaugurated relativistic cosmology in the famous paper which introduced the finite but unbounded universe which is now named after him. Curiously enough this was a false start because the Einstein universe is static, self gravitation being overcome by the repulsive effect of the rather artificial cosmological term which Einstein added to his original field equations of 1915. No doubt it was natural to think in terms of a static universe in 1917, yet in that same year Einstein wrote another famous paper in which he discussed the thermodynamics of radiation in quantum theory, and introduced the A and B coefficients. Had he applied these considerations to his cosmological model he would have seen immediately that the existence of hot stars separated by cold stretches of interstellar space is not compatible with an infinitely old static system (Olbers' paradox). In the event de Sitter showed in the same year that the amended field equations admitted a nonstatic solution, and in 1922 the Russian meteorologist Friedmann discovered that the original field equations led to a range of possible expanding and contracting models. The contracting models can also be ruled out by thermodynamic considerations, leaving just models which, at least at the present time, must be expanding. Further work on these solutions was carried out by Weyl, Lemaitre, Eddington, Robertson, Tolman, Milne, McCrea and Walker, and by the mid thirties these homogeneous and isotropic models were well understood.

* Reprinted from *Physics Bulletin*, **19**, 329 (1968), by kind permission of *The Institute of Physics and The Physical Society.*

Parallel with this development came the observational discovery of the extragalactic nature of the spiral nebulae and the large red shifts in their spectra (large by comparison with stellar red shifts). Again the origins of this great discovery were somewhat confused. The first spiral nebula (hereafter called galaxy) to have its radial velocity measured was the Andromeda galaxy. This was in 1912, when Slipher of the Lowell Observatory found its spectrum to be *blue* shifted by about 200 km s^{-1}. By 1914 Slipher had measured the spectra of 14 galaxies all but two of which he found to be receding at a velocity of from 150 to 300 km s^{-1}. However, it was not until 1924 that it was shown conclusively (by Hubble) that the spiral galaxies lie outside our own Milky Way. Moreover, it was discovered only in 1926–27 (by Lindblad and Oort) that the Milky Way is in rotation, the velocity of the sun around the centre according to present estimates being about 250 km s^{-1}. This motion of the sun must clearly be corrected for, if we are to obtain the velocities of the galaxies relative to the Milky Way as a whole. It was in 1929 that Hubble first announced his linear relation between recession velocity and distance

$$v = \frac{r}{\tau}, \tag{1}$$

and we now know that his value for the Hubble constant τ was too small by a factor of about 5 (the present value is close to 10^{10} years with an unknown uncertainty which could be as large as 50%). In the next few years Humason and Hubble extended the observations out to velocities of about one seventh of the velocity of light and Hubble summarized the situation in his classic book *The Realm of the Nebulae*, published in 1936.

Thus by the mid thirties theory and observation were in satisfactory agreement in the sense that all the homogeneous isotropic world models led to the Hubble law, Eq. (1), in first approximation. The models differed in the next approximation, but not even the 200 in telescope, which came into operation in 1949, led to a reliable determination of the second order term. We do not know, for instance, whether the expansion of the universe will continue indefinitely or will be halted by self gravitation and turned into contraction. It seems fair to say that observational cosmology made very little progress from 1936 until the early sixties, when radioastronomy came to the rescue. On the theoretical side various unorthodox proposals were made, of which the most influential was the steady state theory of Bondi, Gold and Hoyle (1948) with its daring suggestion of the continual creation of matter.

The evidence against this theory is now very strong, but in its time it played an important role in forcing Hoyle and his associates (E. M. and G. R. Burbidge and W. A. Fowler) to devise their theory of the origin of the elements in hot stars. As we shall see, this latter theory may still be largely correct (with the important exception of the origin of helium).

The sterility of observational cosmology ended dramatically in the early and mid sixties. In 1965 two separate and independent discoveries were made which rank with the greatest in astronomy and, from the cosmological point of view, are nearly as significant as the discovery of the expansion of the universe itself. These are the detection of objects with extremely large red shifts and the discovery of the cosmic black body radiation. Needless to say we do not yet know the full implications of these discoveries, but already it is clear that we are experiencing a great renaissance of observational cosmology. For this reason I propose to devote the rest of this article to these developments, to help celebrate the Jubilee of The Institute of Physics and in recognition of the crucial role played by British physicists and radio-astronomers.

THE RADIO SOURCE COUNTS

The first attempt to use counts of radio sources to draw cosmological conclusions was made by Ryle and Scheuer in 1955. They had reason to believe that most of the radio sources which were contained in the second Cambridge catalogue (2C) were extragalactic, so that the distribution of the sources had to do with the structure of the universe as a whole. Ryle and Scheuer came to the conclusion that the counts were incompatible with the steady state theory, and thereby provoked a long, and sometimes violent, controversy, echoes of which can still be heard occasionally today.

The counts themselves consist of the number $N(S)$ of radio sources per unit solid angle whose measured flux density at the operating frequency of the radio telescope exceeds the quantity S. Because of the inverse square law the relation between N and S which would be expected for a uniform distribution of stationary sources has the form

$$N \propto S^{-3/2}.$$

A plot of $\ln N$ against $\ln S$ would then be expected to be a straight line of slope $-3/2$. As we shall see, when the red shift of extragalactic objects is taken into account the quantity $NS^{3/2}$, instead of being independent of S, should decrease with decreasing S. In other words, the ln N/ln S curve should be *flatter* than in the static case. The observed curve is, however, *steeper*.

The anomalous steepness found by Ryle and Scheuer was very marked indeed. For the fainter sources in their analysis the slope of the ln N/ln S curve was -3. We now know that the 2C survey was confusion limited below a relatively large flux density S and that many of the faint sources recorded are actually spurious. Some part of the anomalous steepness is now believed to be due to this effect. Three years later, in 1958, Mills, Slee and Hill used their Sydney catalogue of sources to derive a new slope for the ln N/ln S curve and obtained the value -1.8 (although they regarded their results as compatible with a slope of -1.5). This slope is still anomalously steep, but it has been confirmed by many further surveys, such as those of Scott and Ryle, and of Gower. The most recent and comprehensive ln N/ln S curve,

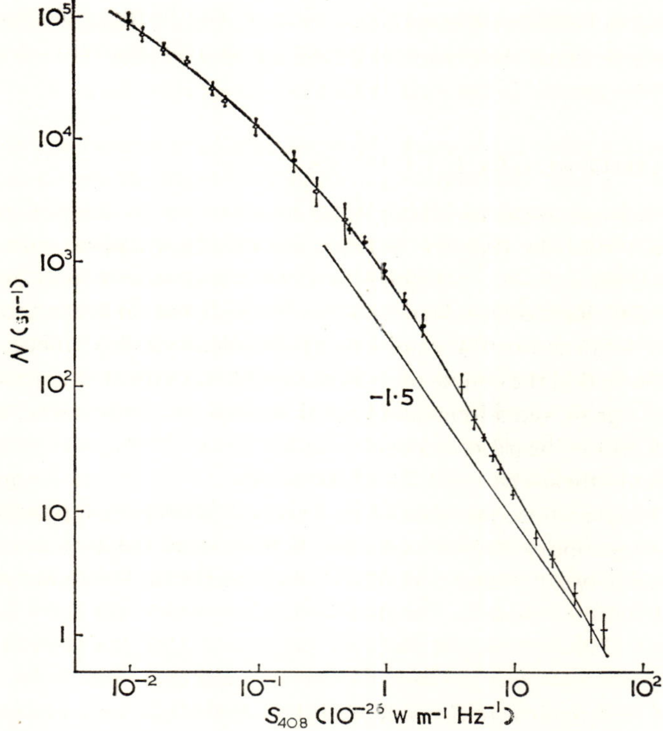

Figure 1. Counts of radio sources[1]. N is the number of sources per unit solid angle whose flux density at 408 MHz exceeds S_{408}

based mainly on Cambridge data, is that due to Ryle and Pooley (Figure 1). It will be seen that for very low values of S the slope has flattened down to about -1.

We now consider the effects of the red shift on the simple theoretical three-halves power law. There are three such effects, all of which are of progressively increasing importance as S decreases:

1) The effective intrinsic intensity of the sources depends on the red shift, since we are observing, in one small frequency band, radiation emitted in a different small frequency band. Allowance must therefore be made for the spectrum of each source.

2) The red shift reduces the apparent brightness of a source over and above the effect of the inverse square law. The distance of a radio source of given S is thus reduced, and so N is reduced.

3) If the red shift is taken to imply an evolutionary universe with no creation of matter, then the expansion of the universe implies that the density of sources was greater in the past. This leads to an increase of N.

In practice effect (1) is small. Most sources have a spectrum of the form $S(\nu) \propto \nu^{-0.7}$, near the frequency ν which is relevant for the Cambridge observations (a few hundred MHz). When we allow for the effect of red shift on the bandwidth, the effective intrinsic luminosity increases with red shift $z\ (=\delta\lambda/\lambda)$ like $(1 + z)^{-0.3}$. Even with z running up to 2 or 3 (see later) this is a fairly weak dependence. By contrast, effects (2) and (3) are very important, and calculation shows that in all reasonable cosmological models (2) is more important than (3) (which is, of course, completely absent in the steady state model). Thus in all likely cosmological models the direct effect of the red shift is to *flatten* the slope of the $\ln N/\ln S$ relation. In this way we arrive at a complete contradiction with the observations.

This contradiction was resolved by Ryle and Scheuer by exploiting the fact that in an evolutionary universe objects with large red shift are being observed at an earlier stage in the development of the universe than are nearby objects of small red shift. The possibility then arises that there has been a significant evolution in the intrinsic properties and distribution of radio sources in the time interval between emission and reception of the radiation. Since we lack a detailed understanding of the origin and development of these sources we are free at this stage to suppose that they have evolved in whatever manner is required to account for the $\ln N/\ln S$ relation. In particular, we would obtain a slope steeper than -1.5 if we assumed that in the past sources had on the average a sufficiently higher intrinsic intensity or a sufficiently higher concentration (over and above the kinematical effects of the expansion) than they have today. Such an explanation is clearly not available to the steady state theory, which requires all intrinsic properties of the sources to have the same average values at all times and in all places in the universe.

Detailed attempts to fit the observations in this way have been made by Davidson and Davies, Longair, and Rowan-Robinson. These attempts suffer from the difficulty that most of the radio sources involved in the counts have not yet been identified optically. In particular we know that two different types of object make substantial contributions to the counts, namely radio galaxies and quasistellar radio sources or quasars. We would clearly like to know the relative importance of these two populations for the anomalously steep slope of the counts. There is in fact preliminary evidence that the quasars are at least in part responsible for the steep slope, and in view of their extraordinary nature, we shall discuss this evidence now, despite its tentative character.

QUASAR COUNTS

The discovery of quasars is an oft-repeated story so we may be brief about it. It begins in 1960 when angular diameters were measured for the brightest 3C sources, thanks mainly to the work at Jodrell Bank. Several of these radio sources had very small angular diameters and so were of special interest. It later turned out that this was a somewhat accidental approach to the discovery of quasars, many of which have in fact substantial radio angular diameters. At any rate in 1960 it was of great interest that 3C48, 3C286, 3C196 and 3C147 had unusually small angular diameters. In September of that year Sandage took photographs with the 200 inch telescope of the regions containing the first three of these sources. These photographs were studied by Matthews who found that in each case the only visible object in the error rectangle of the radio position was what appeared to be a star. In October Sandage obtained a spectrum and photoelectric colours for 3C48. The optical spectrum was very strange, consisting of broad emission lines which could not be identified. Moreover the optical brightness varied appreciably on a time scale comparable with one day. The object was therefore regarded as a star with a puzzling spectrum.

All this was changed early in 1963 when the position of another 3C source, 3C273, was reported by Hazard, Mackey and Shimmins. This position had an unprecedented accuracy (better than 1 second of arc), being derived from observations of a lunar occultation of the source. There was therefore no doubt of its optical identification, which was of a thirteenth magnitude blue star. Schmidt obtained an optical spectrum of this object, which again had broad emission lines with no obvious identification. Then came the historic moment. Schmidt decided to see whether he could interpret the spectrum in terms of a substantial red shift despite the presumption that the object was a star in our galaxy. He was successful. Four of the emission lines fitted very well with the H_α, H_β, H_γ, H_δ lines of hydrogen with a red shift $\delta\lambda/\lambda$ of 0.158 (a fifth hydrogen line in the red being discovered later by Oke), while the other emission lines also had immediate interpretations in terms of this red shift. If this is a Doppler shift the "star" is moving away from us with nearly 16% of the speed of light. This result was published early in 1963, and the quasar era had begun.

It was immediately evident that if the red shift of 3C273 obeys the Hubble law, as the red shifts of radio galaxies appear to do, then this source is exceedingly bright in intrinsic optical power. For its distance would be 5×10^8 parsec, and since it is of thirteenth magnitude its intrinsic optical brightness would be about 100 times greater than that of the brightest known galaxy. This raises profound problems for the astrophysicist, but for the cosmologist the significant inference is that quasars at much greater distances should still be readily detectable and yet have very large red shifts indeed. A first step towards the realization of this was achieved almost immediately. Stimulated by Schmidt's discovery Greenstein and Matthews solved the mystery of the spectrum of 3C48. This source is 3 magnitudes fainter than 3C273, and its spectrum becomes readily understood if it has a red shift of 0.367.

This is a very large red shift by Hubble's standards but it was soon far exceeded. In 1965 Schmidt found a quasar (3C9) with the fantastic red shift of 2.012, a source in which, for the first time, the basic hydrogen Lyman α line (1216Å) was seen from the ground, shifted into the visible at 3666Å. This great result required some intricate argumentation to justify, but so many large red shifts are now known that there is no longer any spectroscopic doubt about the interpretation. If we represent a red shift of 2 in terms of the Doppler formula of special relativity we find a velocity of recession close to 80% of the velocity of light. At last it seemed that Hubble's dream would be realized, that we could observe objects so distant that the linear approximation of Eq. (1) would be insufficient and that it would be possible to distinguish between the different cosmological models.

Unfortunately it has turned out that there is so much spread in the intrinsic properties of quasars, that they are far from being the "standard candles" needed to make the cosmological test. Moreover, if these intrinsic properties vary with the cosmological epoch it will be extremely difficult to extract from the observations the correct model of the universe. The one test we can make is to see whether the steady state model is a possible one, since, as we have seen, this model permits no epoch dependent effects. To make this test one takes the 40 or so quasars in the 3C catalogue whose red shifts are known, since these form a homogeneous sample. One then asks whether the number of quasars of different red shifts is in agreement with the steady state expectation. If I may venture a personal note at this point I would say that I was very much hoping that the steady state theory would survive this test. Alas it did not. My student Martin Rees pointed out to me that there were far too many quasars of large red shift. This result has since been found by several other investigators the most thorough account having been recently published by Schmidt. It is, of course, significant that this discrepancy is in the sense to steepen the ln N/ln S relation for quasars.

The only loophole would be to deny that the red shift of quasars has a cosmological origin. It has been proposed by Terrell and by Hoyle and Burbidge that the quasars may be local, the red shift being either an ordinary Doppler effect unrelated to the expansion of the universe, or a gravitational effect. Along with most astronomers I find this very unlikely. The absence of blue shifts and the restrictions imposed by the known extragalactic radio background would require the quasar cluster nearest to our own local one to be very far away. It then becomes very improbable that we should be in a quasar cluster at all, not merely near the centre of one, as the observed isotropy would require. These are slim grounds on which to save the steady state theory. Moreover we shall meet another powerful argument against this theory when we come to discuss the cosmic black body radiation.

THE INTERGALACTIC MEDIUM

Unless the process of galaxy formation is 100% efficient intergalactic space must contain residual gas. This gas has not yet been detected but it is potentially of great cosmological importance since it may make an appreciable contribution to the mean density of matter in the universe. Indeed many of the relativistic cosmological models lead to a mean density far in excess of that due to the known galaxies. Of these the most attractive in many ways is the so called Einstein–de Sitter model, in which the expansion continues indefinitely—but only just; that is, the velocity of expansion tends asymptotically to zero. In this model the present density ϱ is given by

$$\frac{8\pi}{3} G\varrho\tau^2 = 1,$$

where G is the Newtonian gravitational constant. With the Hubble constant $\tau \sim 10^{10}$ years we have

$$\varrho \sim 2 \times 10^{-29} \text{ g cm}^{-3}.$$

By contrast the mean density ϱ_g contributed by galaxies so far observed is unlikely to exceed 10^{-30} g cm^{-3}. Thus if the Einstein–de Sitter model is even approximately correct most of the matter in the universe is unaccounted for.

The form this missing matter might take has been much discussed. It could be made up of very faint galaxies or intergalactic stars, rocks or neutrinos, without having been detected. However, the most interesting possibility is that it is gaseous, since, as we shall see, it would then be on the verge of detection. We shall also see that its composition would probably be 90% hydrogen, and 10% helium (by number), with an admixture of heavy elements very much smaller than the relative abundance in our galaxy.

Intergalactic *atomic* hydrogen has been searched for by several radio-astronomers who have attempted to detect the hyperfine transition at 21 cm both in emission and absorption. Their results have been negative, and despite some difficulties in interpretation one can say that the density of atomic hydrogen cannot significantly exceed the value 2×10^{-29} g cm^{-3}. A much more stringent limit has recently been obtained by considering the process of absorption at the Lyman α wavelength. Normally this absorption would be undetectable below the atmosphere because it is in the far ultraviolet, but Scheuer, and Gunn and Peterson pointed out that the intergalactic gas near a quasar with a red shift of about 2 would absorb at a wavelength which for a terrestrial observer would be in the visible. Since the Lyman α absorption involves a resonance transition from the ground state this is a very sensitive method for detecting intergalactic atomic hydrogen. Careful inspection of the spectra of quasars with a red shift of about 2 has failed to reveal any absorption shortward of the Lyman α emission line in the quasars (for example *see* Figure 2). The resulting limit on the present intergalactic density of atomic hydrogen is

$$\varrho_{\mathrm{H}} < 10^{-36}\ \mathrm{g\ cm^{-3}}.$$

By a similar analysis Field, Solomon and Wampler have placed an upper limit of 10^{-32} g cm^{-3} on the concentration of molecular hydrogen.

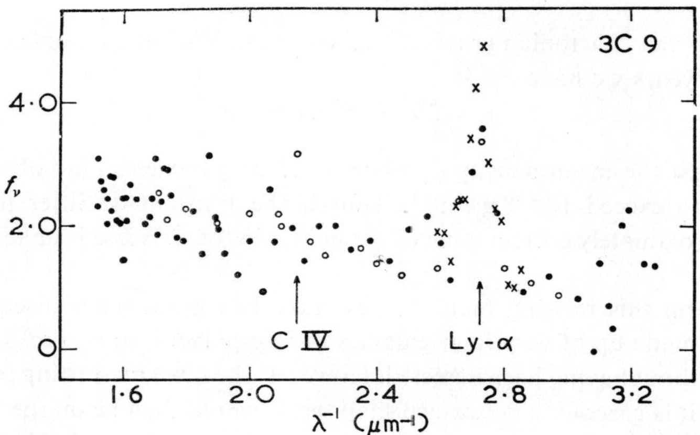

Figure 2. The photoelectric spectrum of the quasar 3C9[2]. Note that the level of the continuum does not change abruptly across the Lyman α-emission line

The likely explanation for these stringent limits is that the hydrogen is highly ionized. If the ionization is collisional in origin, the kinetic temperature of the gas must exceed about 3×10^5 °K if not more than 1 particle in 10^7 is to be neutral. Such a temperature could be achieved by heating processes emanating from galaxies, radio galaxies and quasars. On the other hand the temperature cannot be too high or the gas would radiate X-rays at a rate in excess of the known X-ray background (*see* Figure 3). If the gas density is approximately 2×10^{-29} g cm^{-3} the upper limit on its temperature is about 10^6 °K. If the X-ray observations can be extended out to 50 Å despite the severe galactic absorption at such a wavelength it should be possible to test whether the gas temperature exceeds 3×10^5 °K. The preliminary observations of Bowyer, Field and Mack are in fact compatible with such a temperature, but the experiment is a difficult one, and these observations need to be confirmed.[4]

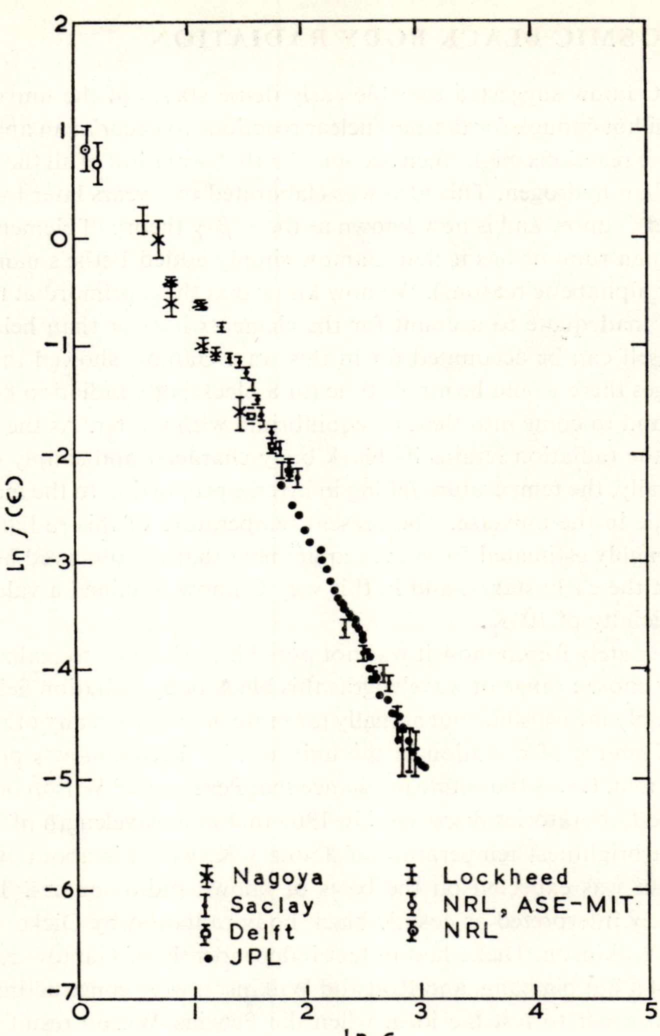

Figure 3. The diffuse X-ray background[3]. ε is the X-ray energy (KeV), and $j(\varepsilon)$ the flux (photons cm^{-2} s^{-1} KeV^{-1})

THE COSMIC BLACK BODY RADIATION

In 1946 Gamow suggested that the early dense stages of the universe may have been hot enough for thermonuclear reactions to occur at an appreciable rate. These reactions might then account for the formation of all the elements heavier than hydrogen. This idea was elaborated two years later by Alpher, Bethe and Gamow and is now known as the α–β–γ theory of element formation (though rumour has it that Gamow simply added Bethe's name to the paper for alphabetic reasons). We now know that these primordial processes are quite inadequate to account for the elements heavier than helium, but helium itself can be accounted for in this way. Gamow showed that in the early stages there would be ample time for a black body radiation field to be built up and to come into thermal equilibrium with matter. As the universe expands the radiation retains its black body character and simply cools off adiabatically, the temperature falling in inverse proportion to the increase of linear scale in the universe. The present temperature of this radiation field can be roughly estimated from the requirement that the observed helium be formed in the early stages, and in this way Gamow obtained a value in the general vicinity of 10 °K.

Unfortunately for Gamow it was not possible at the time to realize that in a suitably chosen range of wavelengths this black body radiation field would be not simply measurable, but actually far more intense than any other extra-terrestrial source of radiation in the universe. In fact Gamow's prediction was forgotten. It was thus quite by chance that Penzias and Wilson of the Bell Telephone Laboratories discovered in 1965 that at a wavelength of 7 cm the sky had a brightness temperature of about 3 °K, which is about 100 times hotter than was expected on the basis of known radio sources. This was immediately interpreted as cosmic black body radiation by Dicke, Peebles, Roll and Wilkinson. Dicke had in fact independently of Gamow conceived the idea of a hot big bang, and Roll and Wilkinson were constructing a 3 cm receiver in order to test the idea, when the Penzias–Wilson result was announced.

The critical step is clearly to check whether the spectrum of the excess radiation is that of a black body. Measurements have now been made at a number of wavelengths in the range from about 60 cm to 0.3 cm, and all these measurements are compatible with a black body temperature of about 2.7 $\pm$ 0.3 °K (*see* Figure 4). In addition there is an independent argument from the observed excitation of instellar CN that the radiation field at

0.25 cm has a similar temperature, which again far exceeds that expected from known sources.

In view of the importance of this question it is necessary to examine these observations very critically. This is not, however, the appropriate place, and

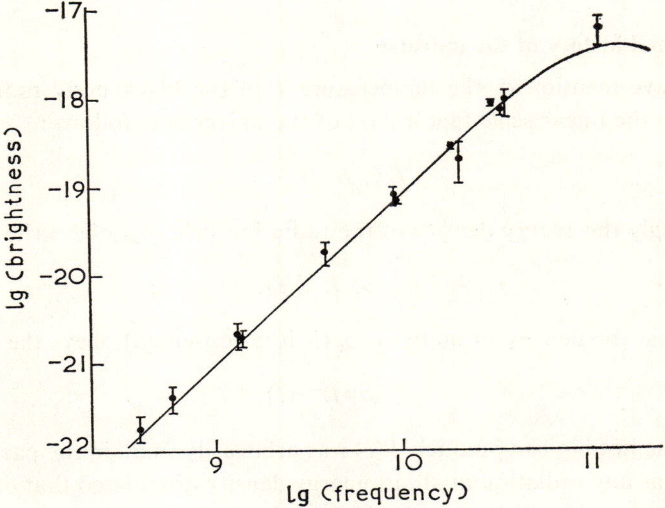

Figure 4. The diffuse microwave background[5]. The curve represents a black body spectrum at 2·7°K

it suffices here to say that the measurements, being absolute in character, are difficult to perform accurately and that relatively large corrections have to be made for extraneous effects such as atmospheric radiation, receiver noise etc. Despite these difficulties the general consensus of opinion seems to be that the measurements can be accepted as genuine, as we shall do for the remainder of this article. Final acceptance must await the results of the rocket measurements now being planned to measure the background in the vicinity of 0.1 cm, where the black body spectrum has its peak.

We shall now discuss the following topics which relate to the existence of the cosmic black body radiation:

1) The thermal history of the universe;

2) The helium problem;

3) Astrophysical effects of the black body radiation;

4) The peculiar velocity of the earth;

5) The isotropy of the universe;

6) The homogeneity of the universe;

7) Singularities in the universe.

The thermal history of the universe

As we have mentioned, the temperature T of the black body radiation is related to the linear scale factor $R(t)$ of the universe as follows:

$$T \propto R^{-1}(t). \tag{2}$$

Accordingly the energy density of the radiation field ϱ_{rad} obeys the relation

$$\varrho_{rad} \propto R^{-4}(t).$$

By contrast the density of matter ϱ_{mat} (if it is conserved) obeys the relation

$$\varrho_{mat} \propto R^{-3}(t).$$

Now in the hot big bang models $R(t)$ was arbitrarily small in the past and so if there was any radiation at all, its energy density dominated that of matter at sufficiently early times. A radiation dominated universe is easy to handle in general relativity with the simple result:

$$R(t) \propto t^{1/2} \quad (t \text{ small}),$$

$$T_{rad} = \frac{10^{10}}{t^{1/2}} \quad (t \text{ small}). \tag{3}$$

At later times two important things happen. The radiation ceases to be strongly coupled to matter when the matter cools down sufficiently so that it can recombine into atomic hydrogen ($T \sim 3000$°K) and the radiation ceases to dominate energetically. If we assume that the matter behaves approximately as a perfect gas then when it is uncoupled from the radiation its temperature obeys the law

$$T_{mat} \propto R^{-2}(t).$$

By comparison with Eq. (2) we see that the matter cools more rapidly than the radiation. Its temperature now should thus be much less than 3 °K, which contradicts the requirement of the last section that it should be about

3×10^5 °K. If this latter requirement is correct the intergalactic gas must have been reheated, presumably by emanations from galaxies, radio galaxies or quasars when they came into being.

It is useful to express the radiation/matter ratio in terms of the entropy per baryon S/n since this quantity is independent of time ($S_{\text{rad}} \propto T_{\text{rad}}^3 \propto R^{-3}(t)$). If the present value of the matter density is 2×10^{-29} g cm^{-3} (Einstein–de Sitter universe) we obtain for the entropy of radiation per baryon the quantity $10^8 k$, where k is Boltzmann's constant. The hot big bang theory in its present form does not specify the processes which produced the observed heat. Either it is a question of the initial conditions at $t = 0$, or processes occurred later which we can legitimately speculate about. For the moment this is an unsolved problem, but we shall mention a possible explanation when we discuss (5), the isotropy of the universe.

The helium problem

Investigation shows that wherever it can be measured spectroscopically or estimated theoretically (in the sun, the stars, the interstellar gas) helium has an abundance by number about 10% that of hydrogen. There are exceptions to this rule in the case of certain old stars, but there seems to be good evidence that these exceptions can be explained away. Now the stars in our galaxy could have manufactured only about 10% of the observed helium in the lifetime of the galaxy. It is therefore attractive to adopt the α–β–γ proposal that most of the helium was formed by thermonuclear reactions in the early stages of the hot big bang.

We see from Eq. (3) that at a time of 1 s after the big bang the temperature throughout the universe was 10^{10} °K. This is beyond the threshold for the creation of electron–positron pairs. Neutrino pairs would also be thermally excited, and the weak and electromagnetic interactions would in fact be strong enough to ensure that thermal equilibrium would prevail between protons, neutrons, electron pairs, neutrino pairs and photons. When the temperature drops somewhat below 10^{10} °K the weak interactions can no longer maintain the neutrons in statistical balance with the protons because the concentration of electron pairs is beginning to drop abruptly. The neutron–proton ratio is then frozen in, until a few hundred seconds have passed and neutron decay begins to be appreciable. This frozen-in ratio, corresponding to thermal equilibrium at a temperature somewhat below 10^{10} °K, is about 15%.

The following nuclear reactions among others now take place:

$$n + p \rightarrow {}^2H + \gamma \tag{4}$$

$$^2H + {}^2H \begin{cases} \nearrow {}^3He + n \\ \searrow {}^3H + p \end{cases}$$

$$^3He + n \rightarrow {}^3H + p$$

$$^3H + {}^2H \rightarrow {}^4He + n$$

The first reaction, Eq. (4), is the slowest, and at temperatures exceeding $10^9\,^\circ K$ there are enough photons to disintegrate the deuterons as soon as they are formed. This is no longer true at $10^9\,^\circ K$ ($t = 100$ s), so this is when the helium gets built up. At this stage the neutrons have their frozen-in abundance and nearly all of them combine with protons to form helium.

This frozen-in abundance depends only weakly on the material density and the entropy per baryon; the main dependence is on the temperature and the properties of the weak interactions. Thus, so long as the material density is great enough for the key reaction (4) to be more rapid than the expansion time, this fixed concentration of neutrons is incorporated into helium nuclei however great the material density may be. Thus the dependence of helium density on S/n has a plateau, which in fact is only appreciably departed from when S/n is so low that the universe was still matter dominated at $10^{10}\,^\circ K$.

There have been many calculations of this plateau abundance. The most accurate were carried out by Peebles in 1966 and by Wagoner, Fowler and Hoyle in 1967, with results in good agreement with the observed relative abundance of 10%. These results would be modified if the 10% reduction that has recently been proposed in the half life of the neutron is correct. For the weak interaction coupling constant would have to be increased, and so the neutrons would remain longer in thermal equilibrium with the protons. This would mean that the frozen-in abundance of neutrons would be reduced and so the resulting abundance of helium would also be reduced. According to Tayler this reduction would be by 10%, which is within the uncertainty of contemporary abundance determinations but is not altogether negligible.

The cosmological theory of the helium formation thus appears to be in good shape. However, the calculations of Wagoner, Fowler and Hoyle, which were very detailed (144 different reactions being included) show clearly that a negligible amount of elements heavier than helium is built up in these primordial reactions (*see* Figure 5). The origin of these heavier elements thus has a different explanation; at the moment it is not clear whether the build-up occurs predominantly in supernovae or whether some form of massive exploding object must be invoked.

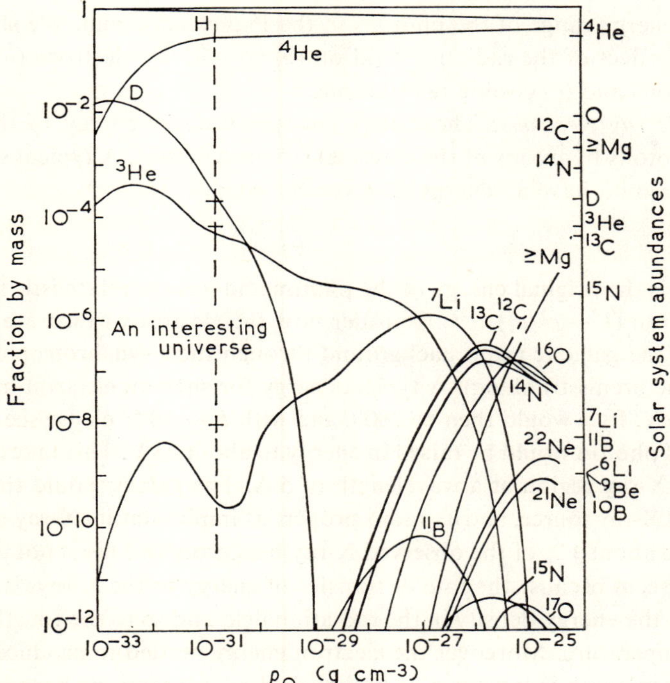

Figure 5. Element production is the hot big bang[6]. The abscissa ϱ_0 is the present mean density of matter in the universe

Astrophysical effects of the black body radiation

From a laboratory viewpoint 3 °K is a low temperature. Indeed to measure it the microwave observers had to use a reference termination immersed in liquid helium. Nevertheless from an astrophysical viewpoint 3 °K is a high temperature. A universal black body radiation field at this temperature contributes an energy density everywhere of about 10^{-12} erg cm^{-3} or 1 eVem^{-3}. This is just the energy density *in our galaxy* of the various modes of interstellar excitation—starlight, cosmic rays, magnetic fields and turbulent gas clouds. In intergalactic space these energy densities probably drop off by a factor of between 100 and 1000, whereas the black body component maintains its energy density at 1 eV cm^{-3}. We may mention that the number density of these photons is about 10^3 cm^{-3} and the mean energy per photon is about 10^{-3} eV. These quantities are useful for making quick estimates of many of the effects of the radiation field, without having to consider in detail the full energy range of the photons in the Planck spectrum. We shall consider the effect of the radiation field on (*a*) cosmic ray electrons (*b*) cosmic ray protons and (*c*) cosmic ray photons.

Cosmic ray electrons: These electrons will transfer energy to the black body photons by means of the (inverse) Compton effect. A typical scattered photon would have an energy E' given by

$$E' \sim \gamma^2 E,$$

where E is the original energy of the photon and γ is the relativistic factor of the electron $(1 - v^2/c^2)^{-1/2}$. Consider now the electrons which are responsible for the galactic radio background through their synchrotron emission (magnetic bremsstrahlung). A typical energy for such an electron might be, say, 1 GeV. Its γ would then be 2000, and with $E \sim 10^{-3}$ eV we see that the scattered photon would be raised in energy to about 4 kV. This takes us right into the X-ray region at a wavelength of 5 Å. The galaxy would thus be an extended X-ray source, and its radio properties imply that its X-ray intensity would be about 1% of the observed X-ray background. This is not as low as it might seem because the rate of transfer of energy to the X-rays is proportional to the energy density in the radiation field and so to the fourth power of its temperature. Moreover the electron energy needed to produce a given X-ray wavelength is less for a radiation field whose photons have a greater mean energy, and there are more electrons of lower energy in the cosmic rays. The net result of all this is that if the black body background had a temperature of say, 10 °K, the X-ray flux from the galaxy would be greater, and the energy drain on the electrons would be very large indeed.

By the same token the X-rays emitted from great distances, at a time in the past when the black body temperature was greater than now, cannot be ignored. Indeed the currently most attractive explanation for the origin of the observed X-ray background is that it is mainly due to inverse Compton processes in distant radio sources, for not only is the radiation density much greater in the past but also, as we have seen in discussing the radio source counts, the concentration of intense radio sources was also much greater in the past. An alternative explanation is that the X-ray background arises from inverse Compton processes in intergalactic space. Whatever its explanation, the X-ray background is likely to be of great cosmological significance.

Cosmic ray protons: From the viewpoint of a cosmic ray proton of 10^{20} eV, which has a γ of 10^{11}, a photon of 10^{-3} eV looks like one of 100 MeV. Such an energetic photon striking a stationary proton would be close to the threshold for producing a pion. This means that from the terrestrial viewpoint a cosmic ray proton of 10^{20} eV can collide with a black body photon, produce a pion, and so be degraded in energy. The importance of this process was first pointed out by Greisen who found that once the threshold is past the proton loses a substantial fraction of its energy in only 3×10^7 years. It is perhaps unlikely that cosmic rays with energies in the range 10^{18} to 10^{20} eV, which are almost certainly not confined to the galaxy by its magnetic field, have a lifetime of less than about 10^{10} years. Greisen therefore proposed that the energy spectrum of cosmic rays would drop very steeply beyond 10^{20} eV. Now it so happens that the spectrum of cosmic rays has been followed out to just about 10^{20} eV, without anything drastic being observed. This would imply that the black body temperature cannot significantly exceed 3°K. Detectors are now being built to extend the spectrum into the range 10^{21} to 10^{22} eV. If our present ideas are correct, not a single event should be detected. This has led some people to suggest that there is no need to build these detectors. In the present state of our knowledge this seems to me a very unscientific attitude.

Cosmic ray photons: If there are high energy cosmic γ-rays then above a threshold at 2.5×10^{14} eV they would be rapidly degraded in energy by interaction with black body photons leading to pair production. Such high energy γ-rays are now being searched for, but there are no definitive results as yet.

The peculiar velocity of the earth

The original measurements of Penzias and Wilson showed that the black body background is isotropic to a precision of a few per cent. Later measurements by Partridge and Wilkinson (Figure 6) and by Conklin and Bracewell (Figure 7) increased the precision to a few tenths of a per cent. Now such an isotropic radiation field defines a rest frame, namely, that frame in which the radiation is observed to be isotropic. An observer moving relative to that frame would, by virtue of the Doppler effect, see an increased intensity in front of him and a decreased intensity behind him. Thus motion relative to the black body radiation, and so to the universe as a whole, can be directly measured. The lack of any observed anisotropy limits the peculiar velocity of the earth to about 300 km s^{-1}, and future measurements should improve this limit or, more likely, actually detect the peculiar velocity. The reason is that in addition to the motion of the earth around the sun at 30 km s^{-1},

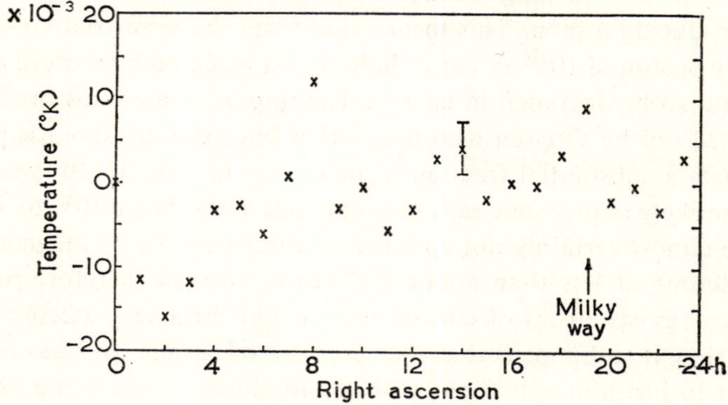

Figure 6. Changes in the temperature of the 3 cm background radiation along a circle parallel to the celestial equator at a declination of $-8°$ [7]

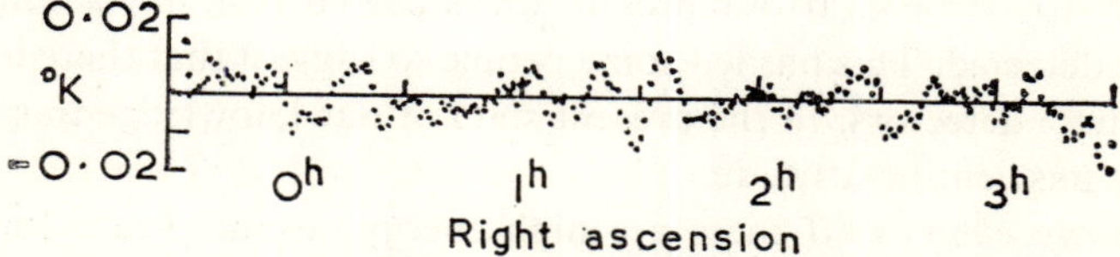

Figure 7. Same as Figure 6 for a different region of the sky[8]

the sun is moving around the centre of our galaxy at about 250 km s^{-1}, the galaxy as a whole is probably moving relative to the local group of galaxies at about 100 km s^{-1}, and the whole local group *may* be moving relative to the local supercluster of galaxies at a few hundred kilometres per second ... It is clear then that a measurement of the net peculiar motion of the earth would be of great importance for our understanding of the hierarchy of irregularities in the universe. It would also link up with Mach's principle, which asserts that local inertial frames are unaccelerated relative to the universe as a whole. We are here on the verge of great clarification.

The isotropy of the universe

The fact that the black body radiation is highly isotropic tells us that the expansion of the universe is highly isotropic too. Can we explain this or must we appeal to the initial conditions at $t = 0$? In an important new theoretical development, worked out most extensively by Misner, it has been proposed that any initial anisotropy may be dissipated away by viscous interactions. Much remains to be done before this proposal can be accepted, but it certainly opens up a new chapter in theoretical cosmology. It may even serve to explain the origin of the black body radiation, since the dissipative processes must produce heat. We may expect rapid developments along these lines in the near future.

The homogeneity of the universe

The high isotropy of the black body radiation limits the inhomogeneity of the universe as well as its anisotropy. Of particular interest for the future is the possibility, pointed out by Sachs and Wolfe, that large scale density fluctuations could affect the black body temperature through the Einstein red shift. In this connection it is intriguing to note that Wilkinson and Partridge have found preliminary signs of a small dip in temperature (see Figure 6) in a direction in the sky which coincides with an apparent cluster of quasars. The possible existence of such a cluster was pointed out by Strittmatter, Faulkner and Walmesley, and related to fluctuations in the black body temperature by Rees and myself. The evidence both for the temperature dip and the quasar cluster is quite uncertain at the moment, but in view of the importance of the problem we may expect the whole sky to be mapped out as precisely as possible in the next few years.[9]

Singularities in the universe

The final use to which we shall put the cosmic black body radiation is perhaps an unexpected one. We can use it to show that, according to general relativity, the universe must have been singular at some time or times in the past. It is well known that the exactly isotropic homogeneous (Friedmann or Robertson-Walker) models of the universe have a point singularity in the past (unless the field equations are modified by the cosmological term). It has often been suggested that this singularity is a consequence of the exact symmetry assumptions of isotropy and homogeneity. However, recently Hawking and Penrose have proved a number of important and powerful theorems which state in effect that even without these symmetry assumptions there must have been at least one singularity in the past, although not necessarily a point singularity. In the formal statement of these theorems a number of assumptions have to be made, most of which are entirely reasonable; for example, one has to exclude matter with negative energy density. However, some of these assumptions, while reasonable, are of a character that would be hard or impossible to check in the actual universe. Hawking, Ellis and Penrose have now been able to show that the presence of the cosmic black body radiation makes the most dubious of these assumptions unnecessary. Oversimplifying slightly we may say that if the universe is causal the gravitational action of the black body radiation ensures that the universe has expanded from one or more singularities, no physically reasonable nonquantum equation of state being able to prevent it. Whether this result must be evaded, and if so how, is not known. We are here at the limits of existing theory.

Acknowledgment

I am grateful to Dr. J. Shakeshaft for his comments on the manuscript.

REFERENCES

1. Published by courtesy of G. G. Pooley and M. Ryle.
2. E. J. Wampler, *Astrophys. J.*, **147**, 1, 1967*.
3. R. J. Gould, *Am. J. Phys.*, 35, **376**, 1967*.
4. The observations have since been confirmed by Henry, Fritz, Meekins, Byram and Friedman, who claim to have actually observed a dense, hot, intergalactic gas. This interpretation of their observations is not the only possible one, however, and the question remains open.
5. J. R. Shakeshaft and A. S. Webster, *Nature*, **217**, 339, 1968**.
6. R. V. Wagoner, W. A. Fowler and F. Hoyle, *Astrophys. J.*, **148**, 3, 1967*.
7. D. T. Wilkinson and R. B. Partridge, *Nature*, **215**, 719, 1967**.
8. E. K. Coklin and R. N. Bracewell, *Phys. Rev. Letters*, **18**, 614, 1967*.
9. Recent measurements at Princeton have shown that the evidence for the temperature drop is not significant.

PAPER 34

Gravitational fields in matter

PETER SZEKERES
King's College, University of London, England

The theory of gravitational radiation in vacuo is by now well understood, particularly with respect to the asymptotic properties far from bounded sources. However very little is known about the actual generation and propagation of gravitational waves in the sources. To discuss this problem it is first of all necessary to pick out a quantity which should represent the gravitational field. Since contractions of the curvature tensor are directly related to the sources by Einstein's field equations

$$R_{ab} - \tfrac{1}{2}Rg_{ab} = -\varkappa T_{ab},$$

a natural candidate for the free gravitational field is the Weyl tensor C_{abcd}, which is formed from the curvature tensor by removal of all traces in such a way that we are left with a tensor having the algebraic properties of a vacuum curvature tensor[1].

In vacuo many of the basic properties concerning the propagation of the gravitational field are consequences of the Bianchi identities which, in a four-dimensional space-time, take on the form

$$C^{;d}_{abcd} = 0.$$

An example of the type of law of propagation which follows is the Goldberg–Sachs theorem[2], which says that a vacuum gravitational field propagates along shear-free null geodesics if and only if it is algebraically special (i.e. if and only if it is "pure radiation"). In the presence of matter this elegant form of the Bianchi identities is upset by the appearance of a source term on the right hand side,

$$C^{;d}_{abcd} = \varkappa J_{abc} \tag{1}$$

20 Kuper/P

where

$$J_{abc} = -(T_{c[a;b]} - \tfrac{1}{3} g_{c[a}T_{,b]}).$$

The result of this is that even for algebraically special gravitational fields the rays of propagation, while still null, are no longer geodesic and shear-free. Typically the amount of shear and departure from geodicity is determined by the kinematic and dynamic properties of the medium[1].

It is tempting at first to regard the departure from geodicity as representing some kind of refraction of the gravitational waves. However this refraction is not anything like refraction as it appears in electromagnetism. For one thing the rays are still null lines and consequently there is no slowing down of the waves as there is for electromagnetic waves in a dielectric. The reason for this discrepancy is not hard to understand. Eq. (1) is in many ways similar to Maxwell's equations

$$F^{ab}_{\ \ ,b} = \frac{4\pi}{c} j^a.$$

Now if we want to treat an electromagnetic field in a material medium, the current j^a varies wildly from point to point as we pass from one atom to another. It is clearly impossible to discuss this situation in detail and it is necessary to consider the average current, averages being taken over volumes encompassing a large number of atoms yet small compared to the large scale fluctuations in the field quantities. The result is an *effective current*

$$(j^a)_{\text{eff}} = (j^a)_{\text{free}} - cP^{ab}_{\ \ ,b} \tag{2}$$

where P^{ab} is a (skew) tensor of polarization arising from the fact that the field induces dipole moments on the individual atoms. For the averaged fields the macroscopic equations are now

$$H^{ab}_{\ \ ,b} = \frac{4\pi}{c} (j^a)_{\text{free}}$$

where

$$H^{ab} = F^{ab} + 4\pi P^{ab}.$$

In a charge-free dielectric we retrieve essentially the vacuum Maxwell equations

$$H^{ab}_{\ \ ,b} = 0,$$

but the typical wave velocity is now $c/\sqrt{\varepsilon\mu}$ where ε is the dielectric constant and μ the permeability of the medium.

For gravitation it should be possible to give an analogous treatment, treating the medium as atomic in structure rather than continuous, with induced quadrupole moments on the atomic constituents. It should be possible then to derive "macroscopic" Bianchi identities,

$$C_{abcd}^{;d} = \varkappa J_{abc} + P_{abcd}^{;d}, \tag{3}$$

and the more familiar refraction effects should follow if suitable constitutive equations are imposed.

I have followed through the gravitational equivalent of the classical electromagnetic treatment[3] in the linearized approximation, for a corpuscular medium with quadrupole moments induced on the individual constituent atoms. The effective energy stress tensor entering into the field equations becomes

$$(T_{ab})_{\text{eff}} = (T_{ab})_{\text{free}} + \tfrac{1}{2}Q_{acbd}^{;cd}$$

where Q_{abcd} is essentially a tensor representing the average quadrupole moment per unit volume. If we define the "electric" component of the gravitational field as

$$E_{\alpha\beta} = C_{\alpha 0 \beta 0}$$

(there are also "magnetic" components $B_{\alpha\beta} = C^{*}_{\alpha 0 \beta 0}$; this splitting is carried out by complete analogy with electromagnetism), and one regards the atoms of the medium to be 3-dimensional harmonic oscillators, it may be shown from the equation of geodesic deviation that the quadrupole moment density induced on the atoms is given by

$$G_{\alpha\beta} = \varepsilon_g E_{\alpha\beta}$$

where

$$\varepsilon_g = 2\pi G m A^2 N / \omega^2,$$

G = Newtonian constant of gravitation
m = mass of atom
A = amplitude of oscillation = radius of atom
ω = frequency of oscillation
N = number of atoms per unit volume.

It is natural to call ε_g the *dielectric constant of gravitation.*

For a plane transverse wave in a homogeneous medium of this type the linearized approximation of Eq. (3) gives rise to a fourth order linear equation. This results in a dispersion relation $\omega = \omega(k)$ connecting the frequency

ω and the wave number k of the waves, and a corresponding group velocity

$$v_g = \mathrm{d}\omega/\mathrm{d}k = 1 - \tfrac{3}{4}\varepsilon_g k^2.$$

In this way we have managed to recover the phenomenon of refraction for gravitational waves in a form more familiar from electromagnetism. The effect will be significant for wavelengths

$$\lambda = k^{-1} \approx \sqrt{\varepsilon_g}.$$

REFERENCES

1. P. SZEKERES, *J. Math. Phys.*, **7**, 751 (1966).
2. J. N. GOLDBERG and R. K. SACHS, *Acta Phys. Polon.*, **22** (Suppl. 13) (1962).
3. J. D. JACKSON, *Classical Electrodynamics*, Wiley, New York, N.Y., 1962.

PAPER 35

Gravitational radiation experiments*

J. WEBER**

Institute for Advanced Study, Princeton, New Jersey, U.S.A.

ABSTRACT

Gravitational radiation experiments have been carried out by observing coincidences of sudden increases in amplitude on detectors 1000 kilometers apart, at *Argonne National Laboratory* and the *University of Maryland.* The detectors measure the Fourier transform of the Riemann curvature tensor.

A brief history is given and followed by discussion of the statistics and experiments to rule out electromagnetic, seismic, and cosmic-ray particle interactions. These data support a conclusion that gravitational radiation is being observed.

The exact theory of the antenna directivity is given. The earth rotates the antenna, therefore the time of each coincidence is some measure of the direction of the source. The largest peak of the intensity pattern is in the direction of the galactic center.

INTRODUCTION

My research on gravitational radiation began in 1956. Among the first papers I read were those of Rosen, and Einstein and Rosen. In 1958 it occurred to me that it would be reasonable to start a search for gravitational radiation. Suppose we imagine what it would have been like to try and discover radio-astronomy in 1880, on the basis of a general theory of electromagnetism. The antenna and technology would have had to be developed and absolutely nothing was known about sources.

---

* Supported in part by the *U.S. National Science Foundation.*

** Permanent address: *University of Maryland, College Park, Maryland, U.S.A.*

The gravitational radiation antenna was invented and analyzed rigorously using Einstein's field equations to develop equations of motion[1,2]. The idea is to employ an elastic solid to measure the curvature tensor. An elastic body is deformed in a curved space. A time-dependent curvature tensor excites normal modes. Observation of the oscillation amplitude of the normal modes enables calculation of the Fourier transform of the curvature tensor.

EXPERIMENTAL PROGRAM

Since nothing was known about sources and calculations were not optimistic it was decided to develop the technology, then wait for radioastronomy to suggest sources. Apparatus could then be built specifically for these sources. Present designs can be extended upwards in volume, mass and time integration to detect pulsars[3] at distances up to 1000 parsecs.

Our first detectors were designed to operate near 1660 Hertz because this led to a convenient size and because this frequency is swept through in a supernova collapse at a time when large amounts of energy are being radiated away.

Dr. J. Sinsky[4] carried out a high-frequency Cavendish experiment to test the detector. This produced a Riemann tensor from the Coulomb gravitational fields, with the required time dependence. Thermal fluctuations[5] imply a mean squared end-face displacement $\langle x^2 \rangle$ for a mass m, with

$$\tfrac{1}{2} m\omega^2 \langle x^2 \rangle \approx \tfrac{1}{2} kT. \tag{1}$$

For our detectors $m \approx 2 \times 10^6$ gram, and (1) implies possibility of detecting end face displacements of the cylinder of 10^{-14} cm, i.e. strains of a few parts in 10^{16}. Such strains were produced and measured by Dr. Sinsky, with a precision of four per cent.

The output of a detector is noise. For the past 6 years the Einstein theory of the Brownian motion has been used far more in these experiments than general relativity. With a random noise source, arbitrarily large amplitudes can be achieved if we wait long enough. An initial experiment searched for diurnal effects in the noise and found none. It appeared that on rare occasions there were pulse-type signals which were probably not statistical fluctuations. A second detector was developed and a coincidence experiment carried out, on a baseline of about 3 kilometers.

My definition of an event is that the rectified output of a detector crosses a certain threshold. If two or more detectors cross threshold within a small time interval this is called a coincidence. Roughly one coincidence every 6 weeks was observed, during this early experiment[6,7], with amplitudes sufficiently large to rule out the possibility that the coincidence was accidental.

To explore the coincidences further, larger detectors were developed, with better sensitivity. One of these was placed at *Argonne National Laboratory*, near Chicago, Illinois. The others are at *College Park, Maryland.* We have thus a baseline of 1000 kilometers. A telephone line transmits the output of the Argonne detector to Maryland.

Over 130 coincidences have been observed up to July 18, 1969 with amplitudes large enough to conclude that they were probably not accidental. Other experiments make it very unlikely that the coincidences are caused by seismic, electromagnetic, or cosmic-ray particle excitation. I will now discuss these issues in detail.

STATISTICS FOR TWO DETECTORS

Let us consider a long period T. For each coincidence the amplitudes are measured in each channel. We wish to compute the probability that the coincidence could have been accidental. Each coincidence is recorded on charts by an on-line computer which has observed that both detector channels have crossed their thresholds from below within some small time interval τ. The statistics are done in the following way. From chart records we observe that in channel A the amplitude at coincidence is exceeded N_A times and in channel B the amplitude at coincidence is exceeded N_B times, during T. The probability that a coincidence with amplitudes exceeding the observed coincidence amplitudes was accidental is

$$P_{AB} = 2\tau^2 N_A N_B / T^2 . \tag{2}$$

The factor 2 in P_{AB} takes account of the fact that channels A and B may cross threshold in either order. Let n_{AB} be the number of accidental coincidences to be expected in M days. Choose T to be the length of the day. n_{AB} is obtained by multiplying (2) by the number of intervals in M days, which is MT/τ. Therefore we have

$$n_{AB} = 2N_A N_B \tau M/T . \tag{3}$$

Expression (3) depends on the product $N_A N_B$, but each coincidence has two observed quantities N_A and N_B associated with it.

It is convenient in the analysis to ask how many accidental coincidences $\tilde{n}_{AB}$ are expected in M days with the given product $N_A N_B$, for which some number N_s is not exceeded by either channel. Consider the two-dimensional space of N_A and N_B. The locus of points $N_A N_B$ = constant is a hyperbola. $\tilde{n}_{AB}$ is calculated by integration under the hyperbola with limits defined by N_s. The result is

$$\tilde{n}_{AB} = 2N_A N_B \tau M \,[1 + \ln (N_s^2/N_A N_B)]/T \,. \tag{4}$$

According to (3) we should expect one accidental coincidence with amplitudes exceeding those implied by N_A and N_B every K_{AB} days with

$$K_{AB} = T/2N_A N_B \tau \,. \tag{5}$$

According to (4) we should expect one accidental coincidence with amplitudes exceeding those implied by the product $N_A N_B$ with N_A, N_B less than some number N_s every $\tilde{K}_{AB}$ days, with

$$\tilde{K}_{AB} = T/[2N_A N_B \tau \,(1 + \ln (N_s^2/N_A N_B))]. \tag{6}$$

It is meaningful to apply (5) and (6) to each coincidence with definite values of N_A and N_B. Statisticians prefer to set up a classification scheme and predict how many accidental coincidences will occur in each class. Proof that there are correlations then consists of showing that within certain classes the observed number of coincidences exceeds the expected number of accidental coincidences.

Expression (3) was confirmed by experiment using counters to count numbers of crossings in various channels, and coincidences. It was discovered that some frequency effect of the pulses occurs with the result that τ changes somewhat for different values of the ratio N_A/N_B. τ was 0.44 second for $N_A = N_B$ and $\tau \to 0.22$ second for $(N_A/N_B) \ll 1$ and for $(N_A/N_B) \gg 1$. For the first 81 days of 1969, $\langle \tau \rangle = 0.35$. During this 81-day period the gravitational radiation detector array was operational for only 56 days. For the data of the June 1969 *Physical Review Letter*[8] the following classification schemes may be employed.

Using Expression 3

N_A	N_B	Number of accidental coincidences expected	Number of coincidences observed	Period for accidental coincidence
$\leqslant 1$	≤ 5	0.0023	1	67 years
$\leqslant 1$	≤ 24	0.011	2	14 years
$\leqslant 150$	≤ 6	0.41	6	137 days

Using Expression 4

$N_A N_B$	N_s	Number of accidental coincidences expected	Number of coincidences observed	Period per accidental coincidence
480	$\leqslant 48$	0.56	8	100 days
528	$\leqslant 110$	0.99	11	57 days

For the period January 1, 1969–July 18, 1969 the data are:

Using Expression 3

N_A	N_B	Number of accidental coincidences expected	Number of coincidences observed	Period per accidental coincidence
$\leqslant 1$	$\leqslant 5$	0.0074	3	67 years
$\leqslant 75$	$\leqslant 75$	8.4	56	22 days

Using Expression 4

$N_A N_B$	N_s	Number of accidental coincidences expected	Number of coincidences observed	Period per accidental coincidence
$\leqslant 6$	$\leqslant 6$	0.0245	4	20 years
$\leqslant 24$	$\leqslant 24$	0.148	5	3.3 years
$\leqslant 100$	$\leqslant 48$	0.61	9	0.81 years
$\leqslant 1000$	$\leqslant 150$	6.0	29	30 days
$\leqslant 6000$	$\leqslant 300$	32.4	116	5.6 days

ELECTROMAGNETIC AND SEISMIC EXCITATION

Analysis has shown that the response of the gravitational radiation detector to the Riemann tensor may be calculated from the equivalent electromagnetic circuit of Figure 1. One detector employed for L_2 a superconducting coil so that the relaxation time of the electromagnetic degree of freedom was large, about 40 seconds. The mechanical system had a relaxation time of about 50 seconds. The normal mode excitations corresponding to normal mode frequencies ω_1 and ω_2 are $I_B = (L_1/L_2)^{1/2} I_A$ and $I_B = -(L_1/L_2)^{1/2} I_A$. The response to a Riemann tensor delta function for $I_A = I_B = 0$ for $t \leqslant 0$ and the delta function at $t = 0$ is then

$$I_A = R\,[\cos \omega_1 t + \cos \omega_2 t] \tag{7}$$

$$I_B = 2R\,[L_1/L_2]^{1/2}\,\{\sin [\tfrac{1}{2}(\omega_1 + \omega_2)\,t] \sin [\tfrac{1}{2}(\omega_1 - \omega_2)\,t]\}. \tag{8}$$

We only observe I_B, and I_B builds up in a time Δt given by

$$\Delta t = (\omega_1 - \omega_2)^{-1} \pi \tag{9}$$

after excitation. For the one detector with superconducting L_2, $\Delta t \approx 11$ seconds. For three other detectors the electromagnetic degree of freedom has the very short relaxation time $\approx 10^{-4}$ seconds and for these the delay (9) is absent. We expect therefore that for gravitational excitation the delay (9) will be present for the detector with superconducting L_2 and absent for the

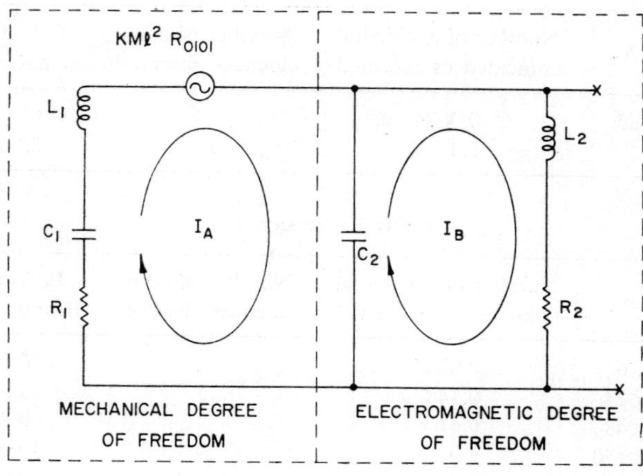

Figure 1. Gravitational radiation detector equivalent circuit for piezoelectric coupling

Figure 2. Delayed response of gravitational radiation detector with long—relaxation-time electromagnetic degree of freedom

other detectors with electromagnetic degrees of freedom with short relaxation time. On the other hand electromagnetic excitation will result in I_B building up without the 11 second delay as can be shown by solving the network of Figure 1 for a delta function source in the electromagnetic degree of freedom loop. Figure 2 is an example of a coincidence in which the eleven second delay was observed, showing that the excitation originated in the mechanical portion of the system. The long arc line is the coincidence marker, for coincidences of the two short relaxation time detectors, and the recording is of the third detector with long relaxation time.

SEISMIC RESPONSE

A seismic array operates at the Maryland gravitational radiation detector site. Earthquakes might cause coincidences as a result of earth motion exciting the detectors or as a result of gravitational radiation associated with the moving earth mass. Coincidences are not registered unless the leading edges of pulses occur within the time τ. This time τ is very short compared with the time required for seismic waves to propagate over distances of 1000 kilometers. Conceivably, seismic disturbances could originate in a zone of the earth with differences of arrival time roughly τ for Argonne and Maryland. Such a zone has a very small fraction of the total earth volume—less than 1 part in 2000. The remainder of the earth would then be a source of a large background at each detector and this is not observed. Figure 3 D shows the response of a long-period vertical seismograph to the great underground nuclear explosion of April 26, 1968, in Nevada. The abscissa is time and the ordinate is the response. The two spikes represent the propagation through the earth via a direct path and a longer, indirect path. Figure 3 C is the response of a horizontal seismograph and Figure 3 B is the response of a high frequency seismometer tuned precisely to the detector frequency. Figure 3 A is the lack of response of the gravitational radiation detecor. All time scales are the same.

ELEMENTARY PARTICLE INTERACTIONS

The detector can be excited[9] as a result of charged particles coupling energy into the electromagnetic degree of freedom and as a result of neutral particles coupling energy into the lowest frequency compressional mode of the cylinder.

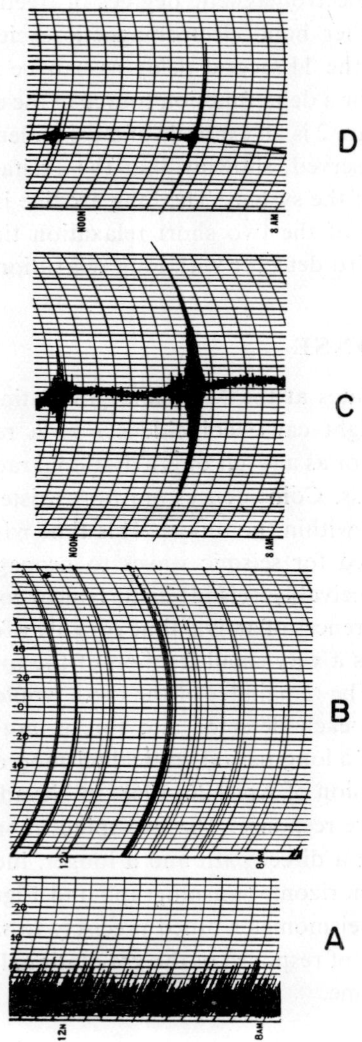

Figure 3. Response of seismic array and lack of response of gravitational radiation detector to underground nuclear explosion

Let the displacement of some element of volume of the cylinder be ξ and expand ξ in normal acoustic modes. Let $\mathbf{k}_{lmn}$ be the wavenumber for a mode designated by integers l, m, n.

$$\xi = \Sigma A_{lmn} \, \mathrm{e}^{\mathrm{i}\omega_{lmn}t} f(\mathbf{k}_{lmn} \cdot \mathbf{r}) \tag{10}$$

$$\dot{\xi} = \Sigma \, \mathrm{i}\omega_{lmn} A_{lmn} \, \mathrm{e}^{\mathrm{i}\omega_{lmn}t} f(\mathbf{k}_{lmn} \cdot \mathbf{r}). \tag{11}$$

Present detectors are instrumented to measure A_{100}. Multiplying (11) by $\mathrm{e}^{\mathrm{i}\omega_{100}t} f(k_{100}x)$ and integrating over the detector volume gives, in terms of an incident particle momentum p and detector mass M

$$A_{100} = \frac{\int \dot{\xi} f(k_{100}x) \, \mathrm{d}^3x}{\mathrm{i}\omega} = \frac{\langle \dot{\xi} \rangle}{\mathrm{i}\omega} = \frac{p_x}{\mathrm{i}\omega M}. \tag{12}$$

In (12) I have omitted the time exponential and $\langle \dot{\xi} \rangle$ is the volume average of the velocity amplitude. From (12) the required momentum p_x to excite A_{100} above the thermal fluctuation limits is then calculated for the most favorable case where a particle enters the detector roughly parallel to an axis and comes to rest within the detector. For this case 10^{18} electron volts are required, but this does not include all possibilities. Much less energy is required for situations where charged particles enter the electromagnetic degree of freedom either directly or as a result of a particle entering the cylinder and scattering a charged particle into the electromagnetic degree of freedom. We expect cosmic-ray particles to affect the background[9], but not to produce coincidences over distances of 1000 kilometers. The observation of the 11 second delay in response of the detector with superconducting degree of freedom is proof that all of the coincidences are not due to charged particles entering the detector.

ANISOTROPY AND POLARIZATION

Without a very directive antenna array, and without good signal-to-noise ratio, identification of the sources is extremely difficult. However even a simple quadrupole has some directivity, which is affected through use of square-law detectors and the fact that even a slight diminution in intensity affects the coincidence rate.

Gravitational radiation antenna polarization response

Consider Figure 4 and employ normal coordinates. The cylinder axis is assumed to coincide with the x axis at the origin and responds to the component R_{0101} of the curvature tensor. The source of the radiation is assumed to be at spherical coordinates ϕ and θ with coordinates represented by $x'y'z'$ as shown. One polarization can be represented by the components R'_{1010}

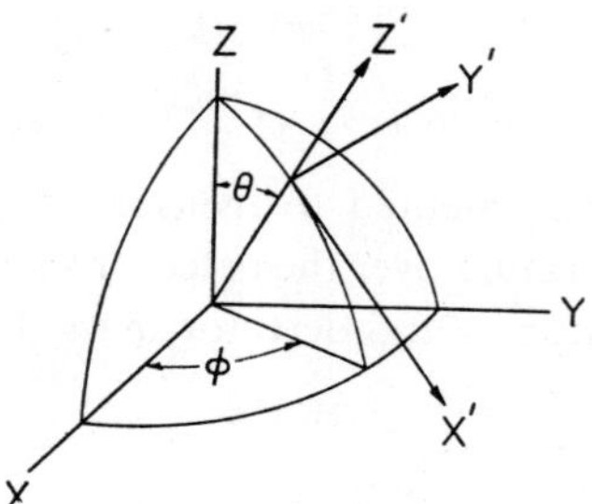

Figure 4. Coordinates for calculation of antenna directivity

and R'_{2020}. The transformation properties of the Riemann tensor lead to

$$R_{1010} = R'_{1010} \frac{\partial x'^1}{\partial x^1} \frac{\partial x'^1}{\partial x^1} + R'_{2020} \frac{\partial x'^2}{\partial x^1} \frac{\partial x'^2}{\partial x^1}. \tag{13}$$

Einstein's field equations for the vacuum require

$$R'_{\mu\nu} = R'_{1010} + R'_{2020} = 0. \tag{14}$$

The partial derivatives in (13) may be evaluated from the direction cosines and give the result

$$R_{1010} = R'_{1010} [(1 + \cos^2 \theta) \cos^2\phi - 1]. \tag{15}$$

Thus for the principal polarization the magnitude of (15) is a maximum for $\phi = \pi/2, 3\pi/2$, corresponding to a source on the meridian for the x axis pointing east and west. The second independent state of polarization is described by the Riemann tensor components R'_{1020} and R'_{2010}. An analysis similar to that for the first kind of polarization leads to

$$R_{1010} = -R'_{1020} \cos \theta \sin 2\phi. \tag{16}$$

Eq. (16) is 0 for a source on the meridian and its magnitude is a maximum for $\phi = \pm\pi/4, \pm(3\pi)/4$.

It is very interesting that the directivity response of the antenna is derived exactly, using only Einstein's field equations and the transformation properties of the Riemann tensor. The theory of the measurement of the Riemann tensor by these antennas also involves no approximations[1,2] other than neglect of the gravitational field of the antenna itself.

Experimental results

The intensity pattern requires for different polarizations a suitable squared combination of (15) and (16). There are then 8 extrema with values dependent on the polarization mixture. There can be two equal large peaks, two equal small peaks and four equal minima. Another possibility is four equal large peaks, 2 equal shallow minima[10] and 2 deep minima.

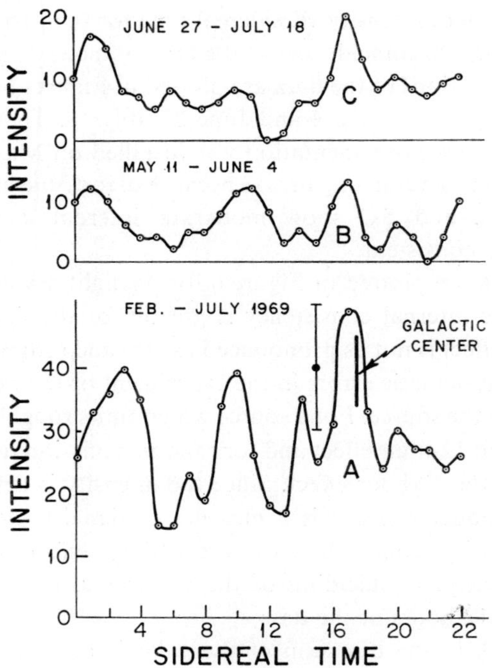

Figure 5. Intensity vs. sidereal time

Intensity is difficult to measure because the signal-to-noise ratio is extremely low. Thus a given amplitude at coincidence may be the result of the detector output being low and then being excited by a large signal. Or the detector output might be high—just below threshold—then excited above threshold by a small signal. Because of the special role played by the product $N_A N_B$, I have assigned intensities to the various coincidences on the following basis:

$N_A N_B$	Intensity
0–500	5
500–1000	4
1000–1500	3
1500–3000	2
3000–6000	1

Figure 5 is a plot of intensity vs. sidereal time for the period February 18–July 18, including 120 coincidences of the 66 centimeter Argonne–Maryland detectors. Two portions of the data are plotted in Figures 5B and 5C covering the periods May 10–June 4 and June 27–July 18. These portions were selected because new instrumentation was installed on May 10. The period June 4–June 27 had several long breaks because of instrumentation problems. Figures 5A, 5B, and 5C show moderate internal consistency. Other stretches are less consistent.

The same data are plotted in Figure 6 for daylight saving time at Maryland. Clearly the internal consistency is greater for the sidereal time plots.

The 12-hour effect is not as pronounced as (15) and (16) predict. This may be the result of systematic errors in the experiment or the random character of emission from the source. For a source which emits roughly every 24 hours there would be no 12-hour effect and for random emission times fluctuations might attenuate the 12-hour effect. Since the intensity is determined statistically by the product $N_A N_B$ it is a measure of signal-to-noise ratio rather than signal intensity alone. Therefore the 12-hour effect might also be attenuated by inadequate shielding of the apparatus, resulting in a higher background for 12 hours of the day.

The large peak in the direction of the galactic center suggests that the center of the galaxy is the source. The apparent consistency in the sidereal time plots is evidence that the array is responding to effects from outside the solar system.

CONCLUSION

Gravitational radiation detectors on a baseline of 1000 kilometers are responding to a common excitation which appears to have a galactic origin. Experiments indicate that seismic, electromagnetic, and cosmic-ray particle

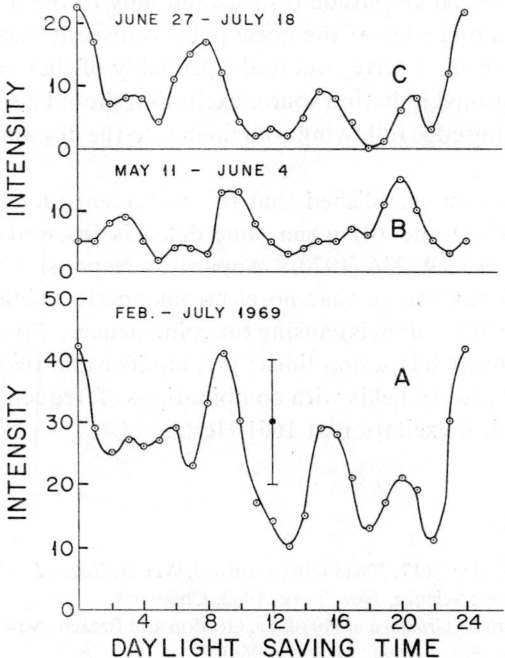

Figure 6. Intensity vs. daylight savings time

origins are very unlikely. These data justify the conclusion that gravitational radiation has been discovered.

It is important that the theory of the antenna directivity, and the theory of the use of the antenna to measure the Fourier transform of the Riemann tensor, are exact. If the galactic center is the source and patterns with multiple peaks are verified by future observations, the spin 2 and transverse character of gravitational radiation will be established experimentally.

Added in proof:

Improved methods have given data showing anisotropy with two peaks, one in the direction of the galactic center and the second one twelve hours away. Histograms showing the 12 hour symmetry with sidereal time anisotropy exceeding six standard deviations have been published *(Phys. Rev. Letters* **25,** 180, 1970). These histograms have been repeated during the ensuing six months. Improved histograms resulted from improved methods for processing the data. The principal improvement was the following: For each coincidence the amplitude is measured only to the point at which the slope of the leading edge of the noise pulse is discontinuous. At the point a new excitation must have occurred—probably of thermal origin. Therefore the gravitational radiation source excitation should be inferred from the part of the leading edge following coincidence to the discontinuity in leading edge slope.

Experiments have established that the coincidence rate decreases to the value given by (3) and (4) when time delay is inserted in one channel (*Phys. Rev. Letters* **24,** 276 (1970)). Apparatus responsive to local electromagnetic fields has shown that no electromagnetic excitation of the mechanical degree of freedom is causing the coincidences. This electromagnetic monitor equipment has a non linear preamplifier and responds to fields at 1661 Hertz and also to fields with combinations of frequencies which might lead to mechanical excitation at 1661 Hertz.

REFERENCES

1. J. Weber, *Phys. Rev.*, **117**, 306 (1960); see also J. Weber, *General Relativity and Gravitational Waves*, Interscience, New York, 1962, Chapter 8.
2. J. Weber, *Relativity Groups and Topology*, Gordon and Breach, New York, 1964, p. 875.
3. J. Weber, *Phys. Rev. Letters*, **21**, 395 (1968).
4. J. Sinsky and J. Weber, *Phys. Rev. Letters*, **18**, 795 (1967); J. Sinsky, *Phys. Rev.*, **167**, 1145 (1968).
5. J. Weber, *Phys. Rev. Letters*, **17**, 1228 (1966).
6. J. Weber, *Phys. Rev. Letters*, **18**, 498 (1967).
7. J. Weber, *Phys. Rev. Letters*, **20**, 1307 (1968).
8. J. Weber, *Phys. Rev. Letters*, **22**, 1320 (1969).
9. B. L. Beron and R. Hofstadter, *Phys. Rev. Letters*, **23**, 184 (1967).
10. I thank Dr. H. U. Schmidt for correcting an earlier statement that there may be six peaks.

PAPER 36

Light propagation in a time dependent gravitational field

P. J. WESTERVELT

Brown University

Providence, Rhode Island, U.S.A.

I have shown[1] that the passage of a light pulse with momentum p at a distance of closest approach d to the sun, imparts a velocity to the sun perpendicular to, and directed towards, the light path. The magnitude of this velocity is

$$v = 4kpc^{-2}d^{-1} \tag{1}$$

in which k is the gravitational constant and c is the speed of light. This result, valid for a sun initially at rest, yields the well known formula for the angle of deflection[2]

$$\phi = 4kmc^{-2}d^{-1}, \tag{2}$$

with m the mass of the sun.

In case the sun has an initial perpendicular velocity $\pm u$ towards the light path, the kinetic energy of the sun will be altered by the passage of the light pulse in the amount

$$\Delta T = \tfrac{1}{2}m\,[(v \pm u)^2 - v^2] \approx \pm muv. \tag{3}$$

This energy change must be balanced by an opposite energy change in the light pulse, since it can easily be demonstrated that the gravitational field energy (and momentum) bilinear in the gravitational field of the sun and the gravitational field of the light pulse vanishes with increasing time.

Applied to a photon the above analysis predicts an additional general relativistic Doppler correction given by

$$\Delta h\nu = \mp muv, \tag{4}$$

with h Planck's constant and ν the frequency.

Combining Eq. (1), (2) and (4) and the relation $p = h\nu c^{-1}$, we obtain finally

$$\frac{\Delta\nu}{\nu} = \mp u c^{-1} \phi. \tag{5}$$

This additional Doppler shift is precisely that which was obtained by I. I. Shapiro[3] in connection with his studies of interplanetary radar time delays.

The above analysis and the work reported in ref. 1 has been put on a completely rigorous basis using the Landau pseudotensor densities and this will be the subject of a more lengthy publication.

This work was performed under the auspices of the *U. S. Atomic Energy Commission* (Report No. NYO-2262TA-208).

REFERENCES

1. P. J. WESTERVELT, *Acta Phys. Polon.*, **35**, 203 (1969).
2. R. ADLER et al., *Introduction to General Relativity*, McGraw–Hill, New York, 1965, p. 188.
3. M. J. TAUSNER, *General Relativity and Effects on Planetary Orbits and Interplanetary Observations*, Technical Report 425, Lincoln Laboratory, M.I.T., Cambridge, Mass., 1966.

PRINTED IN EAST GERMANY